Fractal Geometry *in* Biological Systems

—*An Analytical Approach*—

Edited by
Philip M. Iannaccone
and Mustafa Khokha

CRC Press
Boca Raton   New York   London   Tokyo

Marsha Baker: Acquiring Editor
Denise Craig: Cover Designer
Becky McEldowney: Marketing Manager, Direct Marketing
Susie Carlisle: Marketing Manager, Marketing Management
Carlos Esser: PrePress
Sarah Fortener, Gerry Jaffe: Project Editors
Sheri Schwartz: Manufacturing Assistant
Stephen Iannaccone: Cover computer graphic augmentation

Cover art © 1995 M. C. Escher/Cordon Art, Baarn, Holland. All rights reserved.

Library of Congress Cataloging-in-Publication Data

Fractal geometry in biological systems, an analytical approach / [edited by] P.M. Iannaccone
  and M.K. Khokha.
     p.   cm.
   Includes bibliographical references and index.
   ISBN 0-8493-7636-X
   1. Medicine--Mathematics.   2. Biology--Mathematics.   3. Fractals.
  I. Iannaccone, P. M. (Philip M.)   II. Khokha, M. K. (Mustafa K.)
  R853.M3F67   1996
  574'.01'51474--dc20   96-13895
                                                                                          CIP
                                                                                           AC

   This book contains information obtained from authentic and highly regarded sources. Reprinted material is quoted with permission, and sources are indicated. A wide variety of references are listed. Reasonable efforts have been made to publish reliable data and information, but the author and the publisher cannot assume responsibility for the validity of all materials or for the consequences of their use.
   Neither this book nor any part may be reproduced or transmitted in any form or by any means, electronic or mechanical, including photocopying, microfilming, and recording, or by any information storage or retrieval system, without prior permission in writing from the publisher.
   All rights reserved. Authorization to photocopy items for internal or personal use, or the personal or internal use of specific clients, may be granted by CRC Press, Inc., provided that $.50 per page photocopied is paid directly to Copyright Clearance Center, 27 Congress Street, Salem, MA 01970 USA. The fee code for users of the Transactional Reporting Service is ISBN 0-8493-7636-X/96/$0.00+$.50. The fee is subject to change without notice. For organizations that have been granted a photocopy license by the CCC, a separate system of payment has been arranged.
   CRC Press, Inc.'s consent does not extend to copying for general distribution, for promotion, for creating new works, or for resale. Specific permission must be obtained in writing from CRC Press for such copying.
   Direct all inquiries to CRC Press, Inc., 2000 Corporate Blvd., N.W., Boca Raton, Florida 33431.

© 1996 by CRC Press, Inc.

No claim to original U.S. Government works
International Standard Book Number 0-8493-7636-X
Library of Congress Card Number 96-13895
Printed in the United States of America 1 2 3 4 5 6 7 8 9 0
Printed on acid-free paper

Preface

The Destruction of Civil Order in Flatland

"... I have been assuming — what perhaps should have been laid down at the beginning as a distinct and fundamental proposition — that every human being in Flatland is a Regular Figure, that is to say of regular construction. By this I mean that a Woman must not only be a line, but a straight line; that an Artisan or Soldier must have two of his sides equal; that Tradesmen must have three sides equal; Lawyers (of which class I am a humble member), four sides equal; and, generally that in every polygon, all the sides must be equal.

"If our sides were unequal our angles might be unequal. Instead of its being sufficient to feel, or estimate by sight, a single angle in order to determine the form of an individual, it would be necessary to ascertain each angle by the experiment of Feeling. But life would be too short for such a tedious groping. The whole science and art of Sight Recognition would at once perish; Feeling, so far as it is an art, would not long survive; intercourse would become perilous or impossible; there would be an end to all confidence, all forethought; no one would be safe in making the most simple social arrangements; in a word, civilization would relapse into barbarism."*

Edwin Abbott painted a strange picture of dimensionality in the world of A. Square, lawyer and mathematician, inhabitant of Flatland. Flatland, as we all know, is a two-dimensional universe where only shapes exist. Indeed, only polygons and lines exist (lines being women), and the more sided the polygon, the higher the social position of the figure. In order to ascertain the social position of an individual encountered by chance a Flatlander could "feel" the angle closest to him. The more acute the angle, the more dangerous the encounter! In some individual figures the acquired skill of seeing the angle was possible since the terminal points of the polygon's sides had lights and the relative luminosity of the lights diminished with distance. Thus, the more sided the polygon the less difference there was between the forward-most point and the two nearest terminal points.

One day A. Square was visited by a sphere from Spaceland shortly after he himself had imagined a trip to Lineland. The sphere tried to use analogy to instruct A. Square in the truth of three dimensions. After all, to A. Square the sphere appeared seemingly out of nowhere as a circle of changing diameter! So taken was our Flatland

---

* Edwin A. Abbott, *Flatland: A Romance of Many Dimensions*, Barnes and Noble, New York, 1963. (Reprinted from the second and revised edition, 1884.)

mathematician with this encounter he began to evangelize the population of Flatland concerning the greater universe they had not imagined. This resulted in the immediate incarceration of the hapless mathematician. What occurred next was not suggested by Edwin Abbott.

A. Square was so fervent in his belief that when eventually he was released from prison he attempted by various means to attract attention to himself in the hopes that his visitor from Spaceland would appear. He did succeed in attracting attention but from an unusual visitor, not the well behaved sphere of Spaceland that he expected. Indeed, our new visitor had a new message for A. Square, a message that dimensionality is not as simple as one, two, three, and that well behaved polygons with no irregularities were not the only forms that were to be the population of Flatland.

This new stranger brought new dimensions to Flatland; he was infected with a virus that was highly contagious and had bizarre properties. Flatland women, reprehensibly poorly regarded by Abbott, were lines, a potentially dangerous situation for other flatland individuals. For example, if a flatland woman approached from the end on she was hard to see, appearing as a point. If she approached too rapidly she could slice into another figure, inflicting grave injuries. But this strange new visitor caused an amazing new problem. The virus acted as an initiator and resulted in the repetitive removal of the middle third of a unit line. The individual Flatland female first lost the middle third of herself, then the remaining lines of herself lost their middle third, and so on. So, instead of a line, the infected Flatlander became a cohesive set of dots and dashes, which was reduced to dust in places. This alarming condition immediately raises the question of the victim's dimension ... i.e., is the figure a line with a dimension of 1 or a point with the dimension of 0? It is neither. We will show in Professor Vicsek's chapter that the dimension is 0.6309. We were introduced to equilateral triangles in Abbott's fanciful land, and their fate when infected with the fractal virus is to see the repetitive alteration of the middle third of each side of the Euclidean triangle into an upward bump of a new equilateral triangle. This creates a strange star-shaped figure. When the rule is reapplied to this new figure (the middle third of each side is bumped out into a new triangle), the figure becomes more and more complex. After many applications of the rule, this infected figure becomes very complex and looks like a snowflake. The virus continues and doesn't stop; the figure becomes more and more complex. After an infinite application of the rule, the figure has some very odd characteristics. First of all, the figure begins to fill a plane, so is its dimension 1 as was the perimeter of the triangle that was infected? Or is the dimension 2, that of a filled plane? The dimension is between them ... in fact, about 1.26. Although the Koch snowflake that grows out of the infected equilateral triangle is a closed curve with a finite area, the length of its perimeter is infinite, and no known Euclidean formula can specify the points on the curve.

So what chaos has been brought to Flatland by this fractal virus? No longer could individuals identify each other by deducing shape from estimating angles during encounters. No longer could dimension be counted on as a stable "simple" quantity. The predictable straightforward landscape of Euclidean shapes and relationships was destroyed. Flatland suddenly became very messy and much more like the real world. More and more physicists, geographers, and biologists are coming to realize that describing natural phenomena with Euclidean geometry falls short in providing models of real situations. This was first understood by map makers when they tried to define the length of coastlines or the length of rivers. The data available to them seemed at best a gross estimate of reality. As they thought about it they wondered what the reality of these measures actually was. They had to ask themselves over and over by what measure do we know this length. The question actually meant at what scale do we establish this measure. The actual length of a coast, as pointed out by Mandlebrot, depends on whether it is measured from an airplane or by walking along the beach.* Flatlanders are now confronted with the startling realization that, now that they have been infected, length can never be known absolutely; it can only be known for a given scale of observation. Flatland's fictional occupants, confidant in their sense of two dimensions, were also challenged by the meaning of dimension. They now not only had to worry about the heresy of a third dimension in the visitor from Spaceland but also were confronted with noninteger dimensions. Fractal objects appearing in Flatland had dimensions less than a plane, but more than a curve.

By following the chapters in this book the reader will come to know many examples of fractal objects. Fractal objects are complex, they have detail vested within detail, such that more detail (more coastline, for example) becomes apparent the closer one looks. More detail is apparent than that predicted simply by correcting for the scale of observation. The dimension of these objects, determined by measuring a topologic parameter (length, area, or volume) at many scales of observation and plotting the log of the parameter against the log of the scale of observation, is different from the predicted topologic measure. The importance of these objects is that unlike the characters in Abbot's tale (all Euclidean objects) they occur frequently in nature as we know it. They occur so frequently in fact that many are convinced fractals are a result of very fundamental rules in nature. The recent observation of fractal geometry in biologic systems strengthens this view. The following chapters are devoted to some key biological systems in which fractal geometry appears and can be exploited in the interpretation of results, the formulation of theory, or in the design of experiments. The authors and the editors hope that some of the excitement those of us in this new area feel can be transmitted to others and that new and pioneering uses of fractals can be found.

Philip Iannaccone
Chicago

* James Gleick, *Chaos: Making a New Science*, Viking, New York, 1988.

The Editors

Philip M. Iannaccone, M.D., Ph.D., is currently the George M. Eisenberg Professor of Developmental Pediatrics at Northwestern University Medical School. He is the Core Leader of the Developmental Systems Biology Group at the Children's Memorial Institute for Education and Research at Children's Memorial Hospital, which houses his active research lab. Dr. Iannaccone received his baccalaureate degrees from the State University of New York College of Forestry and Syracuse University. He received his M.D. from Upstate Medical Center in 1972, and his Doctor of Philosophy degree from the University of Oxford (Lincoln College) under the supervision of Henry Harris, F.R.S. (Regius Professor of Medicine) in 1977. His initial faculty appointment was at the University of California, San Diego as Assistant Professor. He moved to Northwestern University in 1979 and was appointed full Professor in 1989. Dr. Iannaccone was Director of the M.D., Ph.D. program and the Medical Scientist Training Program of Northwestern University and was the Director of the Markey Program in Developmental Biology of Northwestern University. He is currently serving as Chairman of the Board of Scientific Counsellors of the National Institute of Environmental Health Sciences (1995–1997). Dr. Iannaccone is a member of the International Academy of Pathology, American Association for the Advancement of Science, American Association of Pathologists, American Association for Cancer Research, American Society of Cell Biologists, International Society for Differentiation, Society for Developmental Biology, and the Society for Perinatal Research. He is on the Editorial Board of the *American Journal of Pathology, Pathobiology, and Transgenics*. Dr. Iannaccone's research interests include the study of organogenesis in chimeric mice and rats, stem cell research, and genetic specification of fate by transcription factors.

Mustafa K. Khokha, M.D., is a resident physician studying Pediatrics at St. Louis Children's Hospital. He received his B.Sc. with Highest Distinction in Biomedical Engineering in 1991 from Northwestern University Technological Institute and his M.D. in 1995 from Northwestern Medical School. He is a recipient of the Howard Hughes Medical Student Research Fellowship and a member of the American Academy of Pediatrics.

Contributors

Sergei V. Buldyrev, Ph.D.
Research Associate
Center for Polymer Studies and
  Department of Physics
Boston University
Boston, Massachusetts

Ary L. Goldberger, M.D.
Department of Medicine
Beth Israel Hospital and
  Harvard Medical School
Boston, Massachusetts

Jeffrey Hausdorff, Ph.D.
Department of Medicine
Beth Israel Hospital and
  Harvard Medical School
Boston, Massachusetts

Shlomo Havlin, Ph.D.
Professor
Department of Physics
Bar Ilan University
Ramat Gan, Israel

Philip M. Iannaccone, M.D., Ph.D.
George M. Eisenberg Professor of
  Developmental Pediatrics
Department of Pediatrics
Children's Memorial Hospital
Children's Memorial Institute for
  Education and Research
Chicago, Illinois

Voyko Kavcic, M.S.
Staff Scientist
Computer and Information Services, Inc.
Denton, Texas

Mustafa K. Khokha, M.D.
Department of Pediatrics
St. Louis Children's Hospital
St. Louis, Missouri

Gabriel Landini, Dr. Odont., Ph.D.
Research Fellow
Oral Pathology Unit
School of Dentistry
The University of Birmingham
Birmingham, England

G. David Lange, Ph.D.
Physiologist
Instrumentation and Computer Section
NINDS
National Institutes of Health
Bethesda, Maryland

Houqiang Li, Ph.D.
Professor
Department of Physics
Sichuan University
Chengdu, Sichuan
China

Larry S. Liebovitch, Ph.D.
Associate Professor
Center for Complex Studies
Florida Atlantic University
Boca Raton, Florida

Rosario N. Mantegna, Ph.D.
Research Associate
Center for Polymer Studies and
  Department of Physics
Boston University
Boston, Massachusetts

Mitsugu Matsushita, Ph.D.
Professor
Department of Physics
Chuo University
Tokyo, Japan

Tohey Matsuyama, M.D., Ph.D.
Associate Professor
Department of Bacteriology
Niigata University School of Medicine
Niigata, Japan

Joseph Mietus, B.S.
Department of Medicine
Beth Israel Hospital
Boston, Massachusetts

Marcos N. Novaes, Ph.D.
Staff Scientist
Computer and Information Services, Inc.
Denton, Texas

Chung-Kang Peng, Ph.D.
Department of Medicine
Beth Israel Hospital and
  Harvard Medical School, *and*
Department of Physics and
  Center for Polymer Studies
Boston, Massachusetts

Michael Simons, M.D.
Assistant Professor
Department of Medicine
Beth Israel Hospital and
  Harvard Medical School
Boston, Massachusetts

Thomas G. Smith, Jr., M.D.
Section Chief
Laboratory of Neurophysiology
NINDS
National Institutes of Health
Bethesda, Maryland

H. Eugene Stanley, Ph.D.
Professor
Center for Polymer Studies and
  Department of Physics
Boston University
Boston, Massachusetts

Tamás Vicsek, Ph.D.
Professor of Physics
Department of Atomic Physics
Eötvös University
Budapest, Hungary

Fuquan Wang, B.S.
Associate Professor
Department of Mathematics
Sichuan Teacher's College
Nanchong, China

Bruce J. West, Ph.D.
Vice President for Research
Computer and Information Services,
  Inc., *and*
Director, Center for Nonlinear
  Science and Professor of Physics
Department of Physics
University of North Texas
Denton, Texas

Contents

Section I: Introduction

chapter one
Fractal Geometry 3
Philip M. Iannaccone and Mustafa K. Khokha

Section II: Molecules

chapter two
Scale Invariant Features of Coding and Noncoding DNA Sequences 15
H. Eugene Stanley, S.V. Buldyrev, Ary L. Goldberger, Shlomo Havlin,
R.N. Mantegna, Chung-Kang Peng, and Michael Simons

2.1 Abstract 15
2.2 Long-range power-law correlations 15
2.3 DNA 16
2.4 The "DNA walk" 17
2.5 Correlations and fluctuations 17
2.6 Detrended fluctuation analysis 18
2.7 Coding sequence finder algorithm 19
2.8 Systematic analysis of GenBank database 19
2.9 Generalized Lévy walk model 21
2.10 Mosaic nature of DNA structure 23
2.11 Linguistic analysis of noncoding and coding DNA 23
2.12 Outlook for the future 26
Acknowledgments 27
References 27

chapter three
Ion Channel Kinetics 31
Larry S. Liebovitch

3.1 Introduction 31
3.2 Energy structure of proteins 33
   3.2.1 Energy structure of ion channels 33
   3.2.2 Energy structure of globular proteins 33

- 3.2.2.1 Potential energy functions ... 33
- 3.2.2.2 Molecular dynamics ... 33
- 3.2.2.3 Reaction rates computed from the average structures ... 34
- 3.2.2.4 Distribution of activation energy barriers determined by the time course of reactions ... 34
- 3.2.2.5 Measurement of motions by X-ray diffraction and nuclear magnetic resonance ... 35
- 3.2.2.6 Induced fit ... 35
- 3.2.2.7 Summary of the properties of proteins ... 35
- 3.3 Markov models of ion channel kinetics ... 36
  - 3.3.1 Mathematical assumptions of the Markov model ... 36
  - 3.3.2 Physical interpretation of the Markov model ... 37
  - 3.3.3 Comparison of the physical interpretation of the Markov model to the known properties of proteins ... 37
- 3.4 Fractal analysis of ion channel kinetics ... 38
  - 3.4.1 Fractals ... 38
  - 3.4.2 Fractal form of the current ... 39
  - 3.4.3 Methods of fractal analysis of the open and closed times ... 39
    - 3.4.3.1 Testing for self-similarity ... 39
    - 3.4.3.2 Testing for scaling: the effective kinetic rate constant ... 40
- 3.5 Physical interpretation of the fractal analysis ... 42
  - 3.5.1 The fractal approach ... 42
  - 3.5.2 Two physical interpretations: structural and dynamical ... 43
  - 3.5.3 Structural interpretation ... 44
    - 3.5.3.1 Examples of cumulative probability distributions $P(t)$ of open and closed times and their corresponding distributions of kinetic rates $g(k)$ ... 45
    - 3.5.3.2 Examples of distributions of kinetic rates $g(k)$ determined from the probability distributions $P(t)$ of open and closed times of patch clamp data ... 46
      - 3.5.3.2.1 Power-law distributions of open or closed times ... 46
      - 3.5.3.2.2 Single exponential distributions of open or closed times ... 47
      - 3.5.3.2.3 Combinations of distributions of open or closed times ... 48
      - 3.5.3.2.4 Dependence of distributions of open or closed times on voltage, ions, and ligands ... 49
  - 3.5.4 Dynamical interpretation ... 50
  - 3.5.5 Structural or dynamical? ... 51
- 3.6 Other approaches ... 51
  - 3.6.1 Are ion channel proteins little machines? ... 51
  - 3.6.2 How do local interactions produce globular structure in ion channel proteins? ... 52
- 3.7 Response to the fractal approach ... 52
  - 3.7.1 Markov vs. fractal ion channel kinetics ... 52
- 3.8 Summary of the fractal approaach to ion channel kinetics ... 53
- Acknowledgments ... 54
- References ... 54

chapter four
Protein Conformation and Enzymatic Kinetics 57
Houqiang Li and Fuquan Wang

4.1 Introduction: basic concepts and terminology..57
    4.1.1 Fat fractal...58
    4.1.2 Multifractal..59
    4.1.3 The spectral (fracton) dimension 61
4.2 Fractal aspects of protein chain structure..62
    4.2.1 The problem of entanglements...62
    4.2.2 Fractal aspects of polymers..63
        4.2.2.1 Linear polymers 63
        4.2.2.2 Star polymers...64
        4.2.2.3 Branched polymers...65
    4.2.3 Universal aspects of protein structure 66
4.3 The fractal dimension of the chain and the mass dimension..................................67
4.4 Fracton dimension of protein..68
4.5 Fractal analysis of protein chain conformation 71
    4.5.1 The fractal dimension of the main chain and tertiary structure 71
    4.5.2 The dimensional calculation of protein chains...76
    4.5.3 Fractal dimensions and conformational entropy of protein chain...........78
        4.5.3.1 The fractal dimensions of protein chains..79
        4.5.3.2 Calculation of conformational entropy 80
        4.5.3.3 The relationship between fractal dimension
               and conformational energy 82
4.6 The correlative dimension for globular proteins 84
    4.6.1 G-P algorithms of correlative dimension and Kolmogorov entropy 85
    4.6.2 The correlative dimension for globular proteins...86
    4.6.3 Estimation of the Kolmogorov entropy on EEG 87
4.7 Chaos — theoretical analysis for biochemical reactions..88
    4.7.1 Metabolite regulation and biochemical attractor 88
    4.7.2 Nonlinear dynamics in glycolysis...89
    4.7.3 Switching process in metabolite regulation 90
4.8 The fractal dimension of protein surfaces...90
    4.8.1 Fractal behavior of surfaces of enzymes and proteins................................90
    4.8.2 Fat fractals of protein and fractal dimension computed
         by the variation method 91
    4.8.3 The multifractals and mass exponents for protein surfaces.......................93
        4.8.3.1 Multifractals for biological macromolecules 93
        4.8.3.2 Mass exponents for macromolecules..95
4.9 Fractal reactions of enzyme surfaces 99
4.10 Fractal approach for protein and enzyme kinetics 102
    4.10.1 Fractal chemical kinetics 102
    4.10.2 Mechanisms for the allosteric effects of proteins and enzymes 104
        4.10.2.1 One-step conformational change..106
        4.10.2.2 Multistep conformational change...108
        4.10.2.3 The generalization of rate equations based
                on the fractal reaction kinetics..110
4.11 Fractal art of enzyme model design...111
    4.11.1 Fractal principle of enzyme model design..112
    4.11.2 Fractals of hybrid orbitals 112
        4.11.2.1 Stereospecificity of enzyme catalysis..114

     4.11.2.2 Cyclodextrin..114
   4.11.3 Structural parameters and enzyme model catalytic properties............114
     4.11.3.1 The fractal dimension of steroid-metalloporphyrin...............115
     4.11.3.2 Application of the topological indices in enzyme models......116
     4.11.3.3 Application of the structural information index.....................117
4.12 Discussion and conclusion..118
Acknowledgment..120
References..120

Section III: Cells

chapter five
Morphogenesis by Bacterial Cells 127
Tohey Matsuyama and Mitsugu Matsushita

5.1 Introduction..127
5.2 Fundamentals in bacterial colony formation......................................128
   5.2.1 Bacterial strains..128
   5.2.2 Culture conditions..128
5.3 Characterization of colony morphology...132
   5.3.1 Diversity in colony morphology..132
   5.3.2 Examination of fractal property...136
     5.3.2.1 Self-similar fractals and fractal dimension...................139
       5.3.2.1.1 Self-similar fractals...139
       5.3.2.1.2 Fractal dimension..140
       5.3.2.1.3 Determination of fractal dimensions............143
     5.3.2.2 Self-affine fractals and scaling indices.........................145
       5.3.2.2.1 Self-affine fractals..145
       5.3.2.2.2 Determination of self-affine scaling exponents.............147
5.4 Pursuit of random pattern growth...148
   5.4.1 Macroscopic generating process..149
   5.4.2 Microscopic generating process...149
5.5 Experimental approach for elucidation of morphogenic mechanisms............155
   5.5.1 Designs of experiments and revealed mechanisms................155
     5.5.1.1 Nutrients..155
     5.5.1.2 Spatial arrangement...157
     5.5.1.3 Substrate construction..158
     5.5.1.4 Genetic approach..158
     5.5.1.5 Terrestrial factor..159
   5.5.2 Identification of chemical factors...160
5.6 Modeling and population dynamics of colony formation................161
   5.6.1 Population dynamics approach..162
   5.6.2 Experiments for verification of a proposed model................166
   5.6.3 Fractal growth of colonies — revisited...................................168
5.7 Prospects..169
Acknowledgments...169
References..169

chapter six
Fractal Studies of Neuronal and Glial Cellular Morphology 173
Thomas G. Smith, Jr. and G. David Lange

6.1 Introduction 173
    6.1.1 Fractal geometry 173
        6.1.1.1 The fractal dimension...173
        6.1.1.2 Range of D..174
        6.1.1.3 Empirical nature of D..174
        6.1.1.4 Methods of determining D...174
        6.1.1.5 Grey scale to binary images...174
        6.1.1.6 Length methods..174
        6.1.1.7 Mass method...176
        6.1.1.8 Factors affecting D 177
6.2 Uses of fractal geometry and D 177
    6.2.1 Quantification and classification 178
    6.2.2 Growth and differentiation 179
6.3 Other studies 182
    6.3.1 Vertebrate central nervous system cortical pyramidal neurons 182
    6.3.2 Bergmann glial cells 182
    6.3.3 Surface-to-volume ratio..182
    6.3.4 Miscellaneous research 182
6.4 Lacunarity and moments of mass distributions 183
6.5 Discussion and conclusions 184
References 185

Section IV: Tissues

chapter seven
Mosaic Pattern in Tissues from Chimeras 189
Mustafa K. Khokha and Philip M. Iannaccone

7.1 Mammalian development 189
7.2 Proposed mechanisms of pattern formation...191
7.3 Mosaic pattern...193
7.4 Models of organogenesis...197
7.5 Mosaic pattern in chimeras...200
7.6 Implications 201
7.7 Further studies 203
Acknowledgment 203
References 203

chapter eight
Applications of Fractal Geometry in Pathology 205
Gabriel Landini

8.1 Current problems in histopathology..205
8.2 More than 1 but less than 2 206
8.3 What is fractal geometry good for?..207
    8.3.1 Quantification and modelling of complex structures...................210
        8.3.1.1 Cell morphology 210

    8.3.1.2 Tumor shape 213
      8.3.1.2.1 Fractal approach 215
      8.3.1.2.2 Multifractal approach 215
    8.3.1.3 Retinal vasculature analysis..217
    8.3.1.4 Spread of herpes simplex virus 218
    8.3.1.5 Periodontal disease modelling...221
  8.3.2 Why fractal?..224
8.4 How to estimate dimension..226
 8.4.1 Self-similarity dimension..226
 8.4.2 Length-resolution methods 226
    8.4.2.1 Yardstick method 226
      8.4.2.1.1 Extension to surface analysis 227
    8.4.2.2 Interpolated yardstick variant 228
    8.4.2.3 Censored intercept method 228
    8.4.2.4 Coordinate skipping method 228
    8.4.2.5 Equally spaced test-lines method................................228
    8.4.2.6 Dilation method 228
 8.4.3 Box-counting method...230
 8.4.4 Area-perimeter relation...230
 8.4.5 Mass-radius relation 231
    8.4.5.1 Radius of gyration 231
    8.4.5.2 Pair correlation function 231
 8.4.6 Fragmentation dimension..232
 8.4.7 Other methods...233
    8.4.7.1 Fourier analysis 233
    8.4.7.2 Root-mean-square fluctuation......................................234
    8.4.7.3 Hurst's rescaled range analysis 235
    8.4.7.4 Semivariogram method...236
    8.4.7.5 Other proposed methods for estimation of dimension...........236
 8.4.8 Fractal harmonics..236
 8.4.9 Multifractals and local dimensions...236
 8.4.10 Lacunarity 238
 8.4.11 Implementation details 238
    8.4.11.1 Image digitization/binarization 239
    8.4.11.2 Dimension reduction, projection, and zerosets........239
 8.4.12 How fractal are biological fractals? 239
    8.4.12.1 Arbitrary range...240
    8.4.12.2 Local slope...240
    8.4.12.3 Equally spaced data points 240
    8.4.12.4 Extended fractal models 240
      8.4.12.4.1 Asymptotic fractal model 241
      8.4.12.4.2 Log-logistic approach 242
Acknowledgment 242
References 242

Section V: Organs

chapter nine
Fractals and the Heart 249
Ary L. Goldberger, Chung-Kang Peng, Jeffrey Hausdorff, Joseph Mietus, Shlomo Havlin, and H. Eugene Stanley

9.1 Introduction 249
9.2 Fractal geometry of the heart 250
9.3 Human heartbeat dynamics...251
      9.3.1 Increment of interbeat interval 254
      9.3.2 Interbeat interval time series: detrended fluctuation analysis 258
9.4 Normal vs. pathologic time series 259
      9.4.1 Crossover phenomena 261
      9.4.2 Clinical applications: preliminary results 261
9.5 Heart rate dynamics: conclusions 263
Acknowledgments...264
References 264

chapter ten
Fractal Probability Density and EEG/ERP Time Series 267
Bruce J. West, Marcos N. Novaes, and Voyko Kavcic

10.1 Introduction 267
      10.1.1 Measures of EEG/ERP times series 268
      10.1.2 Fractal dimensions and learning 270
      10.1.3 Fractal dimensions and performance 270
      10.1.4 Technical overview 271
10.2 Attractor reconstruction technique: a critique 272
      10.2.1 The ART of Takens 275
      10.2.2 Grassberger-Procaccia algorithm 279
      10.2.3 What is wrong with GPA? 280
      10.2.4 Singular value decomposition 283
      10.2.5 Re-embedding technique 285
      10.2.6 Alternative to Takens 285
      10.2.7 Overcoming the limitations: the distribution function 286
      10.2.8 Summary 289
10.3 Processing brain wave data 290
      10.3.1 GPA applied to EEG 292
      10.3.2 EEG time series 292
      10.3.3 Distribution approach to EEG 294
      10.3.4 Distribution and MART applied to EEG 295
10.4 Measuring the event-related potential 298
      10.4.1 Average ERP 302
      10.4.2 Point estimate of ERP fractal dimension 307
      10.4.3 New technique for ERP fractal dimension 307
10.5 Summary and plans for the future 309
Acknowledgment 311
References 311

Section VI: Advanced Topics in Fractal Geometry

chapter eleven
Fractal Geometry 317
Tamás Vicsek

11.1 Introduction 317

11.2 Definitions 318
    11.2.1 Fractals as mathematical and biological objects 318
    11.2.2 Definitions 320
    11.2.3 Useful rules 321
11.3 Types of fractals 322
    11.3.1 Deterministic and random fractals 322
    11.3.2 Self-similar fractals 323
    11.3.3 Self-affine fractals 326
    11.3.4 Fat fractals 329
11.4 Multifractals 330
    11.4.1 Definitions 330
    11.4.2 Multifractal formalism 333
    11.4.3 Recursive multifractals 335
    11.4.4 Mass multifractality 338
11.5 Methods for determining fractal dimensions 339
    11.5.1 Experimental methods for measuring fractal dimensions 339
    11.5.2 Evaluation of numerical data 340
References 342

Index 347

Dedication

To Judy
Because of her library

Section I
Introduction

chapter one

Fractal Geometry

Philip M. Iannaccone and Mustafa K. Khokha

> *Mariners use geometry, and so do builders, but in ordinary circumstances its use is avoided, except for the metrics of driving distances, and cooking.*
>
> **S. Barr**

A snowflake in Vermont, the coastline of Italy, the Nile River: all of these share a characteristic that is very common in nature. They all have geometric complexity. The boundary of the snowflake is difficult to define in geometric terms. Indeed, it must have a very long perimeter, but it is a very small structure. The coastline of Italy is a difficult length to measure. Our geography books tell us how long it is, but different sources provide varying answers. The Nile River is very long indeed, and it appears to have a beginning and an end; yet, how is its length defined? All of these natural examples provide, with a little reflection, a crisis of definition. If we define a geometric measure as the determination of a quantifiable scalar such as length or area, then the geometric measures of physical characteristics of these examples are hard to establish. Indeed, the measures could only be approached on an operational level. That is, if one wants to measure the length of the perimeter of the snowflake, one would have to know by what means to measure it. The measurement of the coastline of Italy will mean one thing to someone in an airplane, but quite another thing to someone walking along the sea coast. The length of the Nile River is an indeterminate measurement which would, no matter how it was measured, depend on how many tributaries of the Nile one wished to measure. If no tributaries are considered part of the river, then its length depends on which channel is called the river. The measurement of the length of a perimeter of a snowflake or a country or the determination of the length of a river also depends on how close one is when the measurements are made. This is because these things have great complexity and display details nested within details. As one gets closer to a snowflake, the geometric complexity becomes apparent; closer yet and one can see more facets of the perimeter; closer yet and a whole new set of edges not previously apparent are seen. If one looks at a map of Italy, there is little doubt how long its coastline is. The little bays and coves seem well defined. With map in hand an airplane ride over the coastline at 30,000 feet would confirm general impressions obtained from the map, but new features such as smaller bays and coves would be apparent. In an airplane at 8000 feet, the geography so assuredly obvious from the map would be an oversimplification of what was actually seen. Any calculation of the length of the coastline would generate different lengths in each of these cases. If one then walked along the coast, whole new suites of features would be obvious, and the length calculated from such an exercise would be much greater. Imagine crawling along the sea shore with a small crab. At times it would be hard to know where the coastline was with our attention focused on the minute details surrounding us, and excursions around rocks or other obstacles along the coastline would create a great and unpredicted length,[1] not predicted on the basis of what was observed on the map.

While exploring the Nile River, the same progression of observed lengths would present themselves, and the same progression of increasing levels of detail in the geography would astonish us. The Nile River, the coastline of Italy, and the snowflake all belong to a class of objects known as *fractals*. Geometric measures cannot be known absolutely in a fractal; they can only be known at a given scale of observation. This attribute is known as *scaling*. If a geometric measure such as length or area of an object were to be determined at several scales of observation and the log of the measurement were to be plotted against the log of the scale of observation, a non-integer slope would be obtained in the case of the fractal objects. The slope of such a line relates to the fractal dimension and there most often is a fractional component to it. Further, one can keep the scale of observation unchanged and use different sized rulers, dividers, etc. as "measuring instruments". In this case of different sized "measuring instruments", the slope is not the fractal dimension, but the negative of the "excess" from the topological dimension. So, measurements of a curve of dimension 1.3 will have a slope of –0.3 in the log-log plots. If the same operation were performed on an empty circle, an empty square, or line, then the dimension of the object would equal the dimension of the space occupied. Hence, for a line which is a nonfractal shape in a plane, the log-log plot of length vs. ruler size would have a slope of 0. The dimensions, then, are $D = 1$ (slope for a line), 2 (slope for a shape), and 3 (slope for a volume). If the dimension of the object studied is different than the predicted topological dimension, the object is a fractal. On occasion, a fractal object occupying a higher space gives an integer value but of the lower order space. That is, a fractal volume (such in the case of the Sierpinski pyramid discussed below) with a predicted dimension of 3 might instead have a dimension of 2 and therefore be a fractal.

The determination of the perimeter length of a snowflake, the coastline of Italy, or the length of the Nile River cannot be achieved without knowing the scale of the observation at the beginning. Importantly, the measurement would be different at different scales of observation. This is not true of the circle or the rectangle or the line. If a snowflake were a pure Euclidean circle, if the coast of Italy formed a pure rectangle, or if the Nile River were a straight line, then the progression of detail with increasing scale of observation would not happen. At any distance from a circle or a rectangle or a line, one can predict absolutely the length or area at any other distance. No matter how close one gets to a circle one does not see more features, only an arc filling more and more of the field of view. The same is true of the rectangle and of the line. There are no more details to be observed by closer examination of these things, no matter how close one gets. The snowflake, the coast of Italy, and the Nile River are all examples of natural fractal objects. The circle, the rectangle, and the line are Euclidean objects fully characterized by classical geometry.

In each of these fractal objects, complexity of form leads to another fascinating property. At every scale of observation new details are revealed, yet these details are reminiscent of details elsewhere in the structure of the object or in the same part of the object, but at different scale. This property is known as *self-similarity* and implies that looking at one part of the object offers the same information as looking at another part of the object. Natural objects have an imprecise self-similarity sometimes called *statistical self-similarity*.

Everywhere one looks in nature, fractal characteristics are obvious: in mountains, in clouds, in the surf splashing on a fractal coastline. The commonness of this geometry implies that some fundamentally important natural attribute is at play in the geometry. How widespread is this geometry? Does it arise or is it always present in nongeographic natural settings? One of the goals of this book is to convince the reader by example that fractal geometry can arise and is widely present in nongeographic natural settings, specifically in biological settings. The forces that give rise to fractal geometry can be deduced, and the geometry may offer an important analytical approach to those forces.

In biological systems fractal geometry abounds, in many ways highly comparable to that seen in physical systems. The geometry of the snowflake can be seen in bacterial colonies, as described in Chapter 5; the coastlines can be seen in the mosaic pattern observed in chimeras made by amalgamating the genetically distinguishable embryos described in Chapter 7; the river bed can be observed in the neuronal outgrowths described in Chapter 6. The biological systems, likewise, can be treated in a quantitative fashion, and fractal dimension can be calculated in a number of ways to reveal the underlying fractal nature of the biological entities. What is the consequence of this geometry or what is the significance of observing it other than that it is there? In order to answer that question, we will consider some *un*natural fractals and begin to look at common themes in the way they are produced and see if those themes are resonant in the biological structures we are interested in.

In the early 20th century, Helge von Koch proposed the formation of a truly bizarre curve to be generated by repetitive application of simple rules to a straight line. Koch proposed to take a straight line and alter its middle third to form an equilateral triangle, making a bump in the line. The straight line has come to be known as an *initiator* and twisting the line to form a bump as a *generator*. Koch now argued that if this procedure, raising the middle third of each of the straight lines into a bump formed by an equilateral triangle, were repeated many, many times, then a highly complex curve would result. Koch's interest in the curve was that while everywhere continuous it was nowhere differentiable. The generator also made a line longer than the initiator; thus, the curve would quickly grow to very large lengths. The curve is also very hard to draw after only a few repetitions of the basic rule used to make it. Other continuous but nondifferentiable functions had been invented in the 19th century and were considered variously as curiosities or monstrosities, depending on the point of view of the mathematician discussing them.[2-4]

If, instead of beginning with a straight line, one begins with an equilateral triangle as the initiator and the middle third of each of the straight sides of the triangle are bumped up into a new equilateral triangle, a six-pointed star results after applying the generator. As the rule to bump up each middle third of each straight line in the structure is applied over and over, an object that looks like an idealized snowflake begins to emerge. As a polygon it is a strange figure indeed; its perimeter becomes infinitely long as the number of repetitions of the simple rule to make the generator increases. Yet, the perimeter encloses a defined area. In fact, as we mentioned before, the very act of measurement of the perimeter of this object cannot be done in terms usual in our experience. Length only has meaning at a given scale of observation of the Koch snowflake, and it cannot fully characterize the perimeter of the object. The mathematician Felix Hausdorff suggested a method that is useful. He argued that if the perimeter of a fractal could be measured by little sticks of known length, the number of little sticks necessary to cover the length of the perimeter times the length of the little sticks would not measure the perimeter, no matter how short the little stick was made. This was true because the Koch snowflake (and all fractal objects) had detail nested within detail and no matter how short the little stick was made it still failed to adequately cover the new edges of the snowflake. Instead, he said that the complexity of the object was more adequately described by a statistic now known as the *Hausdorff dimension*. If the number of little sticks necessary to cover the perimeters is N and the length of the little stick is ε, then the Hausdorff dimension D_H is given by the equation

$$\lim_{\varepsilon \to 0} \frac{\log N}{\log 1/\varepsilon}$$

We see in Chapter 11 that this is equal to 1.26 for the Koch snowflake.

The concept of dimension here is somewhat tricky. We all take for granted that normal topological dimensions are fixed as 1 for lines, 2 for surfaces or planes, and 3 for volumes. Yet these dimensions are rather arbitrary in their own regard;[1,5] if one makes a line twisty enough, it can fill a plane. Nevertheless, in the case of the Koch snowflake, the complex line has a dimension greater than 1 and less than 2. As fractal lines become more complex and begin to fill the plane that they occupy, their fractal dimension[6,7] tends toward 2, the topological dimension of a plane. If the same calculations described in the preceding paragraph were applied to a straight line, the fractal dimension would be 1. There are many definitions of dimensions applied to the description of fractal objects. These are sometimes all considered the same statistic when in fact they are not. It is important to have a clear idea of which dimension is being discussed and to be aware of the fact that different dimensions have different implications. This in turn means that an analytical approach to fractals needs to be consistent in its use of dimension. For the most part, if the term "fractal dimension" is used, it is meant to imply one of a group of possible statistics that can be applied to the characterization of the topology of an object. Some fractal dimensions can imply an extent of complexity in the object. Most dimensions used are variations of the Hausdorff dimension, and virtually all use a scale of observation with some form of topological measure (length, area, or volume). Chapter 8 provides an extensive discussion of dimensions as well as practical advice on their implementation.

If one takes an equilateral triangle with a unit side length and removes an upside down equilateral triangle with a side length 1/3 of the unit length from the center, one will have generated three connected equilateral triangles, each with a side length of 1/3 the unit length. If this procedure is repeated over and over with the remaining triangles a very strange figure called a *Sierpinski gasket* is formed. This structure has many odd mathematical features. The figure is everywhere discontinuous if points of contact at the vertices of the triangles are not considered. The figure at the limit of its iterations is all branch points, i.e., there are more than two points close to any given point along the edges of the Sierpinski gasket. The Sierpinski gasket has rotational symmetry: it can, for example, be rotated 120° and look just like the figure before rotation; its D_H is 1.58. When generalized to three dimensions, the structure becomes a very disconnected pyramid and in this case the D_H is 2, an integer value but less than the topological dimension of the volume, 3. This points out the fact that a fractal can have an integer dimension. Also, dimension may vary from the intuitive topologic dimension by being too large, such as the case of the Koch snowflake where the highly complex curve begins to fill a plane, or it can be too small, such as in the case of the Sierpinski gasket where the removal of sections of the solid triangle reduce the dimension to below that of the predicted value of 2 for a plane. A similar result is found by starting with a unit square, dividing it into nine smaller squares each with a side length of 1/3 the unit side length, and then removing the middle square. This produces a square with a square hole in the middle. If one then removes the middle square in the same way from the remaining solid squares, and repeats the operation over and over again, a very holey figure is obtained. This odd structure is called a *Sierpinski carpet,* and when it is generalized to three dimensions it creates the hard-to-draw Menger sponge. After an infinite number of cubes are removed as described, this sponge has no volume but an infinite surface![8]

If one takes a line of unit length and divides it into three equal parts, one has the initiator of another odd fractal. The generator of this structure is the removal of the middle third of the line. This is repeated over and over on the lines which remain. You can easily satisfy yourself that this leads quickly to the formation of a dust-like figure called *Cantor dust* or the *Cantor set*. This structure has a D_H of 0.63, and, although the set at its limit is infinitely large, it has 0 length.

In what ways do these mathematical curiosities reflect the real world? The following chapters will attempt to define a set of answers to that question with some precision. The systems described are all analytical and in most cases have provided insight or experimental approaches to long-standing problems. Here we will provide a number of concrete examples taken from the chapters to illustrate several characteristics of biological fractals and their use.

Scaling behavior can be seen in very fundamental biological structures. Even DNA demonstrates fractal properties in the distribution of sequence information. In Chapter 2, power–law correlations of DNA sequences are examined using important new approaches. Long-range correlations in sequences across thousands of base pairs in noncoding regions offer a practical approach to distinguishing coding and noncoding sequence. This analysis shows that noncoding DNA sequences have an increasing complexity with evolution. The noncoding correlations and increasing complexity may indicate that these sequences have important functions.

The study of homeostasis in the cornea, maintained by osmotic integrity provided by the corneal epithelium, reveals another form of biologically relevant fractal behavior. In Chapter 3, the role of ion channels in the corneal epithelium is explored. We see that ion channels are thought to be in either an open or closed conformation but that standard models fail to adequately explain experimental data from studies of the dynamic equilibrium between these states. The concept of molecular dynamics is introduced, and, in particular, time course analysis of conformational state change is shown to be self-similar over many scales of time of observation (as opposed to many scales of distance). So, in this example, we see fractal behavior of critical physiological function temporally. This approach provides a new way of interpreting distributions of open and closed times of ion channels which is more consistent with known properties of proteins than has previously been possible. Previously, classical Markov modeling required different descriptive equations to fit data at different time scales, implying at least three different conformational states of the ion channel. These states may not actually exist but might be the consequence of forcing data fits to unrealistic models. Chapter 3 demonstrates how scaling functions may unify the data from short closed states to long closed states in a model with physical significance. The data then can be fit using two adjustable parameters rather than the 12 required to fit the same data with classical models. Importantly, experimental data from ligand binding studies suggest that, using this approach, a ligand does not induce one specific conformational change in the channel but rather may add a missing piece to the channel machinery which allows the ion channel to open over many pathways corresponding to a range of time scales. This leads to a dynamical interpretation of energy barrier change between closed and open states in time. The results imply that a channel protein in a new conformational structure will continue to become more stable in that conformation over time, making it more difficult to escape into a new conformation. As in any useful new interpretation, the fractal approach suggests new testable hypotheses and suggests new experiments.

Proteins have been shown to have fractal surfaces; in fact, they are fat fractals, that is, an object with a fractal surface but a nonzero volume. This concept and the analysis of protein surfaces and enzyme kinetics are presented in Chapter 4. Data demonstrates that the complexity of the surface can be described with multifractal analysis. This implies that the detail of the surface may be "screened" from view by, say, an external probe. Further, this screened surface is a much greater portion of the surface than the unscreened part. The implication presented is that since so many biological processes imbued in proteins are started by surface contact with diffusing molecules, then the unscreened portions of the surface may be important regulatory domains, and their geometry is a legitimate object of study. Interestingly, it is possible to analyze the surface of the active sites in the same way, and often fractal dimensions of active sites are larger than those of the global surface. This increased complexity in the surface of the active sites undoubtedly influences the efficiency with which substrates are absorbed and reacted. Often substrates arrive at active sites in the molecule following diffusion across the surface of the enzyme. This diffusion process can be treated like Brownian motion or a random walk and will be affected by the complexity of the surface of the enzyme in a manner analogous to diffusion-limited aggregation (DLA). The efficiency of the diffusion is related to substrate concentrations which are related to the amount of substrate bound to the enzyme surface, but the relationship between substrate concentration and the amount bound is very different for fractal and nonfractal surfaces. This, in turn, will effect the efficiency with which the substrate arrives at the active site of the enzyme. The amount of substrate on the fractal surface is lower than on a nonfractal surface, there is a higher restriction of movement, and the perimeter is cleared more slowly. This results in a hindrance of further surface binding of substrate, and the fractal surface shows significantly higher substrate inhibition than the nonfractal surface. Additionally, the previous models of ligand binding based on the Hill equation and noninteger values of the Hill coefficient have been confusing. Fractal analysis leads to a new interpretation of the Hill coefficient. It is generally not constant for a given molecule but is shown in Chapter 4 to reflect the fractal properties of the protein or enzyme and not the number of binding sites on each molecule.

The behavior of the dimension can be taken to indicate a number of important aspects of a cellular system, as well. We will explore the morphology of bacterial colonies in Chapter 5. The dimension of this process may tell us something about the way bacteria grow. Similar behavior is displayed in mycelia morphology.[9] In the past, fermenters have had the problem of mechanical disadvantage when the organisms (such as yeast) grew in a filamentous manner. It is necessary to make adjustments for this form of growth as a consequence. It has been shown that the fractal dimension of the colonies in the fermenter are a predictor of changes toward filamentous growth, and this has been used to create a system which measures dimension and then makes adjustments in the fermenter automatically to account for alteration in growth pattern.

In Chapter 6 we explore the growth of neurite extensions in the developing neuron. These structures are fractals, and their dimensions can rise as growth proceeds, leading to new theories of the manner of their formation. Results presented in this chapter suggest two possible fractal growth mechanisms for neuronal growth: L-system-type growth and diffusion-limited aggregation. L-systems were introduced by Aristid Lindenmayer as a way to quantify branching morphogenesis and led to the discovery of numerous algorithms utilizing repetitive application of simple rules to create branched fractals.[10] Percolation theory gave rise to DLA theories, as equations were developed to explain how molecules (e.g., oil) percolated or diffused through porous fields (e.g., soil). It was noticed that DLA or L-systems were good models of growth as well. Such growth systems are shown to increase their fractal dimension as the number of iterations of the growth rules increase, i.e., as development proceeds the structure becomes more complex. Moreover, in the vertebrate central nervous system (CNS), studies have suggested that the degree of morphological complexity correlates with the number of individual functions of a neuron. Similar results are presented for glial cells, and, interestingly, the fractal dimensions for glial cells seem higher in "lower" species, implying higher levels of glial function in these species. In aggregate, the data presented add to the existing literature, which argue that fractal branched structures are optimal functional designs.

In embryonic development, organs form very rapidly and with high fidelity. After parenchyma mass develops, higher orders of organization, genetically determined, are induced and lead to functionality of the organ. This induction is thought to be the result, in part, of mesenchyme–parenchyma interactions involving structures such as blood vessels or extracellular matrix on the parenchyma cells of the organ. There needs to be a way in which the organ can produce the mass of parenchyma cells in a predetermined (genetic) manner that allows the higher levels of organization to occur. This process for most visceral organs is the product of cell proliferation, and in Chapter 7 we explore the way in which cell proliferation can be seen to be stereotypical for several visceral organs and that the repetitive application of simple cell division rules provides a way by which deterministic process of parenchyma proliferation could be represented in very simple genetic terms. Parenchyma proliferation appears at first to have no pattern but when observed properly fractal geometry is revealed. The amalgamation of genetically distinguishable embryos can result in an animal with organs that comprise mixtures of progeny cells from both of the distinguishable embryos. When examined microscopically, the mixtures are apparent as mosaic patterns. The patterns are the net result of cell movement, cell division, and cell death. The pattern was generated in computer simulations of cell divisions, leading to the conclusion that iterating cell division programs could generate self-similar geometrically complex and biologically realistic patterns which did not require cell movement.[11-13] The patterns in real animals were proven to be fractal, and in aggregate the results lead to a new interpretation of mosaic pattern in organ development.[14]

A persistent problem in histopathological diagnosis (that is, the recognition of a disease state by microscopic examination of tissue biopsies) is uniformity of diagnosis. Inter- and intra-observer variation can be high in many settings. The procedures involved are time consuming and related to observer experience. This type of problem is theoretically solvable by artificial intelligence with expert systems operating using image processing techniques. Fractal geometry in pathological states might provide both practical diagnostic additions to the tool box required to achieve artificial intelligence-based diagnosis and insight into some disease processes.[15,16] Chapter 8 looks in depth at neoplastic growth, both in terms of cell and nuclear morphology and tissue distortion with tumor shapes. Fractal dimension analysis is applied to the morphology, and results encourage the belief that fractal geometry will help in diagnosis. Two other histopathologic problems are explored in this chapter. First, the fractal geometry of herpes simplex virus-induced corneal ulcers suggests a natural course of progression of the disease. Second, fractals are applied to the problem of multifactoral analysis of periodontal disease. While classical models of this disease were progressive and linear, it is now known that there are intermittent bursts of disease progress followed by remission. This has now been successfully modeled, accounting for the many factors involved, using fractal analysis of time-scale data. The concept of multifractal analysis is introduced and its potential use in the analysis of disease states is discussed.

In Chapter 9 we see that in the cardiovascular system fractal geometry reveals itself in a number of interesting ways. The arterial and venous networks are self-similar and comprise fractal branching. Some cardiac muscle bundles and the His-Purkinje network are branched in a fractal manner. The heart valve connections to papillary muscles show a fractal spatial distribution. One fascinating aspect of fractal behavior in the heart is that it appears in both spatial and temporal analysis. Importantly, the electrical conducting system displays a behavior which reveals another fractal geometry: that of time scale self-similarity. By looking at interbeat interval times, a significant difference in long-range scaling between healthy and diseased states was discovered. There is more to the potential of this analytical technique than just diagnostic or prognostic resolution, as important as those are. The heart rate dynamics share important characteristics with other biological systems, including some described in this book (i.e., ion channel conformational state changes as in Chapter 3). As discussed in Chapter 9 this implies that fractal behavior may provide an organizing principle for processes which generate fluctuations of physiological significance, and it may help prevent the restriction of functional responsiveness.

Brain wave activity has been used for many years as an entry into the complex domain of brain function and higher order activity such as learning. Brain wave activity behaves as a fractal, and in Chapter 10 an analysis of time series data indicating brain activity is used to formulate new approaches to learning. An important measure of brain activity is the event-related potential, and a method for determining the changing fractal dimension of this waveform is presented. The data presented here show an ordering of the tasks measured and the magnitude of the fractal dimension. The results suggest that learning evokes nonlinear dynamical electrical activity and can be explained as a chaotic process, meaning that fluctuations in brain activity are likely deterministic and not the result of random system noise. In this decade of the brain, an analytical approach to cognitive chaos is welcome indeed. In the future these techniques may prove useful for cognitive disorders, either diagnostically or by developing an understanding of higher orders of misadventure as in Alzheimer's disease.

The analysis of fractal geometry in mathematical forms goes back to the last century, although the explicit description of these forms as fractal is credited to Mandelbrot and is fairly recent.[5] The past several decades have provided a rich literature at the interface of mathematics and physics which describes natural phenomenon in the same terms. This literature suggests practical approaches to the use of fractal geometry, has provided formal proofs, and has described new fractal sets that may be exploited by biologists as well as other natural scientists. In Chapter 11, T. Viscek, one of the foremost mathematicians in this area, provides an elegant mathematical view of the world of fractal geometry.

In conclusion, we have looked at a number of mathematical forms which remind us of structures or functions in important biological systems. The authors have provided a glimpse of the fractal nature of systems at the forward edge of this research. As the principal characteristic of fractals is their scaling properties, it seems fitting that we can see fractals at many biologically relevant scales — in molecules, in cells, in tissues, in organs, and in organisms. The value in this line of research lies in the ability to make dependable predictions, the ability to create new hypotheses about important processes which can be tested, and the ability to provide a view of unifying and fundamental forces operating in biological systems.

All of the fractals described mathematically share key characteristics. They have infinitely increasing detail, with detail hidden at one scale of observation revealed at higher scales of observation. They are self-similar, so looking at one part of the structure reveals aspects of other parts of the structure. Topological measures cannot adequately describe the geometry of the structure, and, in fact, cannot even be known except at a particular scale of observation. The details of structure near any given point are complex, involving branch points or discontinuities. Many curves cannot be integrated; they have no tangents at specific points. There are often counterintuitive mathematical outcomes related to the topology of the structures: for example, the Cantor set, with no length, or the Menger sponge, which has no volume. These characteristics are not shared by Euclidean objects such as straight lines, squares, or circles. Even though the structures (or the sets which define the structures) are highly complex, they are simple to define. Finally, they are produced by recursive procedures.

Finding apparent order in random structures, describing predictability in randomness, and realizing that an operationally random structure is still generated in a deterministic way are lessons learned from the analysis of fractal geometry. The surprise comes from the realization that these characteristics can be found in nature, including in biological systems. Of course, there cannot be true fractals in nature, only in mathematics, as a true fractal must scale to infinity. In natural fractals, there is always some lower limit of scale which can be obtained. Nevertheless, natural objects behave sufficiently like true fractals to make it worth examining how true fractals are generated and to then try to find a similar process in the natural setting. This approach may lead to important insights into at least some aspects of the manner of generation of a wide variety of structures. In the end we want to make the seemingly unpredictable, predictable; we want apparent randomness to be understandable; and we want what appears nondeterministic to have deterministic origins — for then mechanistic explanations of biological processes may be achieved through innovative hypotheses.

References

1. Mandelbrot, B. B., *The Fractal Geometry of Nature*, W.H. Freeman, New York: 1983.
2. Schroeder, M., *Fractals, Chaos, Power Laws*, W.H. Freeman, New York, 1991.
3. Falconer, K., *Fractal Geometry, Mathematical Foundations and Applications*, John Wiley & Sons, New York, 1990.
4. Series, C., In *Exploring Chaos*, 3rd ed., Hall, N. (Ed.), W.H. Norton, New York, 1993.
5. Mandelbrot, B. B., *Science*, 155, 636, 1967.
6. Feder, J., *Fractals*, 3rd ed., Plenum Press, New York, 1988.
7. Farmer, J. D., Ott, E., Yorke, J. A., *Physica Part D*, 7, 153, 1983.
8. Gleick, J., *Chaos*, Viking, New York, 1988.
9. Patankar, D. B., Liu, T.-C., Oolman, T., *Biotech. Bioeng.*, 42, 571, 1993.
10. Prusinkiewicz, P., Lindenmayer, A., Hanan, J. S., Fracchia, F. D., Fowler, D. R., De Boer, M. J. M., Mercer, L., *The Algorithmic Beauty of Plants*, Springer-Verlag, New York, 1990.
11. Iannaccone, P. M., *FASEB J.*, 4, 1508, 1990.
12. Iannaccone, P. M., Lindsay, J., Berkwits, L., Lescinsky, G., Defanti, T., Lunde, A., In *Biomedical Modeling and Simulation*, Eisenfeld, J., Witten, J., Levine, D. S. (Eds.), Elsevier, Amsterdam, 1992.
13. Ng, Y. K., Iannaccone, P. M., *Dev. Biol.*, 151, 419, 1992.
14. Khokha, M., Landini, G., Iannaccone, P. M., *Dev. Biol.*, 165, 545, 1994.
15. Losa, G. A., Baumann, G., Nonnenmacher, T. H. F., *Pathol Res. Pract.*, 188, 680, 1992.
16. Cross, S. S., Cotton, D. W. K., *J. Pathol.*, 166, 409, 1992.

Section II
Molecules

chapter two

Scale Invariant Features of Coding and Noncoding DNA Sequences

H. Eugene Stanley, Sergei V. Buldyrev, Ary L. Goldberger, Shlomo Havlin, Rosario N. Mantegna, Chung-Kang Peng, and Michael Simons

> *I coined* fractal *from the Latin adjective* fractus. *The corresponding Latin verb* frangere *means "to break", to create irregular fragments. It is therefore sensible — and how appropriate for our needs! — that, in addition to "fragmented" (as in fraction or refraction),* fractus *should also mean "irregular", both meanings being preserved in* fragment.
>
> B. B. Mandelbrot

2.1 Abstract

We present evidence supporting the idea that the DNA sequence in genes containing *noncoding* regions is correlated, and that the correlation is remarkably long range — indeed, base pairs thousands of base-pairs distant are correlated. We do not find such a long-range correlation in the *coding* regions of the gene. We resolve the problem of the "nonstationarity" feature of the sequence of base pairs by systematically applying many methods, including a new algorithm called *detrended fluctuation analysis* (DFA). We refute the claim of Voss that there is no difference in the statistical properties of coding and noncoding regions of DNA by systematically applying the DFA algorithm, as well as standard fast Fourier transform (FFT) analysis, to all eukaryotic DNA sequences in the entire GenBank database with more than 512 base pairs (33,301 coding and 29,453 noncoding sequences). We describe a simple model to account for the presence of long-range power-law correlations which is based upon a generalization of the classic Lévy walk. Finally, we adapt to DNA the Zipf approach to analyzing linguistic texts, and the Shannon approach to quantifying the "redundancy" of a linguistic text in terms of a measurable entropy function. We systematically compare coding and noncoding regions and find differences, the possible significance of which is a topic of current study.

2.2 Long-range power-law correlations

In recent years long-range power-law correlations have been discovered in a remarkably wide variety of systems. Such long-range power-law correlations are a physical fact that in turn gives rise to the increasingly appreciated "fractal geometry of nature".[1-12] So, if fractals are indeed so widespread, it makes sense to anticipate that long-range power-law correlations may be similarly widespread. Indeed, recognizing the ubiquity of long-range power-law correlations can help us in our efforts to understand nature, since as soon as we find power-law correlations we can quantify them with a critical exponent. Quantification of this kind of scaling behavior for apparently unrelated systems allows us to recognize similarities between different systems, leading to underlying unifications that might otherwise have gone unnoticed.

Traditionally, investigators in many fields characterize processes by assuming that correlations are negligible or that they decay exponentially, as $e^{-r/\xi}$, so that they become negligible for $r \gtrsim 10\xi$. However, there is one major exception: at the critical point, the exponential decay turns into a power law decay:[13]

$$C_r \sim (1/r)^{d-2+\eta}. \tag{1}$$

Many systems drive themselves spontaneously toward critical points.[1,14] One of the simplest models exhibiting such "self-organized criticality" is invasion percolation, a generic model that has recently found applicability to describing anomalous behavior of rough interfaces.

In the following sections we will attempt to summarize some recent findings[15-35] concerning the possibility that — under suitable conditions — the sequence of base pairs or "nucleotides" in DNA also displays power-law correlations. The underlying basis of such power-law correlations is not understood at present, but this discovery has intriguing implications for molecular evolution,[32] as well as potential practical applications for distinguishing coding and noncoding regions in long nucleotide chains.[34] It also may be related to the presence of a language in noncoding DNA.[36]

2.3 DNA

The role of genomic DNA sequences in coding for protein structure is well known.[37] The human genome contains information for approximately 100,000 different proteins, which define all inheritable features of an individual. The genomic sequence is likely the most sophisticated information database created by nature, through the entirely "random" dynamic process of evolution. Equally remarkable is the precise transformation of information (duplication, decoding, etc.) that occurs in a relatively short time interval.

The building blocks for coding this information are called *nucleotides*. Each nucleotide contains a phosphate group, a deoxyribose sugar moiety, and either a *purine* or a *pyrimidine base*. Two purines and two pyrimidines are found in DNA. The two purines are adenine (A) and guanine (G); the two pyrimidines are cytosine (C) and thymine (T). The nucleotides are linked, end to end, by chemical bonds from the phosphate group of one nucleotide to the deoxyribose sugar group of the adjacent nucleotide, forming a long polymer (*polynucleotide*) chain. The information content is encoded in the sequential order of the bases on this chain. Therefore, as far as the information content is concerned, a DNA sequence can be represented most simply as a symbolic sequence of four letters: A, C, G, and T.

In the genomes of higher eukaryotic organisms, only a small portion of the total genome length is used for protein coding (as low as 3% in the human genome). The segments of the chromosomal DNA that are spliced out during the formation of a mature mRNA are called *introns* (for intervening sequences). The coding sequences are called *exons* (for expressive sequences).

The role of introns and intergenomic sequences constituting large portions of the genome remains unknown. Furthermore, only a few quantitative methods are currently available for analyzing information that is possibly encrypted in the noncoding part of the genome.

2.4 The "DNA walk"

One interesting question that is natural for a statistical physicist is, "Does the sequence of the nucleotides A, C, G, and T behave like a one-dimensional ideal gas, where the fluctuations of density are uncorrelated, or do there exist long-range correlations in nucleotide content (as in the vicinity of a critical point) which would result in domains of all size with different nucleotide concentrations?" Actually, domains of various sizes were known for a long time, but their statistical properties were uncharacterized. A natural language to describe heterogeneous DNA structure is long-range correlation analysis, borrowed from the theory of critical phenomena.[13]

In order to study the scale-invariant, long-range correlations of a DNA sequence, we first introduce a graphical representation of DNA sequences, which we term a *fractal landscape* or *DNA walk*.[15] For the conventional one-dimensional random walk model,[38,39] a walker moves either "up" [$u(i) = +1$] or "down" [$u(i) = -1$] one unit length for each step i of the walk. For the case of an uncorrelated walk, the direction of each step is independent of the previous steps. For the case of a correlated random walk, the direction of each step depends on the history ("memory") of the walker.[40-42]

One definition of the DNA walk is that the walker steps "up" if a pyrimidine (C or T) occurs at position i along the DNA chain, while the walker steps "down" if a purine (A or G) occurs at position i. The question we asked was whether such a walk displays only short-range correlations (as in an n-step Markov chain) or long-range correlations (as in critical phenomena and other scale-free "fractal" phenomena). A different kind of DNA walk was suggested by Azbel.[43] There have also been attempts to map DNA sequences onto multidimensional DNA walks;[16,44] however, recent work[34] indicates that the original purine-pyrimidine rule provides the most robust results, probably due to the purine-pyrimidine chemical complementarity.

The DNA walk allows one to visualize directly the fluctuations of the purine-pyrimidine content in DNA sequences: positive slopes correspond to high concentration of pyrimidines, while negative slopes correspond to high concentration of purines. Visual observation of DNA walks suggests that the coding sequences and intron-containing noncoding sequences have quite different landscapes.

2.5 Correlations and fluctuations

An important statistical quantity characterizing any walk[38,39] is the root-mean-square fluctuation $F(\ell)$ about the average of the displacement of a quantity $\Delta y(\ell)$ defined by $\Delta y(\ell) \equiv y(\ell_0 + \ell) - y(\ell_0)$, where

$$y(\ell) \equiv \sum_{i=1}^{\ell} u(i) \ . \tag{2}$$

If there is no characteristic length (i.e., if the correlation is "infinite" in range), then fluctuations will be described by a power law:

$$F(\ell) \sim \ell^\alpha \qquad (3)$$

with $\alpha \neq 1/2$.

Figure 1a shows a typical example of a gene that contains a significant fraction of base pairs that do not code for amino acids. It is immediately apparent that the DNA walk has an extremely jagged contour which corresponds to long-range correlations.

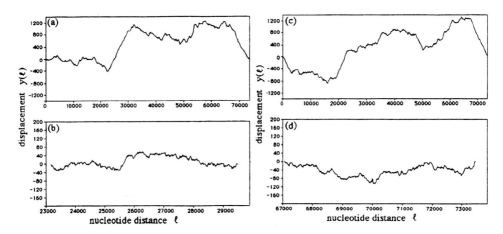

Figure 1 DNA walk displacement $y(\ell)$ (excess of purines over pyrimidines) vs. nucleotide distance ℓ for (a) HUMHBB (human beta globin chromosomal region of the total length $L = 73{,}239$); (b) the LINE-1c region of HUMHBB starting from 23,137 to 29,515; (c) the generalized Lévy walk model of length 73,326 with $\mu = 2.45$, $l_c = 10$, $\alpha_0 = 0.6$, and $\varepsilon = 0.2$; and (d) a segment of a Lévy walk of exactly the same length as the LINE-1c sequence from step 67,048 to the end of the sequence. This subsegment is a Markovian random walk. Note that in all cases the overall bias was subtracted from the graph such that the beginning and ending points have the same vertical displacement ($y = 0$). This was done to make the graphs clearer and does not affect the quantitative analysis of the data.

The fact that data for intron-containing and intergenic (i.e., noncoding) sequences are linear on this double logarithmic plot confirms that $F(\ell) \sim \ell^\alpha$. A least-squares fit produces a straight line with slope α substantially larger than the prediction for an uncorrelated walk, $\alpha = 1/2$, thus providing direct experimental evidence for the presence of long-range correlations.

On the other hand, the dependence of $F(\ell)$ for coding sequences is not linear on the log–log plot: its slope undergoes a crossover from 0.5 for small ℓ to 1 for large ℓ. However, if a single patch is analyzed separately, the log–log plot of $F(\ell)$ is again a straight line with the slope close to 0.5. This suggests that within a large patch the coding sequence is almost uncorrelated. The function $F(\ell)$ was also considered for DNA sequences by Azbel.[45]

2.6 Detrended fluctuation analysis

The initial report[15] on long-range (scale-invariant) correlations only in noncoding DNA sequences has generated contradictory responses. Some[16,18,20,46] support our initial finding, while some[17,22,25,27] disagree. However, the conclusions of Reference 18 and References 17, 22, 25, and 27 are inconsistent *with one another*, in that References 17 and 27 doubt the existence of long-range correlations (even in noncoding sequences) while Reference 18 and References 22 and 25 conclude that even coding regions display long-range correlations ($\alpha > 1/2$). Prabhu and Claverie[22] claim that their analysis of the putative coding regions of the yeast chromosome III produces a wide range of exponent values, some larger than 0.5.

The source of these contradicting claims may arise from the fact that, in addition to normal statistical fluctuations expected for analysis of rather short sequences, coding regions typically consist of only a few lengthy regions of alternating strand bias — and so we have nonstationarity. Hence, conventional scaling analyses cannot be applied reliably to the entire sequence but only to subsequences.

Peng et al.[33] recently have applied the "bridge method" to DNA and have also developed a powerful method specifically adapted to handle problems associated with nonstationary sequences which they term detrended fluctuation analysis (DFA).

The idea of the DFA method is to compute the dependence of the standard error $F_d(\ell)$ of a linear interpolation of a DNA walk on the size of the interpolation segment ℓ. The method takes into account differences in local nucleotide content and may be applied to the entire sequence which has lengthy patches. In contrast with the original $F(\ell)$ function, which has spurious crossovers even for ℓ much smaller than a typical patch size, the detrended function $F_d(\ell)$ shows linear behavior on the log–log plot for all length scales up to the characteristic patch size, which is of the order of a thousand nucleotides in the coding sequences. For ℓ close to the characteristic patch size the log–log plot of $F_d(\ell)$ has an abrupt change in its slope.

The DFA method clearly supports the difference between coding and noncoding sequences, showing that the coding sequences are less correlated than noncoding sequences for length scales less than 1000, which is close to characteristic patch size in the coding regions. One source of this difference in correlation properties is the tandem repeats (sequences such as AAAAAA…), which are more frequent in noncoding sequences but are absent in coding sequences.

2.7 Coding sequence finder algorithm

To provide an "unbiased" test of the thesis that noncoding regions possess, but coding regions lack, long-range correlations, Ossadnik et al.[34] analyzed several artificial uncorrelated and correlated "control sequences" of size 10^5 nucleotides using the GRAIL neural net algorithm.[47] The GRAIL algorithm identified about 60 putative exons in the uncorrelated sequences, but only about 5 putative exons in the correlated sequences.

Using the DFA method, we can also measure the local value of the correlation exponent α along the sequence (see Figure 2) and find that the local minima of α as a function of a nucleotide position usually correspond to noncoding regions, while the local maxima correspond to noncoding regions. Statistical analysis using the DFA technique of the nucleotide sequence data for yeast chromosome III (315,338 nucleotides) shows that the probability that the observed correspondence between the positions of minima and coding regions is due to random coincidence is less than 0.0014. Thus, this method — which we called the "coding sequence finder" (CSF) algorithm — can be used for finding coding regions in the newly sequenced DNA, a potentially important application of DNA walk analysis.

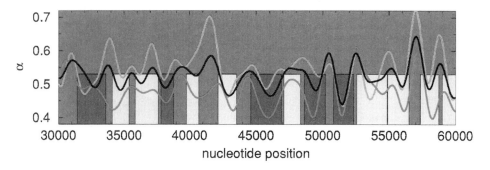

Figure 2 Analysis of section of yeast chromosome III using the sliding box coding sequence finder (CSF) algorithm. The value of the long-range correlation exponent α is shown as a function of position along the DNA chain. In this figure, the results for about 10% of the DNA are shown (from base pair number 30,000 to base pair 60,000). Shown as vertical bars are the putative genes and open reading frames; denoted by the letter "G" are those genes that have been more firmly identified (March 1993 version of GenBank). Note that the local value of α displays minima where genes are suspected, while between the genes α displays maxima. This behavior corresponds to the fact that the DNA sequence of genes lacks long-range correlations (α = 0.5 in the idealized limit), while the DNA sequence in between genes possesses long-range correlations (α ≈ 0.6). The different curves correspond to different "rules" for what constitutes an "up" step.

2.8 Systematic analysis of GenBank database

An open question in computational molecular biology is whether long-range correlations are present in both coding and noncoding DNA or only in the latter. To answer this question, Buldyrev et al.[35] recently analyzed all 33,301 coding and all 29,453 noncoding eukaryotic sequences — each of length larger than 512 base pairs (bp) — in the present release of the GenBank to determine whether there is any statistically significant distinction in their long-range correlation properties.

Buldyrev et al. find that standard fast Fourier transform (FFT) analysis indicates that coding sequences have practically no correlations in the range from 10 bp to 100 bp (spectral exponent β ± 2 SD = 0.00 ± 0.04). Here β is defined through the relation $S(f) \sim 1/f^\beta$, where $S(f)$ is the Fourier transform of the correlation function, and β is related to the long-range correlation exponent α by β = 2α − 1 so that α = 1/2 corresponds to β = 0 (white noise).

In contrast, for noncoding sequences, the average value of the spectral exponent β is positive (0.16 ± 0.05), which unambiguously shows the presence of long-range correlations. They also separately analyzed the 874 coding and 1157 noncoding sequences which have more than 4096 bp and found a larger region of power-law behavior. Buldyrev et al. calculated the probability that these two data sets (coding and noncoding) were drawn from the same distribution and found that it is less than 10^{-10}. Buldyrev et al. also obtained independent confirmation of these findings using the DFA method, which is designed to treat sequences with statistical heterogeneity such as the known mosaic structure ("patchiness") of DNA arising from nonstationarity of nucleotide concentration. The near-perfect agreement between the two independent analysis methods, FFT and DFA, increases the confidence in the reliability of the conclusion that long-range correlation properties of coding and noncoding sequences.

From a practical viewpoint, the statistically significant difference in long-range power-law correlations between coding and noncoding DNA regions that we observe supports the development of gene finding algorithms based on these distinct scaling properties. A recently reported algorithm of this kind[34] is especially useful in the analysis of DNA sequences with relatively long coding regions, such as those in yeast chromosome III.

Very recently Arneodo et al.[46] studied long-range correlation in DLA sequences using wavelet analysis. The wavelet transform can be made blind to "patchiness" of genomic sequences. They found the existence of long-range correlations in noncoding regimes and no long-range correlations in coding regimes, in excellent agreement with Buldyrev et al.[35]

Finally, we note that although the scaling exponents α and β have potential use in quantifying changes in genome complexity with evolution, the current GenBank database does not allow us to address the important question of whether unique values of these exponents can be assigned to different species or to related groups of organisms. At present, the GenBank data have been collected such that particular organisms tend to be represented more frequently than others. For example, about 80% of the sequences from birds are from *Gallus gallus* (the chicken) and about 2/3 of the insect sequences are from *Drosophila melanogaster* (fruit flies). The results indicate the importance of sequencing not only coding but also noncoding DNA from a wider variety of species.

2.9 Generalized Lévy walk model

Although the correlation is long range in the noncoding sequences, there seems to be a paradox: long *uncorrelated* regions of up to thousands of base pairs can be found in such sequences as well. For example, consider the human beta-globin intergenomic sequence of length $L = 73,326$ (GenBank name: HUMHBB). This long noncoding sequence has 50% purines (no overall strand bias) and $\alpha = 0.7$ (see Figure 1a). However, from nucleotide number 67,089 to 73,228, there occurs the LINE-1 region (defined in Reference 48). In this region of length 6139 base pairs, there is a strong strand bias with 59% purines. In this noncoding subregion, we find power-law scaling of F, with $F \sim l^\alpha$, with $\alpha = 0.55$, quite close to that of a random walk.

Even more striking is another region of 6378 base pairs, from nucleotide number 23,137 to 29,515, which has 59% pyrimidines and is uncorrelated, with remarkably good power-law scaling and correlation exponent $\alpha = 0.49$ (Figure 1b). This region actually consists of three subsequences complementary to shorter parts of the LINE-1 sequence.

These features motivated us to apply a generalized Lévy walk model (see Figures 1c, 1d, and 2) for the noncoding regions of DNA sequences.[30] We will show in the next section how this model can explain the long-range correlation properties, since there is no characteristic scale "built into" this generalized Lévy walk. In addition, the model simultaneously accounts for the observed large subregions of noncorrelated sequences within these noncoding DNA chains.

The classic Lévy walk model describes a wide variety of diverse phenomena that exhibit long-range correlations.[49-52] The model is defined schematically in Figure 3a: a random walker takes not one but l_1 steps in a given direction. Then the walker takes l_2 steps in a new randomly chosen direction, and so forth. The lengths l_j of each string are chosen from a probability distribution, with

$$P(l_j) \propto (1/l_j)^\mu \tag{4}$$

where $\sum_{i=1}^{N} l_i = L$, N is the number of substrings, and L is the total number of steps that the random walker takes.

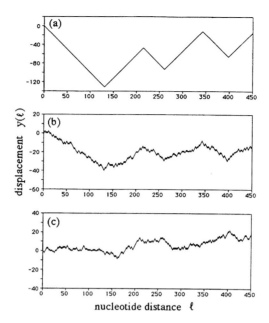

Figure 3 Displacement $y(\ell)$ vs. number of steps for (a) the classical Lévy walk model consisting of six strings of l_j steps, each taken in alternating directions; (b) the generalized Lévy walk model consisting of six biased random walks of the same length with a probability of p_+ that it will go up equal to $(1 \pm \varepsilon)/2$ [$\varepsilon = 0.2$]; and (c) the unbiased uncorrelated random walk. Note that the vertical scale in (b) and (c) is twice that in (a).

We consider a generalization of the Lévy walk[42] to interpret recent findings of long-range correlation in the noncoding DNA sequences described above. Instead of taking l_j steps in the *same* direction, as occurs in a classic Lévy walk, the walker takes each of l_j steps in *random* directions, with a fixed bias probability:

$$p_+ = (1 + \varepsilon_j)/2 \qquad (5)$$

to go up and

$$p_- = (1 - \varepsilon_j)/2 \qquad (6)$$

to go down, where ε_j gets the values $+\varepsilon$ or $-\varepsilon$ randomly. Here $0 \leq \varepsilon \leq 1$ is a bias parameter (the case $\varepsilon = 1$ reduces to the Lévy walk). Figure 3b shows such a generalized Lévy walk for the same choice of l_j as in Figure 3a.

As shown in Reference 30, the generalized Lévy walk — like the pure Lévy walk — gives rise to a landscape with a fluctuation exponent α that depends upon the Lévy walk parameter μ,[42,50]

$$\alpha = \begin{cases} 1 & \mu \leq 2 \\ 2 - \mu/2 & 2 < \mu < 3 \\ 1/2 & \mu \geq 3; \end{cases} \qquad (7)$$

i.e., non-trivial behavior of α corresponds to the case $2 < \mu < 3$ where the first moment of $P(l_j)$ converges while the second moment diverges. The long-range correlation property for the Lévy walk, in this case, is a consequence of the broad distribution of Equation (4) that lacks a characteristic length scale. However, for $\mu \geq 3$, the distribution of $P(l_i)$ decays fast enough that an effective characteristic length scale appears. Therefore, the resulting Lévy walk behaves like a normal random walk for $\mu \geq 3$.

2.10 Mosaic nature of DNA structure

The key finding of this analysis is that a generalized Lévy walk model can account for two hitherto unexplained features of DNA nucleotides: (1) the long-range power law correlations that extend over thousands of nucleotides in sequences containing noncoding regions (e.g., genes with introns and intergenomic sequences), and (2) the presence within these correlated sequences of sometimes large subregions that correspond to biased random walks. This apparent paradox is resolved by the generalized Lévy walk, a mechanism for generating long-range correlations (no characteristic length scale), that with finite (though rare) probability also generates large regions of uncorrelated strand bias. The uncorrelated subregions, therefore, are an anticipated feature of this mechanism for long-range correlations.

From a biological viewpoint, two questions immediately arise: (1) What is the significance of these uncorrelated sub-regions of strand bias? and (2) What is the molecular basis underlying the power-law statistics of the Lévy walk? With respect to the first question, we note that these long uncorrelated regions at least sometimes correspond to well described but poorly understood sequences termed "repetitive elements", such as the LINE-1 region noted above.[48,53] There are at least 53 different families of such repetitive elements within the human genome. The lengths of these repetitive elements vary from 10 to 10^4 nucleotides.[48] At least some of the repetitive elements are believed to be remnants of messenger RNA molecules that formerly did code for proteins.[53-55] Alternatively, these segments may represent retroviral sequences that have inserted themselves into the genome.[56] Our finding that these repetitive elements have the statistical properties of biased random walks (e.g., the same as that of active coding sequences) is consistent with both of these hypotheses.

Finally, what are the biological implications of this type of analysis? Our findings clearly support the following possible hypothesis concerning the molecular basis for the power-law distributions of elements within DNA chains. In order to be inserted into DNA, a macromolecule should form a loop of a certain length l with two ends, separated by l nucleotides along the sequence, coming close to each other in real space. The probability of finding a loop of length l inside a very long linear polymer scales is $l^{-\mu}$.[57,58] Theoretical estimates of μ made by different methods[58-61] using a self-avoiding random walk model[57] indicate that the value of μ for the three-dimensional model is between 2.16 and 2.42. Our estimate made by the Rosenbluth Monte Carlo Method[61] gave $\mu = 2.22 \pm 0.05$, which yields $\alpha = 0.89$, a larger value than the effective value observed in DNA of finite length. However, the asymptotic value of the exponent α remains uncertain since the statistics of Lévy walks converge very slowly due to rare events associated with the very long strings of constant bias that may occur in the sequence according to Equation (4).

In summary, it is clear that the behavior of DNA sequences cannot be satisfactorily explained in terms of only one characteristic length scale even of about 10^3 to 10^4 base pairs long. The asymptotic behavior of the scaling exponent α and whether it reaches some universal value for long DNA chains must await further data from the Human Genome Project.

2.11 Linguistic analysis of noncoding and coding DNA

Long-range correlations have been found recently in human writings.[62] A novel, a piece of music, or a computer program can be regarded as a one-dimensional string of symbols. These strings can be mapped to a one-dimensional random walk model similar to the DNA walk, allowing calculation of the correlation exponent α. Values of α between 0.6 and 0.9 were found for various texts.

An interesting hierarchical feature of languages was found in 1949 by Zipf.[63] He observed that the frequency of words f as a function of the word order ("rank") R decays as a power-law $f \sim R^{-\zeta}$ (with a power ζ close to 1) for more than four orders of magnitude.

In order to adapt the Zipf analysis to DNA, the concept of word must first be defined. In the case of coding regions, the words are the 64 3-tuples ("triplets") which code for the amino acids, AAA, AAT, ... GGG; however, for noncoding regions, the words are not known. Therefore, Mantegna et al.[36] consider the word length n as a free parameter and perform analyses not only for $n = 3$ but also for all values of n in the range 3 through 8. The different n-tuples are obtained for the DNA sequence by shifting progressively by one base a window of length n; hence, for a DNA sequence containing L base pairs, we obtain $L - n + 1$ different words.

The results of the Zipf analysis for all 40 DNA sequences analyzed are summarized in Reference 36. The averages for each category support the observation that ζ is consistently larger for the noncoding sequences, suggesting that the noncoding sequences bear more resemblance to a natural language than the coding sequences. Moreover, the "words" used in coding and noncoding sequences appear in quite different "orders" — i.e., the most popular words for coding are not the most popular words for noncoding (Figure 4).

Related interesting statistical measures of short-range correlations in languages are the entropy and redundancy. The redundancy is a manifestation of the flexibility of the underlying code. To quantitatively characterize the redundancy implicit in the DNA sequence, we utilize the approach of Shannon, who provided a mathematically precise definition of redundancy.[64,65] Shannon's redundancy is defined in terms of the entropy of a text, or, more precisely, the "n-entropy":

$$H(n) = -\sum_{i=1}^{4^n} p_i \log_2 p_i \tag{8}$$

which is the entropy when the text is viewed as a collection of n-tuple words. Here p_i is the normalized frequency of occurrence of n-tuple i. The redundancy is defined as $R \equiv \lim_{n \to \infty} R(n)$, where

$$R(n) \equiv 1 - H(n)/kn; \tag{9}$$

here $k = \log_2 4 = 2$. Mantegna et al.[36] also calculate the Shannon n-entropy $H(n)$ for $n = 1, 2, \ldots 6$.

For sufficiently high values of n (for example, $n = 4$), we found that the redundancy is consistently larger for the primarily noncoding sequences. Indeed, the redundancy of 6-tuples of real sequences (HUMRETBLAS) is 4.7%, about two times larger than the corresponding random case, 2.64 ± 0.03.

Very recently, we have extended the approach of Mantegna et al. to the entire GenBank. This extension requires care, since the entries in the GenBank are not chosen at random; since the sample is not random, one cannot draw conclusions

chapter two: Scale Invariant Features of Coding and Noncoding DNA Sequences

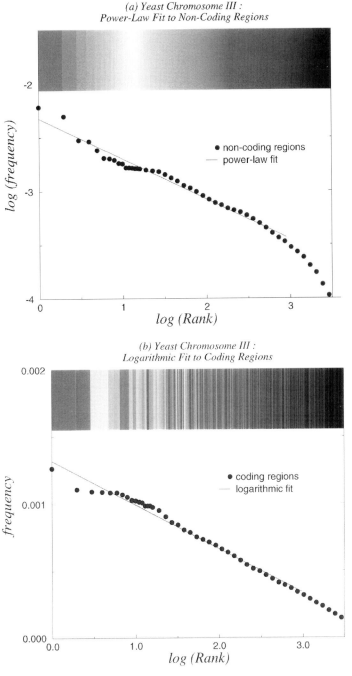

Figure 4 Linguistic features of noncoding DNA. (a) Log–log plot of a histogram of word frequency for the noncoding part of yeast chromosome III (≈315,000 bp). The six-character words are placed in rank order, where rank 1 corresponds to the most frequently used word, rank 2 to the second most frequently used word, and so forth. The straight line behavior provides evidence for a structured language in noncoding DNA. Rainbow color code corresponds to the rank of words in the language of this sequence, which is used as a "reference language" below. (b) Linear–log plot of word frequency histogram for the coding part of the same chromosome. The straight line behavior shows that the coding part lacks the statistical properties of a structured language. The colors are rearranged, corresponding to the rearrangements of their rank with respect to the reference language.

readily. Moreover, one may compare sequences *only* if they have approximately the same percent of GC content, since the redundancy function is roughly a parabola centered around 50% GC content. Preliminary results suggest that the results of Mantegna et al. generalize to the entire GenBank for the classes of plants and invertebrates, while for higher organisms the evidence is rather unclear at the moment. A full report is in preparation.[66]

Linearity of a Zipf plot may be indicative of hierarchical ordering. For example, it is possible that a wide range of systems result in straight-line behavior when subjected to Zipf analysis, and some understanding of the implications of Zipf analysis is now emerging.[67,68] An example that was the subject of some discussion is the remarkable linearity of the Zipf plot giving the annual sales of a company as a function of its sales rank. Bouchaud[69] finds that this plot is linear for European companies, while Stanley[70] finds linearity for American companies. Furthermore, Stanley et al.[71] find a significant deviation from this apparent linearity at rank ≈ 100 and relate this feature to the log-normal distribution of sales (the "Gibrat law").

2.12 Outlook for the future

There is a mounting body of evidence suggesting that the noncoding regions of DNA are rather special for at least two reasons:

1. They display long-range power-law correlations, as opposed to previously believed exponentially decaying correlations.
2. They display features common to hierarchically structured languages — specifically, a linear Zipf plot and a nonzero redundancy.

These results are consistent with the possibility that the noncoding regions of DNA are not merely "junk" but rather have a purpose. What that purpose could be is the subject of ongoing investigation. In particular, the apparent increase of α with evolution[32] could provide insight.

In the event that the purpose is not profound, our results nonetheless may have important practical value since quantifiable differences between coding and noncoding regions of DNA can be used to help distinguish the coding regions.[34]

The results of the systematic and inclusive analysis of GenBank DNA sequences are notable for two major reasons.

1. The GenBank data unambiguously demonstrate that noncoding DNA, but not coding DNA, possesses long-range correlations. This finding is made using two independent, complementary techniques: Fourier analysis and DFA, a modification of root-mean-square analysis of random walks. Indeed, as shown in Tables I and II of Reference 35, the spectral exponent β computed by both techniques for the same sequence is nearly identical.
2. The GenBank data demonstrate an increase in the complexity of the noncoding DNA sequences with evolution. The value of β for vertebrates is significantly greater than that for invertebrates. This finding based on the full GenBank data set supports the suggestion, based upon a systematic study of the myosin heavy gene family, that there is an apparent increase in the complexity of noncoding DNA for more highly evolved species compared to less evolved ones.[32]

Both of these results contradict the report of Voss,[18] who failed to observe any difference in the long-range correlation properties of coding and noncoding DNA and who reported a decrease in the value of the spectral exponent β with evolution.

The ultimate meaning of long-range correlations is still not clear. It is possible that long-range correlations also exist in other systems of biological interest. For example, the idea of long-range correlations has been extended to the analysis of the beat-to-beat intervals in the normal and diseased heart,[72,73] and to the human gait.[74] The healthy heartbeat is generally thought to be regulated according to the classical "equilibrium" principle of homeostasis whereby physiologic systems operate to reduce variability and achieve an equilibrium-like state.[75] We find, however, that under normal conditions, beat-to-beat fluctuations in heart rate display the kind of long-range correlations typically exhibited by physical dynamical systems "far from equilibrium", such as those near a critical point. Specifically, we find evidence for such power-law correlations that extend over thousands of heartbeats in healthy subjects. In contrast, heart-rate time series from patients with severe congestive heart failure show a breakdown of this long-range correlation behavior, with the emergence of a characteristic short-range time scale (see also Chapter 9). Similar alterations in correlation behavior may be important in modeling the transition from health to disease in a wide variety of pathologic conditions.

Acknowledgments

We are grateful to many individuals, including Matsa, M. E., Ossadnik, S. M., and Sciortino, F., for major contributions to those results reviewed here which represent collaborative research efforts. We also wish to thank Cantor, C., DeLisi, C., Frank-Kamenetskii, M., Grosberg, A. Yu., Huber, G., Labat, I., Liebovitch, L., Michaels, G. S., Munson, P., Nossal, R., Nussinov, R., Rosenberg, R. D., Schwartz, J. J., Schwartz, M., Shakhnovich, E. I., Shlesinger, M. F., Shworak, N., and Trifonov, E. N., for valuable discussions. Partial support was provided by the National Science Foundation, National Institutes of Health (Human Genome Project), the Harold, G., and Leila, Y., Mathers Charitable Foundation, the National Heart, Lung and Blood Institute, the National Aeronautics and Space Administration, the Israel–USA Binational Science Foundation, Israel Academy of Sciences, and by an NIH/NIMH postdoctoral NRSA Fellowship (Peng, C.-K.).

References

1. Bunde, A., Havlin, S., Eds., *Fractals and Disordered Systems* (Springer-Verlag, Berlin, 1991); Bunde, A., Havlin, S., Eds., *Fractals in Science* (Springer-Verlag, Berlin, 1994).
2. Vicsek, T., Shlesinger, M., and Matsushita, M., Eds., *Fractals in Natural Sciences* (World Scientific, Singapore, 1994).
3. Garcia-Ruiz, J. M., Louis, E., Meakin, P., and Sander, L., Eds., *Growth Patterns in Physical Sciences and Biology* (Proc. 1991 NATO Advanced Research Workshop, Granada, Spain, October 1991), (Plenum Press, New York, 1993).
4. Grosberg, A. Yu., Khokhlov, A. R., *Statistical Physics of Macromolecules* (translated by Atanov, Y. A.), (AIP Press, New York, 1994).
5. Bassingthwaighte, J. B., Liebovitch, L. S., and West, B. J., *Fractal Physiology* (Oxford University Press, New York, 1994).
6. Barabási, A.-L., Stanley, H. E., *Fractal Concepts in Surface Growth* (Cambridge University Press, London, 1995).
7. West, B. J., Goldberger, A. L., *J. Appl. Physiol.*, 60, 189 (1986); West, B. J., Goldberger, A. L., *Am. Sci.*, 75, 354 (1987); Goldberger, A. L., West, B. J., *Yale J. Biol. Med.*, 60, 421 (1987); Goldberger, A. L., Rigney, D. R., *Sci. Am.*, 262, 42 (1990); West, B. J., Shlesinger, M. F., *Am. Sci.*, 78, 40 (1990); West, B. J., *Fractal Physiology and Chaos in Medicine* (World Scientific, Singapore, 1990); West, B. J., Deering, W., *Phys. Rep.*, 246, 1 (1994); Buldyrev, S. V., Goldberger, A. L., Havlin, S., Peng, C.-K., and Stanley, H. E., in *Fractals in Science*, Bunde, A., Havlin, S., Eds., (Springer-Verlag, Berlin, 1994), pp. 49–83.

8. Vicsek, T., *Fractal Growth Phenomena*, 2nd ed. (World Scientific, Singapore, 1992).
9. Feder, J., *Fractals* (Plenum Press, New York, 1988).
10. Stauffer, D., Stanley, H. E., *From Newton to Mandelbrot: A Primer in Theoretical Physics*, 2nd ed. (Springer-Verlag, Heidelberg, 1995).
11. Guyon, E., Stanley, H. E., *Les Formes Fractales* (Palais de la Découverte, Paris, 1991), (English translation, *Fractal Forms*), (Elsevier/North-Holland, Amsterdam, 1991).
12. Stanley, H. E., Ostrowsky, N., Eds., *Random Fluctuations and Pattern Growth: Experiments and Models*, Proc. 1988 Cargèse NATO ASI (Kluwer Academic, Dordrecht, 1988).
13. Stanley, H. E., *Introduction to Phase Transitions and Critical Phenomena* (Oxford University Press, London, 1971).
14. Stanley, H. E., Ostrowsky, N., Eds., *Correlations and Connectivity: Geometric Aspects of Physics, Chemistry and Biology*, Proc. 1990 Cargèse NATO ASI, Series E, Applied Sciences (Kluwer Academic, Dordrecht, 1990).
15. Peng, C.-K., Buldyrev, S. V., Goldberger, A. L., Havlin, S., Sciortino, F., Simons, M., and Stanley, H. E., *Nature*, 356, 168 (1992).
16. Li, W., Kaneko, K., *Europhys. Lett.*, 17, 655 (1992).
17. Nee, S., *Nature*, 357, 450 (1992).
18. Voss, R., *Phys. Rev. Lett.*, 68, 3805 (1992); Voss, R., *Fractals*, 2, 1 (1994).
19. Maddox, J., *Nature*, 358, 103 (1992).
20. Munson, P. J., Taylor, R. C., and Michaels, G. S., *Nature*, 360, 636 (1992).
21. Amato, I., *Science*, 257, 747 (1992).
22. Prabhu, V. V. and Claverie, J.-M., *Nature*, 357, 782 (1992).
23. Yam, P., *Sci. Am.*, 267(3), 23 (1992).
24. Peng, C.-K., Buldyrev, S. V., Goldberger, A. L., Havlin, S., Sciortino, F., Simons, M., and Stanley, H. E., *Physica A.*, 191, 25 (1992); Stanley, H. E., Buldyrev, S. V., Goldberger, A. L., Hausdorff, J. M., Havlin, S., Mietus, J., Peng, C.-K., Sciortino, F., and Simons, M., *Physica A.*, 191, 1 (1992); Stanley, H. E., Buldyrev, S. V., Goldberger, A. L., Havlin, S., Mantegna, R. N., Ossadnik, S. M., Peng, C.-K., Sciortino, F., and Simons, M., Fractals in biology and medicine, in *Diffusion Processes, Experiment, Theory, Simulations (Proc. Fifth M. Born Symposium)*, Pekalski, A., Ed., (Springer-Verlag, Berlin, 1994), pp. 147–178; Stanley, H. E., Buldyrev, S. V., Goldberger, A. L., Goldberger, Z. D., Havlin, S., Mantegna, R. N., Ossadnik, S. M., Peng, C.-K., and Simons, M., Statistical mechanics in biology: how ubiquitous are long-range correlations?, Proc. International Conference on Statistical Mechanics, *Physica A.*, 205, 214 (1994).
25. Chatzidimitriou-Dreismann, C. A., Larhammar, D., *Nature*, 361, 212 (1993); Larhammar, D., Chatzidimitriou-Dreismann, C. A., *Nucleic Acids Res.*, 21, 5167 (1993); Chatzidimitriou-Dreismann, C. A., Streffer, R. M. F., and Larhammar, D., *Biochim. Biophys. Acta*, 1217, 181 (1994); Chatzidimitriou-Dreismann, C. A., Streffer, R. M. F., and Larhammar, D., *Eur. J. Biochem.*, 224, 365 (1994).
26. Grosberg, A. Yu., Rabin, Y., Havlin, S., and Neer, A., *Europhys. Lett.*, 23, 373 (1993).
27. Karlin, S., Brendel, V., *Science*, 259, 677 (1993).
28. Peng, C.-K., Buldyrev, S. V., Goldberger, A. L., Havlin, S., Simons, M., and Stanley, H. E., *Phys. Rev. E*, 47, 3730 (1993).
29. Shnerb, N., Eisenberg, E., *Phys. Rev. E*, 49, R1005 (1994).
30. Buldyrev, S. V., Goldberger, A. L., Havlin, S., Peng, C.-K., Simons, M., and Stanley, H. E., *Phys. Rev. E*, 47, 4514 (1993).
31. Borovik, A. S., Grosberg, A. Yu., and Frank-Kamenetski, M. D., *J. Biomolec. Struct. Dynamics*, 12, 655 (1994); Pande, V., Grosberg, A. Yu., and Tanaka, T., *Proc. Natl. Acad. Sci. U.S.A.*, 91, 12972 (1994).
32. Buldyrev, S. V., Goldberger, A. L., Havlin, S., Peng, C.-K., Stanley, H. E., Stanley, M. H. R., and Simons, M., *Biophys. J.*, 65, 2673 (1993).
33. Peng, C.-K., Buldyrev, S. V., Havlin, S., Simons, M., Stanley, H. E., and Goldberger, A. L., *Phys. Rev. E*, 49, 1685 (1994).

34. Ossadnik, S. M., Buldyrev, S. V., Goldberger, A. L., Havlin, S., Mantegna, R. N., Peng, C.-K., Simons, M., and Stanley, H. E., *Biophys. J.*, 67, 64 (1994); Stanley, H. E., Buldyrex S. V., Goldberger, A. L., Havlin, S., Peng, C.-K., and Simons, M., Proc. Intl. Conf. on Condensed Matter Physics, Bar-Ilan, *Physica A*, 200, 4 (1993); Stanley, H. E., Buldyrev, S. V., Goldberger, A. L., Havlin, S., Ossadnik, S. M., Peng, C.-K., and Simons, M., *Fractals*, 1, 283–301 (1993); Havlin, S., Buldyrev, S. V., Goldberger, A. L., Mantegna, R. N., Ossadnik, S. M., Peng, C.-K., Simons, M., and Stanley, H. E., *Chaos, Solitons, Fractals*, 6, 171 (1995).
35. Buldyrev, S. V., Goldberger, A. L., Havlin, S., Mantegna, R. N., Matsa, M. E., Peng, C.-K., Simons, M., and Stanley, H. E., Long-range correlation properties of coding and noncoding DNA sequences, *Phys. Rev. E*, 51, 5084 (1995).
36. Mantegna, R. N., Buldyrev, S. V., Goldberger, A. L., Havlin, S., Peng, C.-K., Simons, M., and Stanley, H. E., *Phys. Rev. Lett.*, 73, 3169 (1994); Flam, F., *Science*, 266, 1320 (1994); Pennisi, E., *Sci. News*, 146, 391 (1994); Yam, P., *Sci. Am.*, 272 (3), 24 (1995).
37. Tavaré, S., Giddings, B. W., in *Mathematical Methods for DNA Sequences*, Waterman, M. S., Ed. (CRC Press, Boca Raton, FL, 1989), pp. 117–132; Watson, J. D., Gilman, M., Witkowski, J., and Zoller, M., *Recombinant DNA* (Scientific American Books, New York, 1992).
38. Montroll, E. W., Shlesinger, M. F., The wonderful world of random walks, in *Nonequilibrium Phenomena II. From Stochastics to Hydrodynamics*, Lebowitz, J. L., Montroll, E. W., Eds., (North-Holland, Amsterdam, 1984), pp. 1–121.
39. Weiss, G. H., *Random Walks* (North-Holland, Amsterdam, 1994).
40. Havlin, S., Selinger, R., Schwartz, M., Stanley, H. E., and Bunde, A., *Phys. Rev. Lett.*, 61, 1438 (1988); Havlin, S., Schwartz, M., Blumberg-Selinger, R., Bunde, A., and Stanley, H. E., *Phys. Rev. A*, 40, 1717 (1989); Selinger, R. B., Havlin, S., Leyvraz, F., Schwartz, M., and Stanley, H. E., *Phys. Rev. A*, 40, 6755 (1989).
41. Peng, C.-K., Havlin, M., Schwartz, Stanley, H. E., and Weiss, G. H., *Physica A*, 178, 401 (1991); Peng, C.-K., Havlin, S., Schwartz, M., and Stanley, H. E., *Phys. Rev. A*, 44, 2239 (1991).
42. Araujo, M., Havlin, S., Weiss, H. E., and Stanley, H. E., *Phys. Rev. A*, 43 5207 (1991); Havlin, S., Buldyrev, S. V., and Stanley, H. E., Weiss, H. E., *J. Phys. A*, 24, L925 (1991); Prakash, S., Havlin, S., Schwartz, M., and Stanley, H. E., *Phys. Rev. A*, 46, R1724 (1992).
43. Azbel, M. Y., *Phys. Rev. Lett.*, 31, 589 (1973).
44. Berthelsen, C. L., Glazier, J. A., and Skolnick, M. H., *Phys. Rev. A*, 45, 8902 (1992).
45. Azbel, M. Y., *Biopolymers*, 21, 1687 (1982).
46. Arneodo, A., Bacry, E., Graves, P. V., and Mugy, J. F., *Phys. Rev. Lett.*, 74, 3293 (1995).
47. Uberbacher, E. C., Mural, R. J., *Proc. Natl. Acad. Sci. U.S.A.*, 88, 11261 (1991).
48. Jurka, J., Walichiewicz, T., and Milosavljevic, A., *J. Mol. Evol.*, 35, 286 (1992).
49. Shlesinger, M. F., Klafter, J., in *On Growth and Form: Fractal and Non-Fractal Patterns in Physics*, Stanley, H. E., Ostrowsky, N., Eds., (Martinus Nijhoff, Dordrecht, 1986), p. 279ff.
50. Shlesinger, M. F., Klafter, J., and Wong, Y. M., *J. Stat. Phys.*, 27, 499 (1982).
51. Shlesinger, M. F., Klafter, J., *Phys. Rev. Lett.*, 54, 2551 (1985).
52. Mantegna, R. N., *Physica A*, 179, 232 (1991).
53. Jurka, J., *J. Mol. Evol.*, 29, 496 (1989).
54. Hwu, R. H., Roberts, J. W., Davidson, E. H., and Britten, R. J., *Proc. Natl. Acad. Sci. U.S.A.*, 83, 3875 (1986).
55. Zuckerkandl, E., Latter, G., and Jurka, J., *J. Mol. Evol.*, 29, 504 (1989).
56. Levin, B., *Genes IV* (Oxford University Press, Oxford, 1990).
57. de Gennes, P.-G., *Scaling Concepts in Polymer Physics* (Cornell University Press, Ithaca, NY, 1979).
58. de Cloiseaux, J., *J. Physique (Paris)*, 41, 223 (1980).
59. Redner, S., *J. Phys. A*, 13, 3525 (1980).
60. Baumgartner, A., *Z. Phys. B*, 42, 265 (1981).
61. Birshtein, T. M., Buldyrev, S. V., *Polymer*, 32, 3387 (1991).

62. Schenkel, A., Zhang, J., and Zhang, Y-C., *Fractals*, 1, 47 (1993); Amit, M., Shmerler, Y., Eisenberg, E., Abraham, M., and Shnerb, N., *Fractals*, 2, 7 (1994); Ebeling, W., Neiman, A., *Physica A*, 215, 233 (1995).
63. Zipf, G. K., *Human Behavior and the Principle of "Least Effort"* (Addison-Wesley, New York, 1949).
64. Brillouin, L., *Science and Information Theory* (Academic Press, New York, 1956).
65. Shannon, C. E., *Bell Systems Tech. J.*, 80, 50 (1951); Herzel, H., Schmitt, A. O., Ebeling, W., *Chaos, Solitons, Fractals*, 4, 97 (1994); Herzel, H., Ebeling, W., and Schmitt, A. O., *Phys. Rev. E*, 50, 5061 (1994); Herzel, H., Große, I., *Physica A*, 216, 518 (1995).
66. Mantegna, R. N., Buldyrev, S. V., Goldberger, A. L., Havlin, S., Peng, C.-K., Simons, M., and Stanley, H. E., Systematic analysis of coding and noncoding DNA sequences using methods of statistical linguistics, *Phys. Rev. E*, 52, 2939 (1995); Buldyrev, S. V., Goldberger, A. L., Havlin, S., Mantegna, R. N., Peng, C.-K., Simons, M., and Stanley, H. E., *Phys. Rev. E* (submitted).
67. Czirók, A., Mantegna, R. N., Havlin, S., and Stanley, H. E., Correlations in binary sequences and generalized Zipf analysis, *Phys. Rev. E*, 52, 446 (1995).
68. Havlin, S., Distance between Zipf plots, *Physica A*, 216, 148 (1995).
69. Bouchaud, J.-P., More Lévy distributions in physics, in *Proc. 1993 Intl. Conf. on Lévy Flights*, Shlesinger, M. F., Zaslavsky, G., and Frisch, U., Eds. (Springer-Verlag, Berlin, 1995).
70. Stanley, M. H. R., 1994 Westinghouse Report (unpublished); Stanley, H. E., Buldyrev, S. V., Goldberger, A. L., Havlin, S., Mantegna, R. N., Peng, C.-K., Simons, M., and Stanley, M. H. R., Long-range correlatins and generalized Lévy walks in DNA sequences, in *Proc. 1993 Intl. Conf. on Lévy Flights*, Shlesinger, M. F., Zaslavsky, G., and Frisch, U., Eds. (Springer-Verlag, Berlin, 1995).
71. Stanley, M. H. R., Buldyrev, S. V., Havlin, S., Mantegna, R., Salinger, M. A., and Stanley, H. E., Zipf plots and the size distribution of firms, *Econ. Lett.*, 49, 453 (1995). See also Mantegna, R. N., Stanley, H. E., Ultra-slow convergence to a Gaussian: the truncated Lévy flight, in *Proc. 1993 Intl. Conf. on Lévy Flights*, Shlesinger, M. F., Zaslavsky, G., and Frisch, U., Eds. (Springer-Verlag, Berlin, 1995); Mantegna, R. N. and Stanley, H. E., Scaling behaviour in the dynamics of an economic index, *Nature*, 376, 46–49 (1995); Stanley, M. H. R., Amaral, L. A. N., Buldyrev, S. V., Havlin, S., Leschhorn, H., Maass, P., Salinger, M. A., and Stanley, H. E., Scaling behaviour in the growth of companies, *Nature*, 379, 804–806 (1996); Stanley, H. E., Afanasyeve, V., Amaral, L. A. N., Buldyrev, S. V., Goldberger, A. L., Havlin, S., Leschhorn, H., Maass, P., Mantegna, R. N., Peng, C.-K., Prince, P. A., Salinger, M. A., Stanley, M. H. R., and Viswanathan, G. M., Anomalous fluctuations in the dynamics of complex systems: from DNA and physiology to econophysics (Proc. Intl. Conf. Dynamics of Complex Systems), *Physica A*, 224, 302–321 (1996); Mantegna, R. N. and Stanley, H. E., Stock market dynamics and turbulence: parallels to quantitative measures of fluctuation phenomena, *Nature*, (submitted).
72. Peng, C.-K., Mietus, J., Hausdorff, J., Havlin, S., Stanley, H. E., and Goldberger, A. L., *Phys. Rev. Lett.*, 70, 1343 (1993); Peng, C.-K., Buldyrev, S. V., Hausdorff, J. M., Havlin, S., Mietus, J. E., Simons, M., Stanley, H. E., and Goldberger, A. L., in *Fractals in Biology and Medicine*, Losa, G. A., Nonnenmacher, T. F., Weibel, E. R., Eds. (Birkhauser-Verlag, Boston, 1994). See also Viswanathan, G. M., Afanasyev, V., Buldyrev, S. V., Murphy, E. J., Prince, P. A., and Stanley, H. E., Lévy flight search patterns in animal behavior, *Nature*, 381 (23 May 1996).
73. Peng, C.-K., Havlin, S., Stanley, H. E., Goldberger, A. L., Quantification of scaling exponents and crossover phenomena in nonstationary heartbeat time series (*Proc. NATO Dynamical Disease Conference*, Glass, L., Ed.,) *Chaos*, 5, 82 (1995); Peng, C.-K., Hausdorff, J. M., Mietus, J. E., Havlin, S., Stanley, H. E., and Goldberger, A. L., Fractals in physiological control: from heartbeat to gait, in *Proc. 1993 Intl. Conf. on Lévy Flights*, Shlesinger, M. F., Zaslavsky, G., Frisch, U., Eds. (Springer-Verlag, Berlin, 1995).
74. Hausdorff, J. M., Peng, C.-K., Ladin, Z., Wei, J. Y., and Goldberger, A. L., *J. Appl. Physiol.*, 78, 349 (1995).
75. Cannon, W. B., *Physiol. Rev.*, 9, 399 (1929).

chapter three

Ion Channel Kinetics

Larry S. Liebovitch

> *We [scientists] had built up our concepts (and prejudices) about the characteristic behavior of natural systems by observing artificial systems, systems which were chosen precisely for their regularity!*
>
> **H.-O. Peitgen and P. H. Richter**

3.1 Introduction

We study the cornea, the clear part in the front of the eye. The fluid from the inside of the eye, called the aqueous humor, leaks into the cornea, disrupting the regularity of its structure so that it is no longer transparent. A layer of cells, called the corneal endothelium, lines the inside of the cornea. These cells transport ions such as sodium and bicarbonate. This transport creates local differences in concentration that osmotically forces the fluid out of the cornea, so that it remains dry enough to be transparent. Over the last 15 years we have experimentally measured and theoretically modeled the flow of ions and fluid across the corneal endothelium.[1-18]

One step in the transport of ions through the corneal endothelial cells involves ion channels in the cell membrane; ion channels of different types are ubiquitous in cells.[19,20] In nerve, they generate and propagate the action potential. In muscle, they control calcium levels that initiate contraction. They are involved in signal transduction, such as transforming changes in light intensity to changes in electrical current in the photoreceptors in the retina. They are also involved in the control of concentration in intracellular organelles such as mitochondria.

The lipid cell membrane blocks the passage of ions such as sodium, potassium, and chloride into or out of the cell. Ion channels are large proteins shaped like a pipe with a central hole that span the cell membrane.[21,22] Ions can cross through this hole and enter or exit the cell. The channel protein structure can change so that this hole is squeezed shut. Channels switch between structures that have this hole (and thus are open to flow of ions) and structures that do not have this hole (and thus are closed to the flow of ions). The energy for these conformational switches comes from the binding of ligands, the voltage across the cell membrane, or random thermal fluctuations. In this chapter we will concentrate on the switches caused by random thermal fluctuations.

The switches in channel structure between the open and closed states can be detected by the patch clamp technique, for which Neher and Sakmann were awarded the Nobel prize in physiology or medicine in 1991.[19,20] In this technique, we take a glass micropipette, push it down onto a cell, and suck up a small piece of cell membrane. The patch of membrane seals within the mouth of the micropipette. The patch typically contains a small number of channels, sometimes only one channel. We can even pull the patch off the cell. We put an electrochemical gradient across the patch and record the current through it as a function of time. Figure 1 shows such a current record that we made through a patch containing a potassium channel from a corneal endothelial cell.[7] No current was recorded when the channel was closed, and a small 5-picoamp current was recorded when the channel was open. From the level of the current, we can see that the channel was first closed, then open, then closed, then open again, etc. That is, we can watch a single ion channel protein molecule spontaneously switch between its open and closed conformational structures. Such experiments provide a unique opportunity to study the behavior of an individual biological molecule. In other biophysical or biochemical experiments, the experimentally measured quantity is the sum of a very large number of molecules, each of which may be in a different state. In the patch clamp we measure the sequence of times that one ion channel molecule spends in the open and closed configurations. The study of the timing of the switching from one conformational state to another is called *ion channel kinetics*.

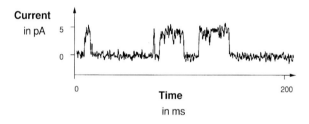

Figure 1 Patch clamp recording of the current through a potassium channel in the cell membrane of a corneal endothelial cell.[7] The current switches between low and high values as the ion channel protein switches between conformational structures that are closed or open to the flow of ions.

It is now possible to inject mRNA of a channel into a frog oocyte which will express the channel protein in its cell membrane, where its kinetics can be measured by the patch clamp technique.[23,24] Site-directed mutagenesis can be used to replace specific amino acid residues. Thus, the effects on the kinetics of the electric charge or the hydrophobicity or the size of a given residue can be determined. These are called "structure-function" studies. Reading such articles one might think that we have a pretty good idea about how ion channels work. In fact, the basic physical properties of ion channels are *not* known.

For example, does the ion channel protein switch between only a few or a very large number of different conformational structures? Are there one or many pathways by which the channel can change from one structure to another? What is the distribution of energy barriers that separate these structures? Are these energy barriers constant or do they change in time?

Is the switching of the channel from one state to another driven primarily randomly by thermal kT fluctuations or is it driven primarily deterministically by the forces in the atomic bonds and electrostatic forces in the channel protein?

Are ion channel proteins in local thermodynamic equilibrium? Systems in which the energy absorbed is equal to the energy dissipated are in a steady state but are not in equilibrium. Such systems form structures in space and time. For example, life on earth is an example of such a system, driven by the energy from the sun. Energy is available to ion channels from voltage gradients and the binding of ligands. Do channel proteins use and dissipate this energy? Does the opening and closing involve local or global changes in protein structure?

In this chapter, we will deal mainly with questions about the number of different conformational structures and the energy barriers between them in ion channel proteins. We have addressed the other questions in other articles.[12,14,16]

3.2 Energy structure of proteins

3.2.1 Energy structure of ion channels

A protein structure is defined by the positions and orientations of the atoms in the protein. For each structure, the energy function is the sum of energies in the atomic and electrostatic interactions between the atoms. This energy function provides a description of the physical properties of a protein. Local minima in the energy function correspond to stable conformational structures. For example, if the energy function has two local minima, then the protein has two stable conformational shapes. In order to go from one stable conformation to another, the protein must increase in energy over an activation energy barrier before it can decrease into the energy well of the other stable conformational state. We can thus characterize a protein by the number of local minima that correspond to the stable conformational states, the distribution of activation energy barriers between these stable states, and how these valleys and mountains of the energy surface change in time.

What is the energy structure of an ion channel? Are there few or many conformational states? Do the energy barriers between states change with time? Because ion channels live partially in the hydrophobic environment of the lipid cell membrane and partially in the hydrophillic environment of the solutions on the inside and outside of the cell, it is very difficult to perform the biophysical experiments needed to directly answer these questions. However, these experiments have been performed on the globular proteins that function in the entirely hydrophillic environment inside of cells. It is reasonable to expect that ion channel proteins have many properties in common with these globular proteins. Thus, we now review the properties of globular proteins revealed by many theoretical and experimental studies over the last two decades.

3.2.2 Energy structure of globular proteins

3.2.2.1 Potential energy functions

Computations of the energies in the atomic bonds and electrostatic interactions for the spatial arrangement and orientation of atoms in proteins have found that proteins have a very large number of shallow, local energy minima.[25,26] Thus, proteins have a very large number of similar, but not identical, conformational substates.

3.2.2.2 Molecular dynamics

Motions within proteins can be computed by molecular dynamics.[25-28] The forces exerted on one atom by all the other atoms in a protein are evaluated. Then, how far that force moves the atom in a time step is computed, and this procedure is then repeated many times. The time steps must be kept very small, typically 10^{-15} s, because the force acting on an atom changes as the structure changes. These studies have shown that protein structure is always changing and shifting between different conformational substates. For example, during 300 ps myoglobin passed through

2000 local energy minima corresponding to different conformational substates.[27] About one third of these motions were variations of structure within the energy well of one conformational substate, and about two thirds were transitions from the energy well of one conformational substate to another.

3.2.2.3 Reaction rates computed from the average structures

The reaction rates of the binding of ligands in proteins can be computed from the energy barriers the ligands must pass through to reach their active site. The average positions of the atoms in the protein, as determined from X-ray crystallography, can be used to determine these energy barriers. These studies found, for example, in the case of oxygen binding to myoglobin, that the predicted reaction rates were far less than those observed experimentally. However, if the fluctuations in the average positions of the atoms were included, then the predicted reaction rates were in approximate agreement with the observed rates. For brief times the protein structure fluctuates into an appropriate structure that makes the reaction happen. The static structure alone does not tell us how proteins work. The reaction rates are proportional to the average of the Boltzmann energy factor $<\exp(-\Delta E/kT)>$, rather than the Boltzmann factor of the average energy, $\exp(-<\Delta E>/kT)$, where ΔE is the activation energy barrier, k the Boltzmann constant, and T the absolute temperature.

3.2.2.4 Distribution of activation energy barriers determined by the time course of reactions

For a chemical reaction $A \Leftrightarrow B$, the rate at which species A is transformed to species B depends on the distribution of activation energy barriers that separate the conformational substates of A and B. The contribution across each pathway to the concentration A(t) is proportional to $\exp(-kt)$ where k is the kinetic rate constant across the energy barrier of that pathway and t is the time since the beginning of the reaction. Summing over all pathways, A(t) is, therefore, proportional to $\int g(k) \exp(-kt) \, dk$, where g(k) is the relative contribution of each pathway.[29] Thus, the time course of the reaction is the Laplace transform of the distribution of kinetic rates of the pathways across the energy barriers. Measurements of the time course of the concentration, A(t), therefore, have been used to determine the distribution of energy barriers. These measurements have been done in many different reaction systems. For example:

1. The visible light emitted when alcohol dehydrogenase was excited by UV was used to measure the fraction of molecules that had unfolded enough to expose an internal tryptophan to oxygen in the solution.[30]
2. The absorption of blue light was used to measure the amount of CO rebound to myoglobin after the MbCO complex was dissociated by a high energy laser pulse.[29]
3. The visible light emitted when nine different proteins were excited by UV was used to measure the rotation of a tryptophan ring.[31,32]
4. The difference in the nuclear magnetic resonance signal between protons and deuterons was used to measure the fraction of lysozyme that had unfolded enough to exchange a deuteron previously loaded into the lysozyme with a proton in the exterior water solution.[33]
5. The rate of the quenching of fluorescent probes by oxygen was used to measure the rate of oxygen penetration from the exterior solution into the protein interior.[34]
6. The rate of decline of radioactivity was used to measure the fraction of tritium previously loaded into the interior of proteins that had exchanged with protons in the exterior water solution.[35]

The analysis of the time course of all these, and many other experiments, showed that proteins have a continuous distribution of activation energy barriers between conformational substates that is very broad and extends over a large range of energies.

3.2.2.5 Measurement of motions by X-ray diffraction and nuclear magnetic resonance

The atoms in a protein bend X-rays into a diffraction pattern that can be used to solve for the spatial structure of the molecule. These patterns are smeared out because the atoms move during the time the diffraction pattern is recorded. The degree of smearing, called the *Debye-Waller factor*, and its dependence on temperature have been used to determine the amount of motion of the atoms in protein structures.[36,37] The interaction between the magnetic moments of nuclei depends on their separation. Thus, nuclear magnetic resonance (NMR) has been used to measure the changing separation between the atoms in a protein.[34,37] Both these techniques have shown that there are significant motions in proteins. The structure of the protein is always changing. These internal motions span an incredible range of time scales from vibrations around bonds as short as 10^{-14} s to local denaturation where the entire protein structure "breathes" over periods of seconds or minutes or hours or days.[25]

3.2.2.6 Induced fit

X-ray diffraction and NMR have been used to determine the structure before and after a protein has bound a ligand. These studies have found that ligand binding changes the structure of the protein. This has led to the concept of "induced fit", namely, that the act of binding is a dynamic interaction between the protein and the ligand that significantly alters the spatial structure of the protein. According to Rini et al., "Analysis of the unliganded structure alone may not be sufficient to define the shape of the antibody combining site with which the ligand binds."[38] An article entitled "Rusting of the Key and Lock Model for Protein-Ligand Binding" concludes that a practical consequence of these structural changes is "the frustration that will often accompany attempts to design drugs by analogy to the structures of flexible, unbound, active substances."[39] Beece et al.[40] have written that "a protein is not like a solid house into which the visitor (the ligand) enters by opening doors without changing the structure. Rather, it is like a tent into which a cow strays."

3.2.2.7 Summary of the properties of proteins

The properties of proteins determined by this extensive theoretical and experimental work over the last 20 years[26,41] are summarized in Figure 2. Each protein has many conformational substates. There are many ways the protein can change its structure to pass from one state to another. Each of these pathways between conformational states passes over an activation energy barrier. Because of this large number of pathways over barriers of different energies, there is a continuous and broad distribution of activation energy barriers between the sets of conformational substates. This results in a continuous distribution of time scales in the motions of the atoms in the protein structure. These motions and the changing energy barriers associated with them are important in how a protein works. The structure that makes a reaction possible may exist only as a rare fluctuation. The structure also changes when a protein interacts and then binds with a ligand. Weber[42] has stated, "Indeed, the protein molecule model resulting from the X-ray crystallographic observations is a 'platonic' protein, well removed in its perfection from the kicking and screaming 'stochastic' molecule that we infer must exist in solution."

> **Many States**
> Very many local energy minima that correspond to very many similar, but not identical, conformational states.
>
> **Many Pathways Between these States**
> Many different scripts of motion that the structure can follow in changing from one structure to another.
>
> **Continuous Distribution of Energy Barriers**
> Since each pathway crosses over a different energy barrier, there is a broad continuous distribution of energy barriers between states.
>
> **Continuous Distribution of Time Scales**
> Transitions over low energy barriers are rapid and transitions over high barriers are slow.
> Thus, there is a very large range of time scales for structural rearrangements from picoseconds to hours.
>
> **Motions and Changing Energy Barriers**
> The structure of a protein fluctuates.
> The structure that makes a reaction possible may exist only as a rare fluctuation.
> The static structure defined by the time average positions of the atoms may not tell us how a protein functions.
>
> **Dynamic Interactions**
> The structure changes when a protein interacts and then binds with a ligand.

Figure 2 Summary of the properties of proteins determined by extensive theoretical and experimental studies over the last 2 decades.

3.3 Markov models of ion channel kinetics

Now that we have reviewed the known properties of proteins, let's see how people have modeled the kinetics of ion channel proteins.[19,20,43,44] A typical model is shown in Figure 3. It consists of a number of closed (C) and open (O) conformational states connected by kinetic rate constants (k). This is called a "Markov model". Markov, a Russian mathematician who died in 1922, did work on integral and differential equations but is best known for his work on probability theory. In 1906 he published the mathematical form that was later used by others to represent ion channel kinetics. The only use of this model to analyze experimental data published by Markov himself was his analysis of the alternations between vowels and consonants in Pushkin's novel *Eugene Onegin*.[45]

3.3.1 Mathematical assumptions of the Markov model

The mathematical assumption of the Markov model is that the channel protein has a small number of discrete, well defined states. The probability per second to switch from one state to another is called the *kinetic rate constant*. These switching probabilities depend only on the current state the channel is in. They do not depend on how long the channel has been in a state. They do not depend on the history of previous states. The kinetic rate constants determine only the probability to switch states; they do not determine exactly when the switch will occur. The switching between states is thought to be "random" in some sense.

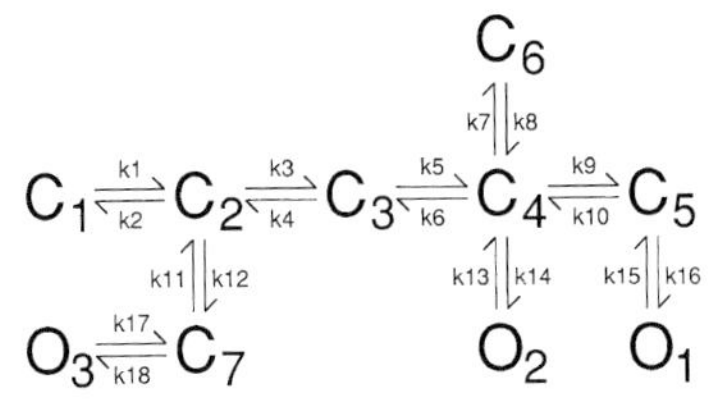

$$C_1 \underset{k_2}{\overset{k_1}{\rightleftarrows}} C_2 \underset{k_4}{\overset{k_3}{\rightleftarrows}} C_3 \underset{k_6}{\overset{k_5}{\rightleftarrows}} C_4 \underset{k_{10}}{\overset{k_9}{\rightleftarrows}} C_5$$

with C_6 connected to C_4 via k_7, k_8; $O_3 \underset{k_{18}}{\overset{k_{17}}{\rightleftarrows}} C_7$ connected to C_2 via k_{11}, k_{12}; O_2 connected to C_4 via k_{13}, k_{14}; O_1 connected to C_5 via k_{15}, k_{16}.

Figure 3 Typical Markov model of ion channel kinetics. The channel protein switches between a small set of closed states (C) and open states (O). The kinetic rate constants (k) are the probabilities per second to switch from one state to another. These switching probabilities depend only the current state. They do not depend on the duration of time that the channel has spent in a state, or the history of previous states.

3.3.2 Physical interpretation of the Markov model

The physical interpretation of the mathematical assumption of the Markov model is that the channel protein has a few, distinct, well defined conformational states. Since these discrete states are well defined, they must be separated by significant activation energy barriers. Since the switching probabilities defined by the kinetic rates are constant, the energy barriers between the states are fixed in time.

3.3.3 Comparison of the physical interpretation of the Markov model to the known properties of proteins

Figure 4 compares the properties of proteins reviewed in the previous section to the physical interpretation of the Markov model given above. Proteins have a very large number of conformational substates, while the Markov model has a small number of states. There are many ways that the atoms of proteins can rearrange to switch the structure from one state to another, while the Markov model has only one unique pathway between states. In proteins the energy barriers are always changing, while in the Markov model they are constant. In proteins there is a continuous, broad distribution of activation energy barrier between states, while in the Markov model there are only discrete activation energy barriers. In proteins, this continuous, broad distribution of energy barriers produces a continuous, broad distribution of time scales of motions, while in the Markov model there is only a small number of discrete time scales determined by the kinetic rate constants. So, there is a real problem here. The physical interpretation of the Markov model is in direct conflict with the known properties of proteins.

These Markov models do fit the open and closed time distributions measured from the patch clamp data. But, this fit requires a large number of adjustable parameters, up to 12 in some models. This makes us worry that the fit to the data is due to this large number of arbitrary parameters and thus these parameters do not provide physically meaningful information about the structure or motions in ion channel proteins.

What we need is a way of interpreting the open and closed time distributions from the patch clamp experiments that is consistent with the known properties of proteins, so that the parameters of those models have physical significance in terms of the structures or motions within ion channel proteins.

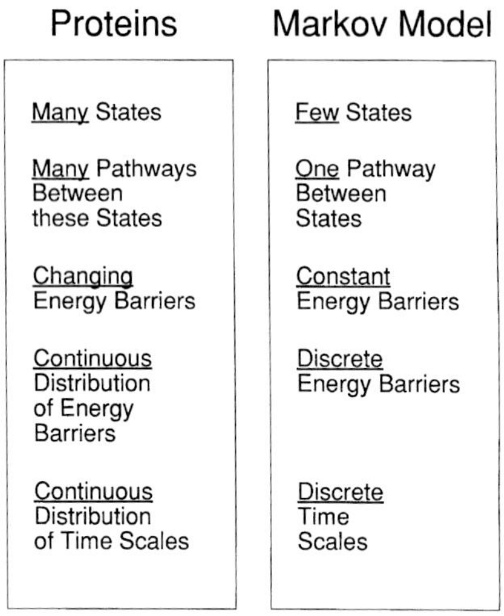

Proteins	Markov Model
Many States	**Few** States
Many Pathways Between these States	**One** Pathway Between States
Changing Energy Barriers	**Constant** Energy Barriers
Continuous Distribution of Energy Barriers	**Discrete** Energy Barriers
Continuous Distribution of Time Scales	**Discrete** Time Scales

Figure 4 Comparison of the known properties of proteins (Section 3.2.2) with the physical interpretation of the mathematical assumptions of the Markov model of ion channel kinetics (Section 3.3.2). The physical interpretation of the Markov model is in conflict with the known properties of proteins.

3.4 Fractal analysis of ion channel kinetics

3.4.1 Fractals

We developed a new way to analyze ion channel kinetics based on fractals. Unlike the Markov analysis, the interpretation of our fractal analysis of experimental data is consistent with the known properties of proteins. The success or failure of our fractal analysis should be judged by whether or not this new approach leads to a better understanding of the structures and motions in ion channel proteins.

Fractals are objects in space or processes in time whose pieces resemble the whole.[46-48] For example, consider a coastline. It looks wiggly. As you observe the coastline at higher spatial resolution you might expect to resolve those wiggles and it should look smooth, but it doesn't. No matter how much you enlarge the coastline, it still looks just as wiggly. It is similar to itself at different spatial scales. This is called self-similarity. More technically, statistical self-similarity means that the distribution of the number of pieces within an object of a certain size has the same mathematical form for pieces of small size as for pieces of large size. Because structures in the coastline at finer spatial resolution are related to structures at coarser spatial resolution, the properties of the coastline as said to scale with the resolution. For example, as you examine the coastline at ever finer resolution, you detect more and more bays and peninsulas. Thus, the value measured for the length of the coastline will be longer when measured at finer spatial resolution. The scaling function describes how a measured property, such as length, depends on the scale at which it is measured.

Many scientists are familiar only with the statistical properties of objects or processes that have a Gaussian or asymptotically Gaussian form. For such objects or processes, as more data is collected, the moments, such as the mean and variance, converge to finite limiting values. However, the statistical properties of fractals are described by a broader class of distributions called *stable distributions*.[48,49] These distributions may not have finite moments. Thus, just as a measurement such as the length of the coastline depends on the resolution used to make the measurement, the values determined for the mean and variance also depend on the resolution or the amount of data analyzed. As more data is collected, the mean and variance may not approach finite limiting values. For example, a fractal signal has ever larger fluctuations that are self-similar reproductions of the small fluctuations. As ever longer data records are analyzed, ever larger deviations will be found. Thus, the variance measured will increase as it is determined from data records of increasing duration.

3.4.2 Fractal form of the current

At a session on ion channels at the Biophysical Society meeting in San Francisco in 1986 each speaker showed ion channel data abruptly changing between periods of frequent and infrequent switches between open and closed states. They suggested these abrupt changes in activity corresponded to changes in the physical properties of the ion channel protein. In the dark conference room with the bright slides on the screen, I suddenly realized that a channel with constant physical properties that produced a current with fractal characteristics would also produce such abrupt changes in activity. For example, if there were hierarchies of bursts within bursts of openings and closings then records made in the center of these hierarchies would show periods of great activity, records made between the hierarchies would show periods of little activity, and records made at the borders of the hierarchies would show abrupt changes in activity, even though the physical properties of the mechanisms that produced the current remained constant.

3.4.3 Methods of fractal analysis of the open and closed times

The qualitative features of the current signal described above strongly suggested that it has a fractal form. We then needed a quantitative method to analyze the data in detail to determine if it was really fractal. The hard part was to decide what properties of the data to analyze and how to do the analysis. We considered a number of different methods which analyze different properties of fractals. We found the methods that better captured the higher order fractal characteristics were more difficult to apply to analyze our type of data. The best compromise was methods that analyzed the statistics of events that depend only on two points in time.[50] For example, we developed the methods described below to analyze the open or closed times which depend on the time between the points where the channel enters and leaves the open or closed state. We are now exploring other methods, such as the Hurst rescaled range analysis, to analyze higher order properties.[51,52]

3.4.3.1 Testing for self-similarity

We used the patch clamp technique to record currents through individual potassium channels on the apical face of corneal endothelial cells.[7] We played back these recordings on an FM tape recorder through an analog-to-digital (A/D) converter and analyzed the durations of the open and closed times on a digital computer (DEC PDP 11/34). (We now store data on a VCR digital interface and analyze it on a Macintosh.) To analyze the data, we determined the histogram of the durations of the closed times. For example, we determined how many closed times lasted between 0 and 20 ms, 20 and 40 ms, etc. To sample the same data at different time scales, we played the same data into the A/D at different sampling rates and then constructed the histograms, which are shown in Figure 5. We know these results are not corrupted by aliasing artifacts because keeping the cutoff frequency of the low pass filter constant or at 1/2.5 times the A/D rate produced the same distributions. Note that

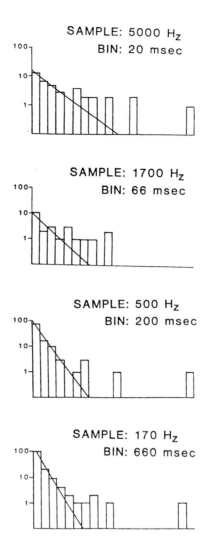

Figure 5 The histogram of closed times determined from the current recording of a potassium channel in a corneal endothelial cell depends on the analog-to-digital (A/D) conversion rate used to sample the data.[7] When the A/D rate is high, short closed times dominate the distribution. When the A/D is low, long closed times dominate the distribution. Yet, the distributions measured at different A/D rates have the same functional form. This is an example of statistical self-similarity characteristic of a fractal process. (From Liebovitch, L. S., Fischbarg, J., and Koniarek, J. P., *Math. Biosci.*, 84, 37–68, 1987. With permission.)

the time axis of each graph in Figure 5 is very different. When we sample rapidly we concentrate on the short closings. When we sample slowly we concentrate on the long closings. Yet, these distributions, measured at different time scales, are similar to each other. Each is nearly a single exponential at short times, with an extended nonexponential tail at long times. This self-similarity in the distributions at different time scales is an example of the statistical self-similarity characteristic of fractals.

3.4.3.2 Testing for scaling: the effective kinetic rate constant
Statistical self-similarity implies that there is a relationship between the short closings and the long closings. This relationship is called the *scaling function*. The simplest scaling function that satisfies self-similarity is a power law.[48] That is, the value of a

property L(x) measured at resolution x has the form $L(x) \propto x^a$. Such power-law scalings are characteristic of fractals. (The more general form of a power law with an additional oscillatory component also satisfies self-similarity.[48])

For example, the length L(x) of a coastline measured at spatial resolution x has the form $L(x) \propto x^{1-d}$ where d is the fractal dimension. Richardson[53] found that d ≈ 1.25 for the west coast of Great Britain. The fractal dimension describes how many additional new segments of an object are found as the object is observed at finer resolution.[48] For example, if N new pieces are revealed when the scale is reduced by a factor F, then the dimension d = log(N)/log(F).

How can we determine if the ion channel kinetics of leaving the closed state has a scaling function with fractal properties? We cannot use the kinetic rate constant, which is the probability per second that a closed channel will open. The channel must be closed long enough for us to detect it as closed. Thus, what we really need is the conditional probability that a channel that has already been closed for at least a certain amount of time will then open. That certain amount of time determines the effective time resolution t_{eff} used to make the measurement. We called this conditional probability the effective kinetic rate constant k_{eff}.[7,60] It is a function of t_{eff}, the time resolution used to make the measurement. The effective kinetic rate constant

$$k_{eff}(t_{eff}) = -\{d[\ln P(t)]/dt\} \text{ evaluated at } t = t_{eff} \qquad (1)$$

where P(t) is the cumulative probability that the channel is closed for a time longer than t. This conditional probability occurs in many other fields, each of which gives it a different name. In renewal theory it is called the age-specific failure rate, and in epidemiology it is called the survival rate.[54]

To analyze the patch clamp data, one $k_{eff}(t_{eff})$ scaling function is determined from the open times and another from the closed times. These scaling functions can be determined in a number of different ways.[7,48,50,55] One method is to construct a series of histograms of different bin sizes. The bin size corresponds to the effective time scale t_{eff}. The value of the parameter b of the function Aexp(–bt) fit to the first few bins of each histogram determines k_{eff} at the value of t_{eff} for that histogram.[7]

The effective kinetic rate constant scaling function determined from our patch clamp experiments of a potassium channel in the apical membrane of corneal endothelial cells is shown in Figure 6. On this log–log plot, the scaling function is a straight line, indicative of a power law that is characteristic of a fractal. That is, the scaling function has the form $k_{eff} \propto t_{eff}^a$. The effective kinetic rate constant k_{eff} is the probability per second that the channel switches states measured at a time resolution t_{eff}. Thus, Figure 6 tells us that the probability to switch states increases as the channel current is examined at finer temporal resolution. That is, the faster we can look, the faster we see the channel open and close. In analogy to the age-specific failure rate of renewal theory, the effective kinetic rate constant k_{eff} is the probability per second to switch states given that the channel has already remained in a state for a time t_{eff}. Thus, Figure 6 also tells us that the longer the channel remains closed, the less the probability per second that it opens. Thus, channel kinetics is not a Markov process. Instead, the channel has memory. The channel remembers how long it has been since the closed state began. The longer it remains closed, the less likely per second that it will open.

The probability P(t) that the channel is closed (or open) for a time greater than t can be determined from the effective kinetic rate constant using Equation (1). The probability density function f(t) that the channel is closed (or open) for a time greater than t and less than t + dt is given by the relationship f(t) = –dP(t)/dt.

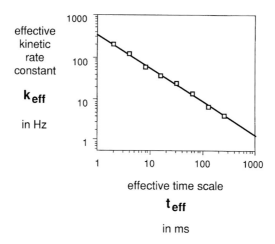

Figure 6 The effective kinetic rate constant $k_{eff}(t_{eff})$ is the probability per second that an ion channel switches states when measured at an effective time resolution t_{eff}. The boxes show the effective kinetic rate constant determined from the closed times of a potassium channel in a corneal endothelial cell.[7] The straight line on this log–log plot indicates that the effective kinetic rate constant for this channel has a fractal scaling which has the power-law form $k_{eff}(t_{eff}) \propto t_{eff}^a$.

Figure 7 shows the histogram of closed times determined from the corneal endothelial potassium channel compared with the probability density function f(t) determined from the fractal scaling function of the effective kinetic rate constant. This probability density function based on the fractal scaling function fits all the closed time durations from those as short as milliseconds to those greater than 1 second long.

Before the fractal approach, this same histogram would have been interpreted in a different way. In the Markov model, every transition between discrete states produces an exponential term, which appears as a straight line on this semi-log plot. Thus, to fit this data might require one straight line at short times, another straight line at intermediate times, and yet another straight line at long times. Hence, it would be concluded that this channel has three real, discrete, well defined conformational states. Such states do not exist. They arise as an artifact of forcing the data to be fit by the functional form of the Markov model, namely, a sum of exponential terms. Rather, what seems to be going on is that there is a scaling function that links the short closings with the long closings. What is the physical significance of this scaling?

3.5 Physical interpretation of the fractal analysis

3.5.1 The fractal approach

The new viewpoint introduced by the fractal analysis of ion channel kinetics is to analyze how a property, such as the effective kinetic rate constant, depends on the time scale at which it is measured. The result of this analysis is a scaling relationship that connects events at different time scales. We can express this scaling relationship in different ways. We can express it as the effective kinetic rate constant k_{eff}, which is the probability per second to switch states, as a function of the time resolution t_{eff} at which it is measured. According to Equation (1), the effective kinetic rate constant, k_{eff}, is related to the probability P(t) that the channel remains in a state longer than time t. Thus, we can equivalently express this scaling as the functional form of the probability distribution P(t). The discovery of the existence of these

chapter three: Ion Channel Kinetics 43

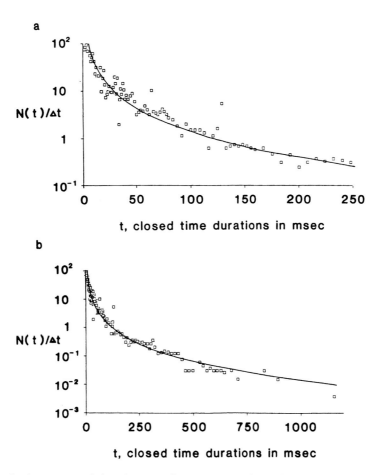

Figure 7 The histogram of closed times of a potassium channel in a corneal endothelial cell (boxes) compared with the probability density function (line) of the fractal analysis determined from the effective kinetic rate constant shown in Figure 6. This probability density function has a stretched exponential form. The fractal approach then drives us to ask what is the physical significance of this functional form. (From Liebovitch, L. S., Fischbarg, J., and Koniarek, J. P., *Math. Biosci.*, 84, 37–68, 1987. With permission.)

scaling relationships now drives us to ask, "What is the physical significance of this scaling relationship in terms of the structure or motions in the ion channel protein?"

3.5.2 Two physical interpretations: structural and dynamical

This scaling relationship can be interpreted in two different ways. If we think of the energy structure of the channel as fixed in time, the scaling relationship gives us information about the number of energy minima that correspond to the conformational substates and the distribution of activation energy barriers between them. We will call this the structural interpretation.[56] If we think of the energy structure of the channel as variable in time, the scaling relationship gives us information about how the values of the energy barriers vary in time. We will call this the dynamical interpretation.[56] These two interpretations are illustrated schematically in Figure 8. These two interpretations may be related to each other. The time-varying dynamical interpretation may be pictured as a set of still frames of a motion picture that is assembled into one fixed energy surface of the structural interpretation. It is not clear

Structural

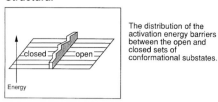

The distribution of the activation energy barriers between the open and closed sets of conformational substates.

Dynamical

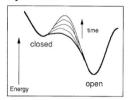

The time dependence of the activation energy barrier between the open and closed states.

Figure 8 The physical significance of the scaling relationships found from the fractal analysis can be interpreted in two different ways. The structural approach interprets the scaling relationships in terms of the distribution of activation energy barriers between the set of open and the set of closed conformational substates. The dynamical approach interprets the scaling relationships in terms of the time variation of energy barrier between the open and closed states.

(to me) if these models can be made entirely equivalent to each other. Perhaps some time-dependent experiments may be able to differentiate between them.[57]

3.5.3 Structural interpretation

The physical properties of the ion channel protein can be characterized by the distribution of activation energy barriers that separate the set of open and the set of closed conformational substates. This distribution can be determined from the histograms of the open and closed times. The cumulative probability P(t) that the channel is closed (or open) for a time longer than t depends on contributions from each pathway over an activation energy barrier. Each contribution is proportional to exp(–kt) where k is the kinetic rate constant across that energy barrier. Summing over all pathways, P(t), therefore, is given by

$$P(t) = \int_0^\infty g(k) \exp(-kt)\, dk \qquad (2)$$

where g(k) is the relative contribution of each pathway.[13] This is analogous to the relationship between the time course of chemical reactions in globular proteins and the distribution of activation energy barriers between their conformational substates[29] that was described in Section 3.2.2.4. The fractal scaling relationship corresponds to the functional form of the cumulative probability distribution P(t). This probability distribution is related to the distribution of the kinetic rate constants g(k). Equation (2) says that P(t) is the Laplace transform of g(k). Using this relationship, we can now interpret the scaling relationship represented by P(t) in terms of the distribution of energy barriers and their associated distribution of kinetic rate constants g(k).

3.5.3.1 Examples of cumulative probability distributions P(t) of open and closed times and their corresponding distributions of kinetic rates g(k)

Liebovitch and Tóth[13] showed that a cumulative probability distribution with the scaling

$$P(t) = \exp(-at^b) \quad \text{where} \quad 0 < b \leq 1 \qquad (3)$$

corresponds to a distribution of kinetic rates g(k) that has the form

$$g(k) = \frac{1}{\pi} \int_0^\infty \left\{ \exp\left[-kx - ax^b \cos(b\pi)\right] \right\} \sin\left[ax^b \sin(b\pi)\right] dx. \qquad (4)$$

We can make more sense out of the complex forms of Equations (3) and (4) by considering the special cases illustrated in Figure 9.

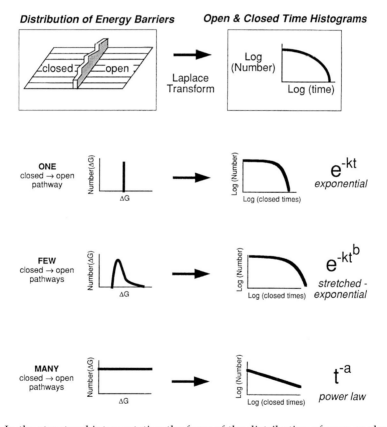

Figure 9 In the structural interpretation the form of the distribution of open or closed times is used to determine the distribution of activation energy barriers between the set of open and the set of closed conformational substates. For example, Equation (3) describes a set of closed time distributions that are parameterized by the single parameter b. As b varies from 1 to 0, the form of the closed time distributions progresses from a single exponential exp(–kt) to a stretched exponential exp(–kt^b) to a power law (t^{-a}). This progression corresponds to an increasing number of closed states, an increasing number of different ways that the channel can open, and an increasingly broad distribution of activation energies barriers between the open and closed states.

For example, consider the distribution of energy barriers for leaving the closed state. If there is only one unique closed state and one pathway out of it by which the channel can open, then g(k) has a single, unique value. (More technically, g(k) is equal to the Dirac delta function $\delta(k - k_o)$ where k_o is the kinetic rate constant across that energy barrier.) This correspond to a probability distribution P(t) of the closed times with a single exponential scaling exp(–kt).

If there are a few closed conformational substates and a few different ways that the channel can open, then there is a small spread in the distribution of activation energy barriers. This correspond to a probability distribution P(t) of the closed times with a stretched exponential scaling $\exp(-kt^b)$.

If there is a very large number of closed conformational substates and a very large number of different ways that the channel can open, then the distribution of energy barriers is very broad. In the limiting case that g(k) is a constant, this corresponds to a probability distribution P(t) of the closed times with a power-law scaling t^{-a}.

Thus, as the spread of the distribution of energy barriers between the set of open and the set of closed conformational substates increases from very narrow to very broad, the histograms of the closed (or open) times observed in the patch clamp recordings progresses from a single exponential through a stretched exponential to a power-law scaling. Hence, the scaling relationships represented by these different forms in the measured histograms tells us about the distribution of energy barriers in the ion channel protein.

3.5.3.2 Examples of distributions of kinetic rates g(k) determined from the probability distributions P(t) of open and closed times of patch clamp data

Figures 10 through 13 show the probability density functions f(t) = –dP(t)/dt for the open or closed times on log–log plots for four different channels. These figures show that all three types of scaling functions represented by single exponential exp(–kt), stretched exponential $\exp(-kt^b)$, and power-law distributions t^{-a} are found in the patch clamp data from different channels.

3.5.3.2.1 Power-law distributions of open or closed times.
Figure 10 shows the distribution of closed times, on a log–log plot, for a potassium channel in cultured neuroblastoma ×glioma hybrid cells.[44] The closed time distribution is approximately a straight line on this log–log plot, indicating that it has a power-law form characteristic of fractals. This power-law form extends impressively over the entire range from which closed times were recorded, over 5 decades in magnitude, from 0.1 ms to 10 s. Equations (2) to (4) imply that this channel has a very large number of closed conformational substates and a very large number of different ways that these closed substates can open. Thus, there is a very broad distribution of activation energy barriers between the set of closed and the set of open conformational substates.

The data in Figure 10 also can be fit a Markov model consisting of the sum of six exponential terms corresponding to six closed states.[44] The fractal approach requires two adjustable parameters, and the Markov approach requires 12 adjustable parameters to fit the same data. One group did a statistical analysis comparing the goodness of fit of different models to this data and concluded that "by such statistical analysis we have provided support for the M[arkov] model of channel gating" and therefore against the fractal interpretation.[59] We showed that their statistical analysis is mathematically flawed,[11] but it is not just a matter of these technical mathematical issues. It is also a matter of common sense. It is important not to be so fixated on

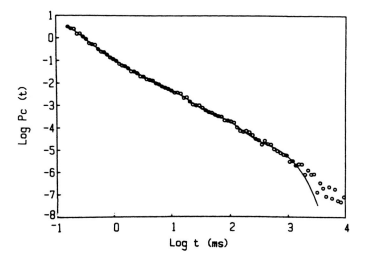

Figure 10 Distribution of closed times, on a log–log plot, from a potassium channel in cultured neuroblastoma × glioma hybrid cells. The approximately straight line on this plot indicates that the scaling relationship for this channel has a power-law form. The structural interpretation of this power-law scaling is that this channel has a very large number of closed states, a very larger number of different ways that the channel can open, and thus a very broad distribution of activation energy barriers between the closed and open states. (From McGee, R., Sansom, M. S. P., and Usherwood, P. N. R., *J. Memb. Biol.*, 102, 21–34, 1988. With permission.)

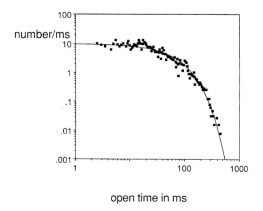

Figure 11 Distribution of open times, on a log–log plot, from a potassium channel channel in cultured hippocampal neurons.[60] The scaling relationship for this channel has a single exponential form. The structural interpretation of this single exponential scaling relationship is that this channel has a single open state, a single way that it can close, and thus a very narrow distribution of activation energy barriers between the open and closed states.

these Markov models that we are unable to see that Figure 10 looks approximately like a straight line.

3.5.3.2.2 Single exponential distributions of open or closed times. Not all channels have open or closed time distributions that are power laws. Figure 11 shows the distribution of open times from a potassium channel in cultured hippocampal neurons.[60] It is a beautiful single exponential, as shown by the solid line. This sharply bent curve is the form of a single exponential on a log–log plot. Equations (2) to (4) imply that this channel has a single open state and single way by which the channel

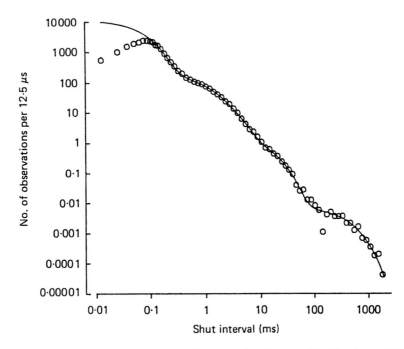

Figure 12 Distribution of closed times, on a log–log plot, from a chloride channel in skeletal muscle. There are different scaling relationships at different time scales. At long times there is a single exponential scaling. At closed times of approximately 1 to 10 ms, there is a scaling relationship that is intermediate between a sharply bent single exponential and a straight line power law. The structural interpretation of this scaling relationship is that there is a process at 1 to 10 ms that has an intermediate spread in the distribution of activation barriers between the closed and open states. (From Blatz, A. L. and Magleby, K. L., *J. Physiol. (Lond.)*, 348, 141–174, 1986. With permission.)

can close. Thus, it has a very narrow distribution of activation energy barriers between the open state and the set of closed conformational substates.

3.5.3.2.3 Combinations of distributions of open or closed times. Not only do different channels have different scalings in distributions of open or closed times, but sometimes the same channel can have different scalings at different time scales. Figure 12 shows the distribution of closed times from a chloride channel in skeletal muscle.[43] At long times there is a clear single exponential scaling. However, at intermediate times, around closed times of approximately 1 to 10 ms, there is a scaling that is not so sharply bent as a single exponential, nor so straight as a power law. That is, at intermediate times, there is a process that has some intermediate spread in its distribution of activation energy barriers.

Many channels have a power-law scaling at short times and a single exponential scaling at long times.[55,61-63] We showed that such a combined scaling would result from a distribution of activation energy barriers that is constant at low energy but has a cutoff at high energies.[13] Activation energy barriers of low energy correspond to short time scales; a closed channel will rapidly cross the energy barrier and open. Activation energy barriers of high energy correspond to long time scales; a closed channel will wait a long time before it crosses the energy barrier and opens. When there is a high energy cutoff in the activation energy barrier distribution it means that no matter how long you wait, a closed channel will never cross those high energy barriers and crank itself open.

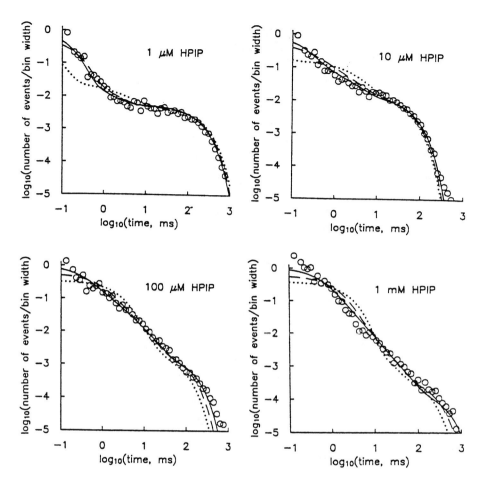

Figure 13 Distribution of closed times, on a log-log plot, from an acetylcholine-activated sodium channel in brain tumor cells at four different concentrations of an analog to acetylcholine (HPIP). At low ligand concentration there is power-law scaling relationship at short times and a single exponential scaling relationship at long times. As the ligand concentration increases, the power-law scaling extends to increasingly longer times. The binding of the ligand does not seem to induce one specific reaction pathway, but rather it seems to allow the channel to open over many different pathways over a broad range of time scales. (From Oswald, R. W., Millhauser, G. L., and Carter, A. A., *Biophys. J.*, 59, 1136–1142, 1991. With permission.)

3.5.3.2.4 Dependence of distributions of open or closed times on voltage, ions, and ligands. The effective kinetic rate constant analysis showed that a potassium channel in fibroblasts has a power-law scaling at short times and a single exponential scaling at long times.[55,61] Variations in the voltage applied across the cell membrane or in the concentration of calcium in the solution outside the cell changes only the time scale at which the scaling changes from a power to a single exponential. These open and closed time histograms also can be described by a Markov model with two open and three closed states.[64] The kinetic rate constants of that Markov model have a complex dependence on voltage and calcium that cannot be given a physical interpretation. However, the scalings found by the fractal approach can be given a physical interpretation. These scalings would be produced by a distribution of energy barriers between the set of open and closed states that is a constant at low energy with an energy cutoff beyond which there are no pathways between the open and closed states.[13] The data indicate that the value of the energy cutoff depends on the voltage applied across the channel and the ambient calcium concentration.

The binding of ligands can alter the form of the scaling. Figure 13 shows the distribution of closed times from an acetylcholine-activated sodium channel in cultured brain tumor cells as the concentration of an analog to acetylcholine is increased.[58] At low ligand concentration there is a power-law scaling at short times and a single exponential scaling at long times. As the ligand concentration is increased, the power-law scaling extends to longer times, and the single exponential scaling decreases in prominence. This is the opposite of what we would have expected. When the ligand binds to the channel, it should specifically drive the channel to a new conformational structure. The specific reaction with the ligand should induce a specific conformational change at a specific time scale. Thus, we would have expected that the single exponential scaling, corresponding to this time scale, would increase with increasing ligand concentration. Instead, as the ligand concentration is increased, the single exponential scaling decreases in prominence and the closed time distribution becomes more like a power law. Such a power-law scaling corresponds to switching between different conformational states at all time scales. Thus, the binding of the ligand does not drive the closed channel over one pathway into the open state; rather, it made possible all pathways at all time scales. It is as if the closed channel is lacking a piece so that it cannot open over a range of time scales. The ligand does not induce one specific conformation change, but rather it adds this missing piece so that the channel can now open over many different pathways corresponding to this range of time scales.

3.5.4 *Dynamical interpretation*

The scalings found from the fractal analysis can be interpreted also in terms of how the energy barriers change in time between the open and closed states. This can be done by using the effective kinetic rate constant $k_{eff}(t_{eff})$ defined in Equation (1).[7,60] In terms of the age-specific failure rate of renewal theory,[54] the effective kinetic rate constant $k_{eff}(t_{eff})$ is the probability per second to switch states given that a channel has already remained in a state for a time t_{eff}. For all the published and unpublished patch clamp data that we have ever seen, the semi-log plot of the cumulative distribution of the open or closed times (that is, ln P(t) vs. t) is concave upward. The absolute value of the slope of ln P(t) vs. t decreases with increasing t. Hence, from Equation (1), we find that $k_{eff}(t_{eff})$ decreases with increasing t_{eff}. In other words, the probability per second that a channel switches states decreases the longer the channel remains in a state. This implies that when a channel enters a state, the activation energy barrier to leave that state continues to increase in time. It is as if once a channel protein settles into a new conformational structure, it continues to become ever more stable in that structure and it becomes ever more difficult to escape from it into another structure.

As shown in Figure 14, when the activation energy barrier and the kinetic rate constant across it are constant in time, then the cumulative distribution of the open or closed times P(t) has a single exponential scaling exp(–kt). When the activation energy barrier increases slowly in time so that the kinetic rate constant associated with it is proportional to t^{b-1} where $0 < b < 1$, then P(t) has a stretched exponential scaling exp(–kt^b). When the activation energy barrier increases as rapidly as ever observed from the patch clamp data, so that the kinetic rate constant associated with it is proportional to t^{-1}, then P(t) has a power-law scaling t^{-a}.

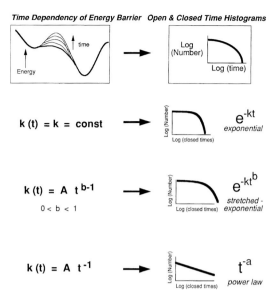

Figure 14 In the dynamical interpretation, the form of the distribution of open or closed times is used to determine how the activation energy barrier between the open and closed state varies in time. For all ion channels studied so far, the probability per second to switch states decreases the longer a channel remains in a state. As the form of the time distributions progresses from a single exponential $\exp(-kt)$ to a stretched exponential $\exp(-kt^b)$ to a power law (t^{-a}), the activation energy barrier rises increasingly rapidly.

3.5.5 Structural or dynamical?

Right now, the reader may be thinking, "Wait a minute, you've just told me over the last two sections that we can interpret the open and closed times from the patch clamp either as distributions of activation energy barriers between conformational substates (structural interpretation) or as the changing value in time of the energy barrier between the open and closed states (dynamical interpretation). Which is it? What is really going on inside the ion channel protein?" The blunt answer is that we simply do not know. As noted by the questions raised at the end of the Introduction, we do not yet understand the basic physical properties of ion channel proteins. At this time, we need to explore as many possible models of ion channel structure and motions as we can.

3.6 Other approaches

This chapter reviews our work on fractal approaches to analyzing and interpreting ion channel kinetics, but we have also used other approaches to ask different questions about the physical properties of ion channel proteins.

3.6.1 Are ion channel proteins little machines?

It used to be thought that experimental data that looks random must be the result of a random process, that is, of many independent influences. We now know that some data that looks random was actually generated by deterministic processes. This phenomena is now called *chaos*.[48] As opposed to its normal usage, the technical mathematical meaning of the word "chaos" here means a nonlinear and often a very simple system, the output of which is so complex that it mimics random behavior. There are now a number of methods to analyze experimental data to determine if seemingly random data could have been generated by a deterministic, chaotic processes.[48]

Ion channels seem to switch randomly from one conformational state to another. Is this switching really random or could it arise from deterministic, chaotic processes?

We have shown that the statistical properties of ion channel kinetics, once attributed only to random processes, can also be produced by deterministic processes.[12,14] If the switching in ion channels is produced in part by the forces in the atomic bonds and electrostatic interactions between the atoms, then perhaps we should think of these molecules more as little mechanical machines made up of balls and springs rather than as small molecules driven by random thermal fluctuations. The relative contributions of deterministic atomic and electrostatic forces vs. random thermal fluctuations in switching a channel protein between different conformational states has not yet been determined.

3.6.2 How do local interactions produce globular structure in ion channel proteins?

How local short range interactions between the atoms in the channel protein sculpt it into a globally organized structure is not understood. A similar phenomenon happens in a computational structure called a *neural network*. A neural network consists of nodes and the connections between them. At each time step, the new value of a node depends on the values of the other nodes and on the strengths of the connections between them. We are beginning to study how the degree of locality of the interactions between the nodes of a neural network affects the statistics of the switching of the network from one state to another. This may help us to find out if the statistical properties of the open and closed times can be used to determine the spatial extent of the mechanism that opens or closes the channel. We are also investigating the usefulness of neural networks to compute the motions in proteins.[18]

3.7 Response to the fractal approach

3.7.1 Markov vs. fractal ion channel kinetics

Some people in the channel field have fit Markov models to ion channel data over the last 15 years and one suspects that they want to continue to do exactly the same work for another 15 years. As they see it, the fractal model is not as good a fit to the experimental data as the Markov model.[59,65-68] They compared the fit of our simplest fractal model with a power-law scaling to that of their most complex Markov models. This fractal model has two adjustable parameters, and their Markov models have as many as 12 adjustable parameters. Thus, in order to compare these models, they used certain statistical procedures to determine the goodness of fit of models that have different numbers of parameters.

We think that the statistical procedures that they used to compare models with different numbers of parameters are not valid.[11] Moreover, their statistical comparisons were done in an odd way that may betray their intentions. When the Markov model did not fit the data sufficiently well, they added more adjustable parameters to improve the fit. When our fractal model did not fit the data sufficiently well, they did not add any additional parameters; instead, they concluded that the fractal approach is "inadequate"[65] or that it "ignores the experimentally observed details."[67] When we extended the simplest fractal, power-law scaling by adding additional parameters so it could be directly compared to a Markov model with the same number of parameters, we found that the fractal scaling was a much better fit to the data than the corresponding Markov model.[9]

As we see it, much more important than these statistical issues is the basic point that the physical interpretation of the Markov model (Section 3.3) is contradicted by extensive theoretical and experimental studies conducted over the last 20 years which have elucidated the biophysical properties of proteins (Section 3.2).[9] This contradiction may not have been given the attention it deserves because electrophysiologists who study ion channel kinetics may not have been aware of the biophysical properties of proteins. This is not surprising. Science is highly compartmentalized with different subfields separated by cultural and language barriers.

The values of the parameters of the Markov model can be adjusted so that the model fits the measured open and closed times. However, since the physical interpretation of the Markov model does not match the physical properties of proteins, the values of these parameters do not have physical significance. What we need is a way of analyzing the experimental data that is consistent with the known properties of proteins so that the values of the parameters determined from fitting the data have physical significance in terms of protein structure or motions. The fractal analysis provides one method to interpret the channel data in ways that are consistent with the known properties of proteins. That is, the scalings found from the fractal analysis can be interpreted in terms of the distribution of activation energy barriers between sets of conformational substates and/or the time dependence of those energy barriers.

3.8 Summary of the fractal approach to ion channel kinetics

Fractals have provided a new approach to analyze and interpret the currents through individual ion channels measured by the patch clamp technique; this new approach is to analyze how a property depends on the time scale at which it is measured. For example, we can determine how the effective kinetic rate constant, $k_{eff}(t_{eff})$, which is the probability per second that a channel switches out of an open or closed state, depends on the time scale t_{eff} used to make the measurement.

The fractal analysis found that there are scaling relationships which can be expressed as the functional form of the effective kinetic rate constant or the distribution of the open or closed times. These scaling relationships can be interpreted in terms of the distribution of activation energy barriers between the set of the open and closed conformational substates or how those barriers change in time. The fractal analysis has made it possible to interpret the patch clamp data in ways that are consistent with the known properties of proteins determined by extensive theoretical and experimental studies over the last two decades.

The fractal analysis suggests that new types of experiments should be done. The fractal approach emphasizes the importance of determining the number of the conformational substates, the distribution of energy barriers between them, and how those energy barriers change in time. Thus, experiments should be done to more clearly measure these properties in ion channel proteins. For example, rather than statistical analysis of patch clamp data, or "structure-function" patch clamp experiments of genetically modified channels, it may be more important to measure the motions of different pieces of the channel with respect to each other. These experiments are beyond what is now technically feasible, but not much beyond. For example, fluorescent or NMR labels on different pieces of the channel could be used to determine the motions, or the statistical distribution of the motions, of different pieces of the channel with respect to each other. Time-resolved circular dichroism could be used to measure the change in helicity as a set of channels gate open in response to a voltage pulse. Spin echo experiments using short-pulse, free-electron lasers could be used to determine whether the vibrational modes within a channel protein are discrete or are coupled together.

The success or failure of our fractal analysis will depend on whether or not this new approach will lead to a better understanding of the structures and motions in ion channel proteins.

Acknowledgments

Our work was supported in part by grants from the Whitaker Foundation, the American Heart Association (Established Investigatorship), and the National Institutes of Health (EY6234).

References

1. Liebovitch, L. S. and Weinbaum, S., A model of epithelial water transport, *Biophys. J.,* 35, 315, 1981.
2. Liebovitch, L. S., Fischbarg, J., and Koatz, R., Osmotic water permeability of rabbit corneal endothelium and its dependence on ambient concentration, *Biochim. Biophys. Acta,* 646, 71, 1981.
3. Liebovitch, L. S. and Fischbarg, J., Effects of passive Na and HCO_3 fluxes on electrical potential and fluid transport across rabbit corneal endothelium, *Curr. Eye Res.,* 2, 183, 1982.
4. Lim, J. J., Liebovitch, L. S., and Fischbarg, J., Ionic selectivity of the paracellular shunt across rabbit corneal endothelium, *J. Membrane Biol.,* 73, 95, 1983.
5. Liebovitch, L. S., Fischbarg, J., and Koniarek, J. P., Optical correlation functions applied to the random telegraph signal: how to analyze patch clamp data without measuring the open and closed times, *Math. Biosci.,* 78, 203, 1986.
6. Liebovitch, L. S. and Fischbarg, J., Membrane pores: a computer simulation of interacting pores analyzed by $g_1(\tau)$ and $g_2(\tau)$ correlation functions, *J. Theoret. Biol.,* 119, 287, 1986.
7. Liebovitch, L. S., Fischbarg, J., and Koniarek, J. P., Ion channel kinetics: a model based on fractal scaling rather than multistate Markov processes, *Math. Biosci.,* 84, 37, 1987.
8. Koniarek, J. P., Lee, H.-B., Rosskothen, H. D., Liebovitch, L. S., and Fischbarg, J., Use of transendothelial electrical potential difference to assess chondroitin sulfate effect in corneal preservation media, *Invest. Ophthalmol. Vis. Sci.,* 29, 657, 1988.
9. Liebovitch, L. S., Testing fractal and Markov models of ion channel kinetics, *Biophys. J.,* 55, 373, 1989.
10. Liebovitch, L. S., Analysis of fractal ion channel gating kinetics: kinetic rates, energy levels, and activation energies, *Math. Biosci.,* 93, 97, 1989.
11. Liebovitch, L. S. and Tóth, T. I., The Akaike information criterion (AIC) is not a sufficient condition to determine the number of ion channel states from single channel recordings, *Synapse,* 5, 134, 1990.
12. Liebovitch, L. S. and Tóth, T. I., A model of ion channel kinetics using deterministic chaotic rather than stochastic processes, *J. Theoret. Biol.,* 148, 243, 1991.
13. Liebovitch, L. S. and Tóth, T. I., Distributions of activation energy barriers that produce stretched exponential probability distributions for the time spent in each state of the two state reaction A ⇔ B, *Bull. Math. Biol.,* 53, 443, 1991.
14. Liebovitch, L. S. and Czegledy, F. P., A model of ion channel kinetics based on deterministic, chaotic motion in a potential with two local minima, *Ann. Biomed. Engr.,* 20, 517, 1992.
15. Liebovitch, L. S., Selector, L. Y., and Kline, R. P., Statistical properties predicted by the ball and chain model of channel inactivation, *Biophys. J.,* 63, 1579, 1992.

16. Liebovitch, L. S., Interpretation of protein structure and dynamics from the statistics of the open and closed times measured in a single ion channel protein, *J. Stat. Phys.,* 70, 329, 1993.
17. Liebovitch, L. S., Single channels: from Markovian to fractal models, in *Cardiac Physiology: From Cell to Bedside,* 2nd ed., Zipes, D. P. and Jalife, J., Eds., W. B. Saunders, Philadelphia, 1995, pp. 293–304.
18. Liebovitch, L. S., Arnold, N. A., and Selector, L. Y., Neural networks to compute molecular dynamics, *J. Biol. Systems,* 2, 193, 1994.
19. Sakmann, B. and Neher, E., Eds., *Single-Channel Recording,* Plenum Press, New York, 1983.
20. Hille, B., *Ionic Channels of Excitable Membranes,* Sinauer Assoc., Sunderland, MA, 1984.
21. Hille, B. and Fambrough, D. M., Eds., *Proteins of Excitable Membranes,* Wiley-Interscience, New York, 1987.
22. Langs, D. A. and Triggle, D. J., Structural motifs for ion channels, in *The Structure of Biological Membranes,* Yeagle, P., Ed., CRC Press, Boca Raton, FL, 1992, chap. 16.
23. Catterall, W. A., Structure and function of voltage-sensitive ion channels, *Science,* 242, 50, 1988.
24. Miller, C., 1990: Annus mirabilis of potassium channels, *Science,* 252, 1092, 1991.
25. McCammon, J. A. and Harvey, S. C., *Dynamics of Proteins and Nucleic Acids,* Cambridge University Press, New York, 1987.
26. Karplus, M. and Petsko, G. A., Molecular dynamics simulations in biology, *Nature,* 347, 631, 1990.
27. Elber, R. and Karplus, M., Multiple conformational states of proteins: a molecular dynamics analysis of myoglobin, *Science,* 235, 318, 1987.
28. Karplus, M. and McCammon, J. A., Dynamics of proteins: elements and function, *Ann. Rev. Biochem.,* 52, 263, 1983.
29. Austin, R. H., Beeson, K. W., Eisenstein, L., Frauenfelder, H., and Gunsalus, I. C., Dynamics of ligand binding to myoglobin, *Biochemistry,* 14, 5355, 1975.
30. Saviotti, M. L. and Galley, W. C., Room temperature phosphorescence and the dynamic aspects of protein structure, *Proc. Natl. Acad. Sci. U.S.A.,* 71, 4154, 1974.
31. Alcala, J. R., Gratton, E., and Prendergast, F. G., Interpretation of fluorescence decays in proteins using continuous lifetime distributions, *Biophys. J.,* 51, 925, 1987.
32. Alcala, J. R., Gratton, E., and Prendergast, F. G., Resolvability of fluorescence lifetime distributions using phase fluormetry, *Biophys. J.,* 51, 587, 1987.
33. Welch, G. R., Ed., *The Fluctuating Enzyme,* John Wiley & Sons, New York, 1986.
34. Gurd, F. R. N. and Rothgeb, T. M., Motions in proteins, *Adv. Protein Chem.,* 33, 73, 1979.
35. Englander, S. W., Englander, J. J., McKinnie, R. E., Ackers, G. K., Turner, G. J., Westrick, J. A., and Gill, S. J., Hydrogen exchange measurement of the free energy of structural and allosteric change in hemoglobin, *Science,* 256, 1684, 1992.
36. Ringe, D. and Petsko, G. A., Mapping protein dynamics by X-ray diffraction, *Prog. Biophys. Molec. Biol.,* 45, 197, 1985.
37. Frauenfelder, H., Parak, F., and Young, R. D., Conformational substates in proteins, *Ann. Rev. Biophys. Biochem.,* 17, 451, 1988.
38. Rini, J. M., Schulze-Gahmen, U., and Wilson, I. A., Structural evidence for induced fit as a mechanism for antibody-antigen recognition, *Science,* 255, 959, 1992.
39. Jorgensen, W. L., Rusting of the lock and key model for protein-ligand binding, *Science,* 254, 954, 1991.
40. Beece, D. Eisenstein, L., Frauenfelder, H., Good, D., Marden, M. C., Reinisch, L., Reynolds, A. H., Sorensen, L. B., and Yue, K. T., Solvent viscosity and protein dynamics, *Biochemistry,* 19, 5147, 1980.
41. Frauenfelder, H., Silgar, S. G., and Wolynes, P. G., The energy landscape and motions of protein, *Science,* 254, 1598, 1991.
42. Weber, G., Energetics of ligand binding to proteins, *Adv. Protein Chem.,* 29, 1, 1975.
43. Blatz, A. L. and Magleby, K. L., Quantitative description of three modes of activity of fast chloride channels from rat skeletal muscle, *J. Physiol. (Lond.),* 378, 141, 1986.

44. McGee Jr., R., Sansom, M. S. P., and Usherwood, P. N. R., Characterization of a delayed rectifier K$^+$ channel in NG108-15 neuroblastoma × glioma cells: gating kinetics and the effects of enrichment of membrane phospholipids with arachidonic acid, *J. Memb. Biol.*, 102, 21, 1988.
45. Youschkevitch, A. A., A. A. Markov, in *Dictionary of Scientific Biography*, vol. 4., Gillispie, C. C., Ed., Charles Scribner's Sons, New York, 1974, pp. 124–130.
46. Mandelbrot, B. B., *The Fractal Geometry of Nature*, W. H. Freeman, San Francisco, 1983.
47. Feder, J., *Fractals*, Plenum Press, New York, 1988.
48. Bassingthwaighte, J. B., Liebovitch, L. S., and West, B. J., *Fractal Physiology*, Oxford University Press, New York, 1994.
49. Feller, W., *An Introduction to Probability Theory and Its Applications*, vol. 2, 2nd ed., John Wiley & Sons, New York, 1971.
50. Liebovitch, L. S., The fractal random telegraph signal: signal analysis and applications, *Ann. Biomed. Engr.*, 16, 483, 1988.
51. Churilla, A. M., Gottschalk, W. A., Liebovitch, L. S., Selector, L. Y., Todorov, A. T., and Yeandle, S., Membrane potential fluctuations of human t-lymphocytes have the fractal characteristics of fractional Brownian motion, *Ann. Biomed. Engr.*, 24, 99, 1996.
52. Nogueira, R. A., Varanda, W. A., and Liebovitch, L. S., Hurst analysis in the study of ion channel kinetics, *Braz. J. Med. Biol. Res.*, 28, 491, 1995.
53. Richardson, L. F., The problem of contiguity: an appendix to *Statistics of Deadly Quarrels*, *General Systems Yearbook*, 6, 139, 1961.
54. Cox, D. R., *Renewal Theory*, Science Paperbacks, London, 1962.
55. French, A. S. and Stockbridge, L. L., Fractal and Markov behavior in ion channel kinetics, *Can. J. Physiol. Pharm.*, 66, 967, 1988.
56. Dewey, T. G. and Spencer, D. B., Are protein dynamics fractal?, *Comments Mol. Cell. Biophys.*, 7, 155,1991.
57. Tian, W. D., Sage, J. T., Srajer, V., and Champion, P. M., Relaxation dynamics of myoglobin in solution. *Phys. Rev. Lett.*, 68, 408, 1992.
58. Oswald, R. E., Millhauser, G. L., and Carter, A. A., Diffusion model in ion channel gating. *Biophys. J.*, 59, 1136, 1991.
59. Sansom, M. S. P., Ball, F. G., Kerry, C. J., McGee, R., Ramsey, R. L., and Usherwood, P. N. R., Markov, fractal, diffusion, and related models of ion channel gating, *Biophys. J.*, 56, 1229, 1989.
60. Liebovitch, L. S. and Sullivan, J. M., Fractal analysis of a voltage-dependent potassium channel from cultured mouse hippocampal neurons, *Biophys. J.*, 52, 979, 1987.
61. Stockbridge, L. L. and French, A. S., Characterization of a calcium-activated potassium channel in human fibroblasts, *Can. J. Physiol. Pharm.*, 67, 1300, 1989.
62. Liebovitch, L. S., unpublished.
63. Teich, M. C., unpublished.
64. French, A. S. and Stockbridge, L. L., Potassium channels in human and avian fibroblasts, *Proc. Roy. Soc. London (B)*, 232, 395, 1988.
65. McManus, O. B., Weiss, D. S., Spivak, C. E., Blatz, A. L., and Magleby, K. L., Fractal models are inadequate for the kinetics of four different ion channels, *Biophys. J.*, 54, 859, 1988.
66. Korn, S. J. and Horn, R., Statistical discrimination of fractal and Markov models of single-channel gating, *Biophys. J.*, 54, 871, 1988.
67. Horn, R. and Korn, S. J., Model selection: reliability and bias, *Biophys. J.*, 55, 379, 1989.
68. McManus, O. B., Spivak, C. E., Blatz, A. L., Weiss, D. S., and Magleby, K. L., Fractal models, Markov models, and channel kinetics, *Biophys. J.*, 55, 383, 1989.

chapter four

Protein Conformation and Enzymatic Kinetics

Houqiang Li and Fuquan Wang

> Since earth's relief is finely "corrugated", there is no doubt that, just like a coastline's length, an island's total area is geographically infinite. But the domains surrounded by coastlines have well defined "map areas".
>
> B. B. Mandelbrot

4.1 Introduction: basic concepts and terminology

A brief introduction to the basic concepts of fractal theory applying to proteins and enzymes will be presented in this section. Fractal geometry[1] is in essence a new mathematical approach which reveals the regularity behind matter with apparently irregular forms, e.g., the shape of clouds in the sky or the coastline on a map, etc. All of the above mentioned have been out of the reach of Euclidean and differential geometries. Since the structure of protein has such a complex form it only can be analyzed appropriately by the new approach of fractal geometry.

A fractal implies a complex pattern with self-similarity and self-affinity. In other words, a fractal is a shape made of parts similar to the whole in some way. The parameter characterizing a fractal feature is called *fractal dimension* (D_f). Many pioneering works have proved time and again that the protein having statistical self-similarity can be handled by the fractal approach.[2] Although iterated fractals that are perfectly self similar can be constructed, the self-similarity of a protein chain is true in common occurrence only in a statistical sense, i.e., the part will not always look exactly like the whole. Another difference between biological macromolecules and ideal objects is that a fractal is not self similar over all length scales. There are both upper and lower size limits beyond which a macromolecule is no longer fractal. The main contribution of fractal theory to modern scientific thinking is not so much through the development of useful mathematical tools, as in helping many scientists to overcome the psychological barriers of treating problems that involve complex geometries. The fractal research into enzymes and proteins is currently an active field of enzymology and biological macromolecular science.[2]

In fractal analysis of proteins, one often seeks a power law of the form $p \sim v^\alpha$, where p is a property; v, the variable, and α, the exponent, may be related to fractal dimensions which can be obtained either theoretically or experimentally. In this chapter, we illustrate this kind of power law for protein conformation and enzymatic kinetics, and several kinds of fractal dimensions have been calculated to describe the features of proteins.

A protein consists of a polypeptide chain made up of residues of amino acids linked together by peptide bonds. The polypeptide chain or backbone forms a linear polymer composed of repeating units that are identical except for the chain termini. Hence, proteins have fractal aspects. There are few covalent cross-link (disulfide) bonds, but the chain is never branched. The types of proteins are decided by the amino acids in the polypeptide chain. There are 20 commonly occurring amino acids that differ in their side-chains, forming a number of types of proteins from the simplest glycine to the most complex tryptophan. Therefore, the fractal analysis of protein chain conformation may be important for the study of protein features.

The primary, secondary, and tertiary structure of proteins have been presented in Reference 2. The primary structure is referred to as the sequence of amino acids and determines the native conformation. The secondary structure elements are the α-helics and β-pleated sheets, while the tertiary structure relates to the overall spatial arrangements of the amino acid residues in proteins. In general, proteins are tightly packed systems, and there is a relatively small number of voids.[2] Thus, fractal dimension may be used to characterize tertiary structures of proteins and enzymes.

For proteins, other properties such as the density of the vibrational frequencies, chaotic behavior, the surface geometry, and kinetics, etc. appear to obey scaling laws in some limiting cases. They can also be described by different fractal models and different kinds of fractal dimensions. So, it is necessary to introduce some special topics of fractal theory such as fat fractal, multifractal, fracton dimension, and so on. These topics will be discussed in this chapter.

4.1.1 Fat fractal

A fat fractal[3] is a set with a fractal boundary, finite Lebesque measure, and a non-zero measure at the same (integer) dimension as the underlying space (d).[4] For the definition of the term fat fractal, we first must note that the main point is that sets with fractal structure but finite (Lebesque) measure have (integer) dimensions that are insensitive to this fractal structure. As a result, the fractal structure of such a set is not well characterized by the dimension of the set, but by a fat-fractal exponent (β). In one dimension, every set is the complement of a disjointed collection of holes (intervals and points) of well defined sizes. If one fills in all of the holes of a size less than ϵ, one obtains a coarse-grained set of measure

$$\mu_\beta(\epsilon) = \mu_0 + F(\epsilon) \qquad (1)$$

where $F(\epsilon)$ is the total size of all of the holes smaller than ϵ. The fat-fractal exponent is given by

$$\beta = \lim_{\epsilon \to 0} \frac{\ln F(\epsilon)}{\ln \epsilon} = \lim_{\epsilon \to 0} \frac{\ln |\mu_\beta(\epsilon) - \mu_0|}{\ln \epsilon} \qquad (2)$$

which lies in the range $0 \leq \beta \leq \infty$, $\lim_{\epsilon \to 0} \mu_\beta(\epsilon) = \mu_0 > 0$. The $|\mu_\beta(\epsilon) - \mu_0|$ value is the fat volume.

According to the above discussion, a protein may be regarded as a fat fractal since the surface of the protein is a fractal but at finite volume.

4.1.2 Multifractal

Multifractals[3] or fractal measures are related to the study of a distribution of physical or other quantities on a geometric support. The support may be an ordinary plane, the surface of a volume, or it could itself be a fractal. In general, a multifractal possesses an infinite number of singularities of infinitely many types. The term "multifractality" expresses the fact that points corresponding to a given type of singularity typically form a fractal subset the dimension of which depends on the type of singularity. The idea that a multifractal may be represented in terms of intertwined fractal subsets having different scaling exponents opens a new realm for the application of fractal geometry to physical systems. The study of multifractals is a rapidly developing field.[5]

In order to describe a multifractal in a more quantitative way, we imagine that a d-dimensional hypercubic lattice with a lattice constant L is put on the fractal and denoted by p_i the stationary probability associated with the jth box of volume ι^d, where $\Sigma_i p_i = 1$. First we are interested in the behavior of p_i as a function of the box size measured in units of the linear size L of the structure. As before, this dimensionless unit will be denoted by $\varepsilon = \iota/L$, and the scaling of various quantities in the limit $\varepsilon \to 0$ will be studied, trivially for a homogeneous structure with a uniform distribution (density) $p_i(\varepsilon) \sim \varepsilon^d$. In the case of a uniform fractal of D with a uniform distribution on it, $p_i(\varepsilon) \sim \varepsilon^D$, since a distribution with a uniform density p_i corresponds to the volume of support in the ith box. In the more complex situation, when a nonuniform fractal with a distribution having infinitely many singularities is considered, we are led to assume a general form

$$p(\varepsilon) \sim \varepsilon^\alpha \qquad (3)$$

where $\varepsilon \ll 1$ and α can take on a range of values depending on the given value of the measure. The noninteger exponent α corresponds to the strength of the local singularity of the measure and is sometimes also called *crowding index* or Hölder exponent. Although α depends on the actual position on the fractal, there are usually many boxes with the same index α. In general, the number of such boxes scales with ε as

$$N_\alpha(\varepsilon) \sim \varepsilon^{-f(\alpha)} \qquad (4)$$

where $f(\alpha)$ is the fractal dimension of the subset of boxes characterized by the exponent α. The exponent α can take on values from interval $[\alpha_\infty, \alpha_{-\infty}]$ and $f(\alpha)$ is usually a single humped function with a maximum

$$\max_\alpha f(\alpha) = D \qquad (5)$$

The $f(\alpha)$ spectrum of an ordinary uniform fractal is a single point on the f-α plane. Thus, a typical multifractal is assumed to be made of interwoven sets of singularities of strength α, each characterized by its own fractal dimension $f(\alpha)$. This fact is the reason for the name *multifractal*, which is frequently used for such fractal measures.

In order to determine f(α) for a given distribution, it is useful to introduce a few quantities which are more directly related to the observable properties of the measure. Then, relations among these quantities and f(α) make it possible to obtain a complete description of a fractal measure.

An important quality which can be determined from the weights p_i is the sum over all boxes of the qth power of box probabilities

$$Z_q(\varepsilon) \equiv \sum_i p_i^q \qquad (6)$$

for $-\infty < q < \infty$. This definition of Z is closely related to the Renyi entropies. For q = 0 Equation (6) gives N(ε), the number of boxes of size ε needed to cover the fractal support (the region where $p_i \neq 0$). Therefore,

$$Z_0(\varepsilon) = N(\varepsilon) \sim \varepsilon^{-D} \qquad (7)$$

where D is the dimension of the support. Since the distribution is normalized, $Z_1(\varepsilon) = 1$.

Because of the complexity of multifractal distributions, the scaling of $Z_q(\varepsilon)$ for $\varepsilon \to 0$ generally depends on q in a nontrivial way:

$$Z_q(\varepsilon) \sim \varepsilon^{(q-1)D_q} \qquad (8)$$

where D_q is the so-called order q generalized dimension. It is also given by

$$D_q = \lim_{\varepsilon \to 0} \frac{1}{q-1} \frac{\ln \sum p_i^q}{\ln \varepsilon} \qquad (9)$$

The distribution of $p_i - \varepsilon$ is extremely inhomogeneous concerning both values and the number of boxes with the same p_i. As a result, when $\varepsilon \to 0$, the dominant contribution to Equation (6) comes from a subset of all possible boxes. This subset forms a fractal with a fractal dimension f(q) depending on the actual value of q. Thus,

$$N_q(\varepsilon) \sim \varepsilon^{-f(q)} \qquad (10)$$

where $N_q(\varepsilon)$ is the number of boxes giving the essential contribution in Equation (6). In addition, all of these boxes have the same $p_i = p(q)$. We shall denote the singularity strength for boxes with probability p(q) by α(q), namely

$$p(q) \sim \varepsilon^{\alpha(q)} \qquad (11)$$

for $\varepsilon \to 0$. The f(q) spectra defined by Equations (10) and (11) provide an alternative description of a multifractal with regard to Equations (3) and (4) given by

$$(q-1)D_q = q\alpha(q) - f(q) \qquad (12)$$

$$\alpha(q) = \frac{d}{dq}\left[(q-1)D_q\right] \qquad (13)$$

Therefore, the spectra D_q and $f(\alpha)$ represent equivalent descriptions of multifractals, since they are Legendre transforms of each other.

The formalism described in the previous material assumes that the knowledge of p_i gives the set of box probabilities. For a real system or computer models, p_i has to be deterministic multifractals constructed by an exact recursive procedure and the box probability distribution can be obtained analytically, just like the fractal dimension, for ordinary deterministic fractals.

4.1.3 The spectral (fracton) dimension

The fractal structures are described by (at least) three dimensions: d, the dimension of embedding Euclidean space; D_f, the fractal dimension; and d_f, the spectral (fracton) dimension.[6] For Euclidean spaces, these three dimensions are equal. The spectral dimension d_f is defined by

$$\rho(\omega) \sim \omega^{d_f - 1} \qquad (14)$$

$$N_t \sim t^{d_f/2} \qquad (15)$$

where ω is the frequency, $\rho(\omega)$ is the density of states, and N_t is the number of distinct sites in the fractal visited by a random walker up to time t. The spectral dimension d_f differs in general from fractal dimension D_f because d_f reflects the topological structure properties of the fractal and D_f reflects the geometrical structure of the fractal. For an example, let us take the Sierpinski gasket[8] in d-dimensional Euclidean space. Its fractal dimension is easily found as $D_f = \ln(d + 1)/\ln 2$, and the spectral dimension $d_f = 2\ln(d + 1)/\ln(d + 3)$. It can readily be shown that d_f is related to D_f by

$$d_f = 2D_f/d_w \qquad (16)$$

where d_w is the exponent connecting the root-mean-square displacement R_w of the random walker on the fractal with the number of steps N_w, $R^{d_w} \sim N_w$, where d_w is called the fractal dimension of the walk, similar to d_w. Furthermore, many other fractal dimensions may be defined.

Because many types of fractal dimensions are used in this chapter, if there is no special statement then D or D_f represents a fractal dimension; d, the dimension of the embedding Euclidean space; D_s, the surface fractal dimension; D_b, the box-counting dimension; D_H, Hausdorff dimension; D_m, mass fractal dimension; d_f, fracton (or spectral) dimension; d_s, self-similar dimension; d_w, the fractal dimension of the walk.

In this chapter we shall deal with the following subjects. The fractal aspects, dimensions, and analysis of protein chains will be discussed in Sections 4.2, 4.3, and 4.5, respectively. Closing a portion of the protein in a sphere of radius R, denoted by S(R), and the R dependence of the mass enclosed in the sphere will be considered in Section 4.3.

How does one describe the density of the vibrational frequencies and protein dynamics? We shall illustrate the fracton dimension of proteins in Section 4.4, the correlative dimension for globular proteins in Section 4.6, chaotic analysis for biochemical reactions in Section 4.7, and the fractal approach for protein and enzyme kinetics in Section 4.10.

The surface roughness of the protein may have a significant effect on ligand binding, ligand diffusion on the surface, and product release. The fractal dimension of the surface, D_s, may be a compact parameter describing the degree of surface corrugation. The surface roughness of the protein may be related to fat fractals and multifractals. In Section 4.8, we will calculate the fractal dimension of the protein surface, and in Section 4.9, the fractal reactions of enzyme surfaces will be considered.

Additionally, in Section 4.11 we shall illustrate the fractal art of enzyme model design. In Section 4.12, we will make our comments and discuss the fractal studies of proteins and enzymes.

4.2 Fractal aspects of protein chain structure

As is well known, proteins have well defined, three-dimensional structures dictated by their amino acid sequence. Recently, many investigations have shown that the protein chain structure has fractal aspects. Because linear polymer molecules containing active chemical groups cross-link or associate irreversibly to form larger macromolecules of a more complex structure closely related to the protein and enzyme, fractal aspects and dynamics of linear polymers have been widely studied by scientists, e.g., Cates,[9] Muthukumar,[10] Dewey,[11] and Adam and Lairez[12] have investigated the fractal comformations of linear polymers, star polymers, and branched polymers in dilute and semidilute solutions in good solvents.

A polymer molecule exhibits fractal behavior, i.e., statistical invariance under dilatation. Cates[9] has stated that the fractal dimension D_f of a polymer can indicate the arrangement of a given polymer or molecular network, but not their intrinsic self-similar connectivity, (e.g., in the ideal limit, $D_f = 2$; in the limit of a swollen coil, $D_f \approx 5/3$) but their internal connectivity is one dimension. The connectivity may be described by the fracton dimension d_f introduced in Section 4.1 (we shall discuss the fracton dimension in Section 4.4). For a linear chain, $d_f = 1$; while for a fractal sol-molecule containing a hierarchy of branches and/or loops on every length-scale, $1 < d_f < 2$, e.g., sol-molecules described by the percolation model have a value of d_f close to 4/3.

"Polymeric fractal" is a class of fractals made of flexible polymer chains at a short length-scale and having self-similar connectivity at a larger length-scale. Therefore, the term can be used to describe aspects of many kinds of polymers, particularly the macromolecules arising by disorderly cross-linking (gelation or aggregation) of flexible chain precursors.

4.2.1 The problem of entanglements

There is an entanglement effect at a high density in polymer molecules. Entanglements should dominate at high enough molecular weights and densities, but for linear chains, there exists a well defined molecular weight below which they can be ignored. For example, an enduring entanglement between linear chains near a concentration C* may have much higher densities than other cases. Linear chains can form long-lasting entanglements at concentrations just above the overlap threshold, but the topological interaction may remain harmless in effective medium (EM) theory; hence, they may be ignored, also.

The effects of entanglements in the vicinity of the overlap threshold may be decreased by the excluded volume repulsion between neighboring molecules. Since an entanglement between a large molecule and a smaller one will resolve on a time-scale that is short compared to the diffusive relaxation time of the larger, polydispersity reduces the effects of entanglements. For example, for branched molecules most entanglements form between side-limbs and can therefore resolve themselves on a shorter time-scale than the overall chain relaxation. From this point of view, it would seem that the entanglement effect should be neglected.

Cates[9] investigated the fractal features of a solution of flexible chain macromolecules of arbitrary self-connectivity, the static configuration, and the ideal chain dimension of polymeric fractals. Cates' main goals were to discuss the dynamic of polymeric fractals associated with fracton dimensions and to calculate various exponents describing the dynamical behavior of a polymer sol-gel system close to its critical point.[9] The investigations indicate that the fractal aspects of polymers (such as linear, star, branched, and/or containing loops, and even simple star-branched molecule chain) will affect polymeric kinetics to varying degrees. So, it is necessary to study the fractal aspects of chain structure.

4.2.2 Fractal aspects of polymers

4.2.2.1 Linear polymers

A linear polymer is an object with a connectivity dimension equal to 1, i.e., a linear polymer is a cord with respect to a walk along its chemical sequence. We will first discuss dilute solutions of linear polymers. This situation corresponds to the case of $d \gg R$, where d is the mean distance between neighboring polymer chains and $R = \langle R_g^2 \rangle^{1/2}$ is the square root of the mean square radius of gyration of the polymer. In some conditions, the polymer conformation can be imagined as a self-avoiding walk (SAW) in three dimensions. Taking the origin anywhere on the polymer, the number n(r) of monomers in a sphere of radius r scales as $n(r) = (r/a)^D$, where "a" is the size of the monomers constituting the polymer. By the Flory theory, D = 5/3, while field theory would result in D = 1.70. Introducing a function g(r) to represent the probability of finding a monomer at a distance from the origin, then

$$\int_0^R g(r) r^2 \, dr = N$$

where N is the number of monomers per chain and $N = (R/a)^D$. For r > R, g(r) = 0; for r < R, a polymer molecule is self similar; and g(r) corresponds to the monomer concentration G(r) in the sphere of radius r and scales as $G(r) = n(r)/r^3 \sim r^{D-3}$ for a < r < R.

Because polymers interact, the osmotic pressure is another important quantity to describe polymers. The relationship between osmotic pressure and concentrations has been investigated.[12] The quantity g(r) and osmotic pressure can be measured by scattering experiments (light or neutron), and quantities N, R, G(R), and particularly the fractal dimension D of polymer chains can be deduced. For different benzene dilute polystyrene samples, Miyaki et al.[13] obtained the relationship $M \sim R^D$ between molecular mass M (in g/mol) vs. radius of gyration R (in Å) with D = 1.67 ± 0.01 using light scattering. In dilute cyclohexane solutions of polyisoprene, Davidson et al.[14] found $C(R) \sim M^{-3/D+1}$ with D = 1.69 ± 0.01 by scattering. From neutron scattering experiments performed on polystyrene samples in dilute and good solvent solutions by Cotton et al., it was found that $I(q) \sim q \sim q^{-D}$ with $D = 1.70 \pm 0.01$,[12] where $I(q)$ is the scattered intensity; $I(q)$ and $g(q^{-1})$ are Fourier transforms of each other; $q = 4\pi\sin(\theta/2)/\lambda$, in which λ is the effective wavelength in the medium, θ is the scattering angle; and $q^{-1} < R$.

Now semidilute solutions of linear polymers are considered. By increasing the concentration in dilute solutions, one reaches a particular value $C^* = C(R)$. The case $C < C^*$ is the dilute solution discussed previously, while $C > C^*$ is the semidilute situation. C^* can be expressed as $C^* = a^{-3}N^{1-3/D}$; hence, C^* may be obtained by digital analysis, e.g., taking $N = 10^4$ and $D = 5/3$ results in $C^* \approx a^{-3} \times 10^{-3}$. So, the value of C in the range $C^* \leq C \leq a^{-3}$ may be regarded as the concentration of semidilute solution. In this case, $d \ll R$ and linear polymers form a transient network with a mesh size ξ. The correlation length ξ is an important concept. $\xi < R$ and ξ decrease as the C increases, while $C(r < \xi)$ remains unaffected. At a length-scale r between "a" and ξ, D is equal to the fractal dimension of a SAW. When $r > \xi$, binary contact between polymer chains screen excluded volume interactions, and the material appears as a homogenous and densely packed set of correlation volumes (called "blobs") ξ^3. Inside a blob, polymer conformation results from the excluded volume interactions (D = 5/3), while at a larger r, polymer conformation is Gaussian (D = 2). ξ depends only on the concentration:

$$\xi \sim C^{1/(D-3)} \qquad (17)$$

The concentration dependence of the osmotic pressure can be expressed as

$$\pi = \pi_0 (C/C^*)^{D/(3-D)} \qquad (18)$$

This scaling law illustrates how physical properties of polymers in semidilutions present a universal nature. The power-law behavior given by Equation (17) is comfirmed by the neutron experiments performed on polystyrene solutions with $1/(D - 3) = -0.72 \pm 0.04$. Light scattering measurements of the osmotic gradient performed on polyisoprene and poly(α-methylstyrene) agree well with the behavior given by Equation (18),[12] where $D/(3 - D) = 1.32 \pm 0.02$. In other cases, except for the Gaussian (D = 2), the experimental fractal dimensions are in good agreement with the experimental one.

4.2.2.2 Star polymers

Linear polymers join together at one end to form star-shaped molecules. Star polymers are depicted as a densely packed set of blobs, the sizes of which increase with the distance from the center of the star. The fractal aspects may be characterized by the scaling law $c(r) = a^{-3}(r/a)^{D-3}f^{(3-D)/2}$, where $c(r)$ represents the internal concentration of monomers at a distance r from the center of the star, f, defined by $\xi(r) = rf^{-1/2}$. If R is the radius of the star, then $R = \int_0^R c(r) r^2 \, dr$ is also equal to the total number N_t of monomers belonging to the star, where

$$N_t = \left[f^{(3-D)/2D} (R/a) \right]^D \qquad (19)$$

This shows that the star polymer has the same fractal dimension as linear polymers; however, a single star fractal dimension depends on the distance from the center, which is due to the symmetry around the center. This implies that $g(r) \neq c(r)$ and leads to a particular scattered intensity profile I(q) different from that obtained for a linear polymer

$$I(q) \sim CN_t F(qR, f) \quad (20)$$

where, for $qR < 1$, $F(qR,f) = 1$; for $1 < qR < f^{1/2}$, and $F(qR,f) = (qR)^{-3}$, and for $f^{1/2} < qR$, $F(qR,f) = f^{(D-3)/2} (qR)^{-D}$. Equation (19) can be verified from light scattering data.[15] A typical neutron scattering from a labeled 12-arm star polyisoprene,[16] shown in Equation (20), agrees well with the experiment results. We think that the star polymer has some self-affinity and multifractal features, which should be studied further.

For the case of semidilute, the fractal investigation is making progress, e.g., at a given concentration one obtains $X = f^{1/2} G^{1/(D-3)}$, where X is the size of the region of space where the star structure is unaffected. However, this remaining star structure affects so few monomers that they cannot be seen by scattering experiments. In this concentration region, the stars look like linear polymers; therefore, the model and ideas which are similar to the linear case are applied to describe a single star polymer within which arms screen each other and theoretical predictions have been verified experimentally.

4.2.2.3 Branched polymers

Branched polymers may be defined as polymers having a connectivity dimension larger than 1. Their fractal aspects and mass distribution may be characterized by a percolation model, in which not only are their fractal comformation and polydispersity described very well, but also the hyperscaling law linking the fractal dimensions of polymers to their mass distribution is revealed.

In the present case, branched polymers correspond to the percolation/gelation clusters and the degree of connectivity corresponds to the probability p. Here, p is defined as the ratio of the number of chemical bonds between monomers and the total number of potential bonds. Near the gelation/percolation threshold $p = p_c$, the arrangement of branched polymers shows self-similarity, as does the conformation of each gelation cluster. The fractal dimension $D_p \approx 2.52$, which is obtained by computer simulations. The fractal conformation of percolation clusters result from the random connection alone as well as the sum of interactions between monomers belonging to the same polymer and those between monomers belonging to different polymers. In experiments, a swollen conformation was found for a single branched $d_s \approx 2$. Apart from this swelling effect, scattering experiments are sensitive to the infinite polydispersity effects of a branched polymer.

The hyperscaling law was proposed by Stanley.[17] Using it, the number of free exponents for a critical phenomenon can be reduced. In the case of gelation, the hyperscaling law of three dimensions becomes $3/D_p = \tau - 1$, where τ is the exponent defined by the formula $n_i = M_i^{-\tau} F(M_i)$, where n_i is the fraction of polymers having a mass M_i and F is a cut-off function with $F(x < 1) = 1$ and $F(x > 1) = 0$. The value of τ is generally obtained by computer simulation applying the Monte-Carlo method, e.g., in a three-dimension lattice, it is found that $\tau \approx 2.20$. The hyperscaling law has very strong implications on the relative arrangement of branched polymers. The neutron experiments show that for a fractionated sample, $d_s = 1.98 \pm 0.03$ and corresponds to the swollen state, while for a polydisperse sample, $d_s(3 - \tau) = 1.59 \pm 0.5$, which agrees well with the expected theoretical value.

Muthukumar[10] studied the viscosity of a solution containing polymeric fractals. Suppressing the entanglement effects, branched polymers containing no loops are considered first. In this case, the branch units have different functionalities and each strand has a given number of Kuhn lengths. Adopting any suitable convention to label the strands in increasing order, every segment i of all n segments can be labeled. Of course, the presence of loops also can be adequately described.

Suppose a solution contains N branched polymeric fractals in a total volume V. According to effective medium theory and for dilute solutions and zero-frequency limits, whether excluded volume effect is present or not, the viscosity change is given by $\delta\eta \sim CM^{(d-D)/D}$, where δ is the Dirac delta function; η, the viscosity; $\delta\eta$, the change in viscosity; C, the polymer concentration; and M, the molecular weight. For a Gaussion chain, D = 2; for the self-avoiding linear chain, D $\approx$ 5/3; and for a percolating cluster, D = 2.5. These values of the molecular weight exponent are in agreement with the experimental values of 0.5, 0.8, and 0.17, respectively.[10] If the frequency $\omega \to \infty$ in a dilute solution, there is a relationship $\delta\eta \sim C\omega^{(D-d)/d}$. The cases of semidilute and dense solutions can be discussed similarly.

The scaling laws also were derived by Muthukumar.[10] Suppose ξ_1 is the hydrodynamic screening length, then the fractal scaling laws are presented as follows: for dilute solutions, $\xi_1 \to \infty$; $\xi_1 \sim C^{1/(\overline{D}-d)}$ and $\xi_1 \sim C^{1/(\overline{D}-d)}$ correspond to semidilute and dense solutions, respectively.

4.2.3 Universal aspects of protein structure

The universal aspects of protein structure may be analyzed by scaling laws related to the radius of gyration, volume, and surface area. Proteins have been described as "collapsed polymers" by the fractal dimensions of protein backbone and surface. Recently, Dewey[18] has presented a polymer collapsed model to address the scaling properties of proteins, and this model is considered in another paper.[11] The investigations for 45 different proteins show the relationship $R_g \sim N^v$, where R_g is the radius of gyration and N, the residue number. The experimental value of v is 0.35 ± 0.03, while the theoretical value is 3/5 for an excluded volume polymer, 1/2 for an ideal polymer, and 1/3 for a collapsed polymer. Taking into account experimental error, the fractal dimension of the backbone D = 1/v is close to 3. This indicates that proteins behave as collapsed polymer and globular structures.

The surface area of protein may be determined from the "ball rolling" algorithm. Since molecular weight is proportional to volume, the molecular weight and surface area data may be used to determine the fractal surface dimension. Dewey's result is D_s = 2.16 for 15 different proteins.[11] It is very close to values obtained for a variety of proteins using a number of different algorithms. Additionally, the surface dimension of proteins can be deduced from the co-dimension principle and the backbone dimension. Dewey[11] obtained the surface fractal dimension of proteins (D_s = 2.14) according to the co-dimensional principle. Dewey thinks that the co-dimensional principle can be used to explain why a fractal backbone dimension of slightly less than 3 gives a surface dimension slightly greater than 2. The global model may be used not only to describe the fractal aspects of proteins, but also to determine the fracton dimension d_f associated with vibrational relaxation in proteins.

A variety of global relaxation models can be used to predict the values of d_f by Equation (16). Constraining the random walk to the fractal, the value of d_w is model dependent, e.g., the polymer has "local" bridges which participate in relaxation, and $d_w = (D + \overline{Z})/2$, where Z = 0.92 for a two-dimensional space and 1.46 for a three-dimensional space. For the global models considered in Dewey's paper,[11] the "local" bridge case appears to be the most appropriate. Not only is this due to d_f = 1.32 being in excellent agreement with the ironsulfur protein, but it also provides Dewey's motivation for investigating the alpha helical model. This model has local bridges due to H-bonding between every fourth residue in the chain. To emphasize the importance of global modes, Dewey also examined the long-range correlations in the Debye-Waller factor using R/S analysis and X-ray data. The results indicate that the fluctuations of atomic displacements are highly correlated along the protein chain and the correlation is higher with backbone atoms than with the side chains. Since the protein chain folds back on itself and consequently will bring residues at distant positions near each other, this long-range correlation may be the effect of the surface on the thermal factors.

4.3 The fractal dimension of the chain and the mass dimension

In this section, we consider the fractal dimension of the chain, D_c, then we will contrast D_c with the mass dimension, D_m. The linear polymer is described by successive jumps between monomers along the chain. Jumps between different portions of the chain which may be close in space but far apart along the chain are not allowed. Thus, the polypeptide chain is modeled by SAW, the "walker" generates the polypeptide chain, and each successive step corresponds to a new monomer.[2] Let R be the position vector of a monomer N, the origin being at a monomer number 0, then R^2 is the square distance between 0 and N. The protein molecule generally has a well defined average structure which is different from the polymer statistic.[2] The structure fluctuations around the average in a stable protein typically are much smaller than the range of statistical fluctuations in a random coil. In order to calculate the average R^2 for a protein molecule, one must adopt a different approach. Divide a protein molecule into segments of length L(R) being (much) smaller than N. Now, let $R \equiv (R^2)^{1/2}$ for a uniformly random walk with some restrictions. Flory found the approximate relationship $L(R) \sim R^{D_c}$, $D_c = 5/3$ in three dimensions.[2] This approximation agrees extremly well with many numerical and experimental tests; however, any attractive forces were ignored in this mode and, hence, deviation will appear due to existing cross-links (e.g., hydrogen) between pieces of the chain which are far apart along the chain but close in space and due to the finite size N of polymer.

Let M(R) represent the mass in a sphere of radius R, denoted by S(R). The mass-radius relation can be writen as

$$M(R) \sim R^{D_m} \tag{21}$$

where D_m is called *mass dimension*. In the protein molecule, M(R) is the number of monomers enclosed in S(R), and the origin is set at monomer zero. Equation (21) indicates that the larger the D_m, the more monomers are in the sphere. For polymeric fractals, Equation (21) has been verified experimentally. For different benzene dilute polystyrene, Equation (21) holds for the molecular mass M vs. radius of gyration R_g with $D_m = 1.67 \pm 0.01$ (see Section 4.2.2). D_m also describes the power-law correlation of the internal monomer concentration c(R), e.g., in dilute cyclohexane solutions of polyisoprene, $C(R_g) \sim M^{-3/D_m+1}$ with $D_m = 1.69 \pm 0.01$ (see also Section 4.2.1).

By expressing R in units of ε, Equation (21) may be rewriten as

$$M(R) = B(R/\varepsilon)^{D_m} \quad \text{for} \quad R \gg \varepsilon \tag{22}$$

where B is a unitless constant and a local quantity that is different from the shape factor. Equation (22) can also be interpreted in terms of the Hausdorff measure: let $H^D(R)$ be the Hausdorff measure of the part of the object that is cut out by S(R). One can deduce that[19]

$$H^D \approx M(R)\varepsilon^D = BR^D \qquad (23)$$

The mass-radius relation is often applied to a series of systems. If the systems have constant shape and variable diameter L, the total mass is denoted by $M_t(L)$. Then[19]

$$M_t(L) \sim L^{D_m} \qquad (24)$$

Applying Equation (24) to polymers, one has

$$R_q \sim (W_m)^v \qquad (25)$$

where W_m represents molecular weight. Substituting the two sides of Equation (25) with L and (M_t) yields $D_m = 1/v$ for the fractal dimension of polymers (see also Section 4.2.3).

The chain dimension D_c and the mass dimension D_m coincide only if all the monomers in the sphere are between monomer zero and monomer M(R). In the evaluation of D_c, only covalent-bond atoms in S(R) are counted, while for D_m, all atoms in S(R) are counted. D_m is therefore larger than D_c. Helman et al.[20] suggested that D_m approaches the same value in the limit $M(R) \to \infty$, but for a finite polymer this is not true. This difference was demonstrated first by Wagner et al.,[21] Colvin and Stapleton,[22] and Stapleton.[23]

Another difficulty in calculating D_m is due to the existence of 20 different possible monomers (amino acids) which differ in their side-chains, while the main backbone is the same. Stapleton and co-workers considered only the C_α of the main backbone to simplify the calculation of D_c and D_m. The importance of the main backbone of the polypeptide chain was emphasized while the contribution of the side-chains was ignored. The estimation of the mass density is inaccurate owing to the larger difference in the side-chain masses, especially for the extreme cases of glycine and tryptophan. When the side-chain was taken into account for myoglobin and the mass density calculated, the numerical result was $D_m = 2.9 \pm 0.2$,[2] the error reflecting boundary effects. Within the accuracy of current calculations, the protein is a three-dimensional object. Colvin and Stapleton[22] obtained $D_m = 1.91$ for C_α of myoglobin. Therefore, it seems that the C_α are not representative of the mass density in the protein. Stapleton[23] obtained $D_c = 1.6$. Hence, D_c and D_m for C_α are not equal to the D_m value of the protein. What the protein chain dimension and D_m of C_α actually mean is not yet clear. Continued investigation should focus on fracton dimension d_f and conformation of the protein chain. Therefore, after we discuss the fracton dimension of proteins in Section 4.4, in the begining of Section 4.5 we shall focus our attention on the four structural classes, e.g., α, α/β, $(\alpha + \beta)$, and β, of protein conformation. In other words, we shall report the results of attempts to apply fractal theory to the analysis of the tertiary structure of proteins having complex irregular form.

4.4 Fracton dimension of proteins[28]

Because the property and function of a protein relate closely to the vibration of interior atoms and chain conformation, researchers have initiated a number of investigations into the protein using vibrating normal modal analysis, molecular kineses, Monte Carlo simulations, etc. Wako et al.[25] have developed a series of fast methods to compute the protein conformational energy. Gō et al.[26] have proposed a dynamical method to study the low frequency vibrational mode of proteins, and they computed the distribution of the normal-modal density of bovine proteins trypsin inhibitor (BPTI). In recent years, we have studied the fractal features of proteins and have proposed some new methods.[3,24] In this section, we shall calculate the fracton (spectral) dimensions of some proteins by the method of normal-modal analysis and discuss the relationship between the fracton dimension and the atomic vibration and macrodynamical behavior of proteins.

In fractal theory, the static structural properties can be described by Euclidian space (embedding space) dimension d and fractal (mass scaling exponent) dimension D, whereas the dynamical properties such as the vibration of an atom and the diffusion of a particle may be described by the fracton dimension d_f. The vibrational nature of fractal structure is characterized by frequency ω. The relationship between the vibrational state densities $\rho(\omega)$ and the frequency ω is denoted by

$$\rho(\omega) \sim \omega^{d_f - 1} \quad \text{for} \quad 0 \leq \omega \leq \omega_{max} \tag{26}$$

where $\rho(\omega)d\omega$ is the number of vibrations in the interval of ω and $\omega + d\omega$, and ω_{max} is the Debye cut-off frequency. The selection of ω_{max} depends upon the modal number of total vibrations being equal to the number of degrees of freedom in the system. Looking upon the dihedral angle of protein conformation as the variable of the function of conformational energy, according to the X-ray diffraction data, one can calculate ω_{max} applying the method proposed by Wako et al.[25] When the frequency is less than ω, the number of the normal mode is given by

$$F(\omega) = \int_0^\omega \rho(\omega) \, d\omega \sim \omega^{d_f} \tag{27}$$

where d_f is the fracton dimension. When the frequency is low, a protein resembles an elastomer and the modal number can be regarded as the number of degrees of freedom of the protein backbone. Within the range of $0 \leq \omega \leq \omega_{max}$, in the plot of $\log F(\omega) - \log \omega$, one can calculate the fracton dimensions. The results are listed in Table 1. Figure 1 indicates the relationship between the modal densities of myoglobin and the frequencies.

Table 1 The Fracton Dimension of Proteins

Proteins	N[a]	m[b]	ω_{max}	d_f	D[c]
Ovomucoid (3rd region)	51	276	73.77	1.527	1.537–1.867
Plastocyanin	99	543	71.19	1.685	1.495–1.798
Bence-Jones protein	114	613	72.53	1.503	1.648–1.855
Ribonuclease A	124	714	73.80	1.718	1.327–1.903
Lysozyme (hen egg-white)	129	747	73.11	1.614	1.302–1.971
Flavodoxin	138	820	69.71	1.739	1.315–1.966
Hemoglobin (deoxy)	141	824	71.48	1.583	1.387–1.976
Myoglobin	153	921	72.16	1.728	1.335–2.099
Dihydrofolate	159	924	68.27	1.644	1.304–1.995
Glyceraldehyde-3-phosphate dehydrogenase	333	2024	70.28	1.847	1.382–1.897

[a] N is the number of residues.
[b] m is the number of degrees of freedom.
[c] D is the Hausdorff dimension.

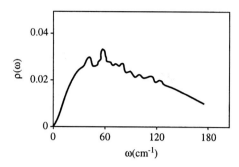

Figure 1 The relationship between the modal densities and frequencies of myoglobin.

From Table 1, we find that the values of ω_{max} are distributed within the range of 68.27 to 73.80 and they are independent of the size of the proteins. There is a linear relationship m = 6.266N − 62.357 (r = 0.9992) between the residues number N and the number of degrees of freedom, m. The values of fracton dimensions are distributed within the range of 1.503 to 1.739, which is in the range of Hausdorff dimensions and coincides basically with the results of Wagner et al.[21] According to the study of Alexander and Orbach,[7] $d_f = 2D_f/d_w$, where d_w is the walk fractal dimension. When $d_w = 2$, $d_f = D_f$. Their study shows that the d_f values are reasonable in the range of D_f values. By the self-avoiding random walk (SAW) model we know that $D_f = (d + 2)/3$. When d = 3, $D_f = d_f \approx 5/3 \approx 1.667$. The fracton dimension calculated by us is about 1.6, which agrees with the fractal dimension in Euclidian space. It shows that the fracton dimension can be replaced by the Hausdorff dimension under some given conditions.

The fracton dimension of protein mainly reflects the vibrational properties and some structural features. The crossing concept of the vibration excitation of fractal structure from long wavelength (low frequency) phonon to short wavelength (high frequency) fracton was first proposed by Alexander and Orbach.[7] Aharony et al.[28] also have defined a scaling method. The relationship between crossing length ξ and the characteristic frequency ω_c is given by $\omega_c \sim \xi^{-D/d_f}$, i.e., the fracton frequency is greater than ω_c and the phonon frequency is less than ω_c. On a protein fractal, the vibration of the high-frequency local quantum will lead to relaxation, which results in the emitting and absorbing states of the local electrons. The relaxation rate is calculated by the probability density. And there are two interactions: that of local electron–single fracton and that of local electron–double fractons. Alexander et al.[29] have further considered the interactions between phonons and fractons: two phonons make up a fracton, and one fracton and one phonon make up another fracton. We think that there are differences between the geometric localization (fracton excitation) of the vibrational module of proteins and Anderson localization (localizational phonon) coming from the irregular scattering. The spatial distribution of proteins is not heterogeneous and the vibrational module is completely different from the vibrational state coming from Anderson localization. The Anderson localizational processing of all phonons is regarded as a weak scattering phenomenon, the state density is invariant, and there is an Anderson localizational edge ω_l, whereas a fracton edge is a crossing frequency ω_c, $\omega_l/\omega_c \approx (1/2\pi)^{1/2}$. There are three scales — wavelength (λ), scattering length (λ_s), and local length (λ_l) — near the Anderson localization edge. But, near the fracton edge, one can consider that $\lambda_l \approx \lambda$ for all frequencies ω; therefore, it is always satisfying the Ioffe-Regel limitation.

Since the vibration of proteins relates to temperature, the fracton dimension is a useful parameter integrating microvibration with macrotemperature. Stapleton[23] pointed out the relationship between Raman relaxation rate T_R and temperature T in the proteins:

$$1/T_R \sim \int_0^{\omega_{max}} \rho^2(\omega)\omega^4 f(\sigma\omega/kT)\,d\omega \tag{28}$$

where $f(z) = e^z/(e^z - 1)^2$, $\sigma = k/2\pi$, σ and k are the Plunk constant and the Boltzmann constant, respectively. Because $\rho(\omega) \sim \omega^p (p = d_f - 1)$,

$$1/T_R \sim (kT/\sigma)^{5+2p} F_{4+2p}(\theta/T) \tag{29}$$

where $\theta = \sigma\omega_{max}/k$. When $T \ll \theta$, $F_{4+2p}(\theta/T)$ is a constant; when $T \gg \theta$, $F_{4+2p}(\theta/T) = T^{-(3+2p)}$. Therefore, one can measure fracton dimensions of proteins by the electron spin resonance (ESR) method. According to the effective medium algorithm (EMA), one can know that the fracton dimension relates to hydrogen bonds in the protein chain. The more hydrogen bonds there are, the less the d_f value will be; conversely, the fewer hydrogen bonds there are, the greater the d_f value will be. Therefore, the value of d_f reflects the properties of the protein chain.

On the other hand, the fracton dimension of proteins or enzymes relates to the macrodynamical behavior.[2] We have studied the allosteric effects of proteins[6] and found that in regard to the multistep conformational change, the relationship between the Hill coefficients and the fracton dimension can be found as follows:

$$h_a = 1 + 2/d_f, \qquad h_b = 2 + 2/d_f \tag{30}$$

For glyceraldehyde-3-phosphate dehydrogenase, $d_f = 1.84$. By Equation (30) we have $h_a = 2.083$ and $h_b = 3.083$; therefore, the mean value $\bar{h} = 2.54$ is almost identical to the experimental value 2.3. For hemoglobin (deoxy), $d_f = 1.583$ and by Equation (30), $h_a = 2.263$ and $h_b = 3.263$. The mean value $\bar{h} = 2.763$ agrees well with the experimental value, $h = 2.8$, of hemoglobin-absorbing oxygen. This shows that some dynamical properties of proteins can be explained by the fracton dimension.

4.5 Fractal analysis of protein chain conformation

4.5.1 The fractal dimension of the main chain and tertiary structure

Since the tertiary structure of a protein molecule is mainly dictated by the primary structure of its sequence of amino acids, it might be quite desirable for our understanding of the morphology of proteins to find a certain structural quantity that reflects the tertiary structure of a protein structure based on its amino acid sequence. Recently, there has been a great deal of interest in studies of the fractal properties of protein molecular chains as a means of achieving the above-mentioned goal. The fractal dimension could be a potential candidate for this structural quantity that reflects information of an amino acid sequence.

Starting from the site of the α-carbons, we can draw a zigzag line using the method proposed by Isogai et al.[30] and Wang et al.[31] The backbone length L(R) of a protein molecule consisting of N residues measured with a scale of R can be written as follows:

$$L(R) = Lz(R) + nLz(R)/RK \tag{31}$$

where Lz(R) is the length of the zigzag line and the second term on the right is the correction term in which n and K are the number of the remaining unconnected residues and the folding number of the zigzag line, respectively. The zigzag line is drawn by connecting the C_α atoms of the protein step-by-step at intervals of R residues starting from the C_α atoms of the N-terminal residue. The length of the molecule with a scale of R is defined as the sum of the length of the zigzag line and a correction term which takes into account the contribution from the residues left unconnected at the C-terminal side. Defining H(R) as the slope of a fractal diagram logL(R) – log(R) obtained by linear regression, a local fractal dimension $D_0(R)$ is given by

$$D_0(R) = 1 - H(R) \qquad (32)$$

Although there are some self-similarities in the polypeptide chains of proteins, rigorous self-similarity is lacking in proteins; hence, the resulting value of the fractal dimension depends, to a certain extent on the choice of the scale R. The results, therefore, are averaged over the number of various choices for the scale R. The mean fractal dimension $D_f(R)$ is defined as

$$D_f(R) = \frac{1}{j} \sum_{R=1}^{j} D_0(R) \qquad (33)$$

where j is the number of local fractal dimensions.

We have found that the results given by Isogai et al.[30] are different from the experimental and simulation data of Stapleton et al.[32] and Wang et al.[31] In addition, when the scale R is considerably large, the fluctuation of the logL(R)–log(R) curve is very significant. We find that there remain some drawbacks with the works of Isogai et al. for a D(L) value greater than 2. The local fractal dimension $D_N(R)$ concept proposed by Havlin et al.[32] is of advantage in characterizing the protein molecular chain. Here, we report some refinement in the method and reasonable computational results. The $D_N(R)$ is defined as

$$\left[\left(\langle R_R^2 \rangle_N\right)^{1/2}\right]^{D_n(R)} = AR \qquad (1 \leq R \leq N-1) \qquad (34)$$

where $\langle R_R^2 \rangle_N$ is the mean square separation of the end points of a segment containing R residues in a protein consisting of N residues; A is a constant. The slope of the curve of the log(R)–log $[(\langle R_R^2 \rangle_N)^{1/2}]$ diagram is the local fractal dimension $D_N(R)$. To obtain a stable value of fractal dimension D, we have calculated the mean value of $D_N(R)$, namely,[27]

$$\begin{bmatrix} D = \dfrac{2}{N} \sum_{R=1}^{N/2} D_N(R) & \text{(N being even)} & \qquad (35a) \\ D = \dfrac{2}{(N+1)} \sum_{R=1}^{(N+1)/2} D_N(R) & \text{(N being odd)} & \qquad (35b) \end{bmatrix}$$

To explore the relationship between the fractal dimension and tertiary structure of a protein, the 39 proteins adopted in this work are selected from the proteins in the database of the protein data bank so as to cover the four structural classes of protein. These are abbreviated hereafter as α, β, (α + β), α/β, respectively. The names of proteins adopted here and their classes are listed in Table 2. The fractal dimension values of three kinds of each protein are calculated by the above-mentioned procedure within the ranges of R ≤ 15, 1 ≤ R ≤ (1/2)N, and 1 ≤ R ≤ (N – 1), and the results are listed in Table 2.

The fractal diagram of superoxide dismutase is not linear but is jagged especially within the range of R > 15.[33] The part of the diagram within the range of R ≤ 15 is smooth and may be regarded as a straight line with negative slope; the slope of the fractal curve does not change considerably. These features found in the case of superoxide dismutase are observed in the other proteins. From these observations it has been determined that the fractal diagram of a protein consists of two components, although the folding of the protein chain should not be thought of as a simple fractal form like the coastline. Nevertheless, the folding of protein seems to retain its fractal nature within certain ranges of R because the diagram within each range can be regarded to be linear in the statistical sense. Since the fractal diagram within the range of smaller R reflects the local folding of the protein backbone and the range of larger R reflects the global folding of the protein backbone, the local and global folding of proteins seems to be dictated by different rules. The mean fractal dimensions within the range of $1 \leq R \leq 15$, $1 \leq R \leq \frac{1}{2}N$, $1 \leq R \leq (N-1)$ are dubbed fractal dimension D_L, D_M, and D_G for short-, middle-, and long-range constructions, respectively. It has been found that the order of fractal dimensions for four classes of proteins is as follows: α > α/β > (α + β) > β. This fact indicates that the fractal dimension may be used to characterize the tertiary structres of enzymes and proteins. From Table 2, we find that the order of fractal dimensions for some protein molecules is as follows: $D_L < D_M < D_G$. The value of fractal dimension D_L reflecting the local conformation of a protein is smaller than the value of fractal dimension D_G reflecting the global conformation of the protein. These results would be helpful for studying the principles of folding and morphology of protein. It has been found that folding of the protein chain does not follow a single rule through the local and global structures. In Table 2, the values of fractal dimension D_L, D_M, and D_G are distributed closely near 1.33, 1.56, and 1.95, respectively. This means that although there are some distinct differences in the primary structure of 39 proteins, the property of their three-dimensional structure tends to be conservative in a certain range of scale R. The magnitude of the mean values of the fractal dimensions for four structural classes increases in the order of β, (α + β), α/β, and α, because there are only α-helix structures in the α class, and there are almost exclusively β-sheet in the β class. The local structure of the β class extended more than that of the α class; hence, the α and β classes show the largest and the smallest values of fractal dimension, respectively, reflecting the extended character of the β structure and the closely packed character of the α-helix; those of the (α + β) and α/β classes are in between. The (α + β) class of α-helix and β-sheet tends to be segregated along the chain and the α/β class tends to alternate along the chain. The values of fractal dimension D(L) calculated by Isogai and Itoh[30] are distributed within the range of 1.5911 to 2.3445, especially D(L) far larger than the fractal dimension (D = 2) of the trajectory of a particle under random Brownian motion. This is an interesting and open question in the works of Isogai and Itoh. In fact, this phenomena is caused by the fluctuation of the fractal diagram when scale R is too large.

Table 2 Fractal Dimensions of Some Proteins

Structural class	Protein	Fractal dimensions D_L	D_M	D_G
α	Cytochrome C (albacore, reduced)	1.416 ± 0.004	1.567 ± 0.006	1.985 ± 0.007
	Hemerythrin	1.433 ± 0.003	1.485 ± 0.005	1.963 ± 0.003
	Myohemerythrin	1.367 ± 0.003	1.433 ± 0.008	1.980 ± 0.008
	Hemoglobin (deoxy)	1.387 ± 0.006	1.480 ± 0.006	1.976 ± 0.004
	Myoglobin (sperm, whale)	1.335 ± 0.005	1.462 ± 0.005	1.985 ± 0.004
	Virus coat protein	1.390 ± 0.003	1.445 ± 0.005	1.874 ± 0.005
β	Rubredoxin	1.326 ± 0.005	1.538 ± 0.007	1.955 ± 0.005
	α-Chymotrypsin A	1.305 ± 0.003	1.473 ± 0.008	1.956 ± 0.006
	Superoxide dismutase	1.317 ± 0.004	1.456 ± 0.003	1.954 ± 0.003
	Immunoglobulin	1.334 ± 0.007	1.373 ± 0.002	1.823 ± 0.008
	Trypsin (pH 8)	1.320 ± 0.002	1.380 ± 0.006	1.908 ± 0.007
	Prealbumin (human)	1.283 ± 0.003	1.355 ± 0.007	1.922 ± 0.006
	Concanavalin A	1.298 ± 0.008	1.355 ± 0.006	1.904 ± 0.005
	Tosylelastase	1.303 ± 0.007	1.377 ± 0.004	1.914 ± 0.004
	Acid protease	1.313 ± 0.004	1.447 ± 0.004	1.892 ± 0.003
α + β	Trypsin inhibitor (bovine pancreas)	1.288 ± 0.005	1.489 ± 0.003	1.998 ± 0.003
	Oxidized high potential iron protein	1.376 ± 0.003	1.433 ± 0.004	1.938 ± 0.002
	Cytochrome B_5	1.309 ± 0.006	1.555 ± 0.008	1.985 ± 0.002
	Ribonuclease A	1.332 ± 0.007	1.421 ± 0.007	1.874 ± 0.004
	Lysozyme (hen egg-white)	1.302 ± 0.007	1.502 ± 0.006	1.971 ± 0.003
	Papain	1.335 ± 0.008	1.442 ± 0.007	1.996 ± 0.002
	Carbonic anhydrase B	1.336 ± 0.008	1.405 ± 0.007	1.969 ± 0.005
	Thermolysin	1.343 ± 0.003	1.391 ± 0.003	1.780 ± 0.002
	Bacteriochlorophy 1-A-protein	1.322 ± 0.007	1.337 ± 0.003	1.953 ± 0.004
α/β	Thioredoxin (*E. coli*, oxidized)	1.367 ± 0.007	1.530 ± 0.003	1.954 ± 0.002
	Flavodoxin	1.343 ± 0.006	1.457 ± 0.004	1.966 ± 0.002
	Dihydrofolate reductase	1.328 ± 0.006	1.431 ± 0.003	1.995 ± 0.002
	Adenylate kinase (porcine muscle)	1.337 ± 0.003	1.401 ± 0.005	1.976 ± 0.003
	Yeast phosphoglycerate mutase	1.344 ± 0.002	1.444 ± 0.008	1.945 ± 0.004
	Triose phosphate isomerase	1.326 ± 0.003	1.487 ± 0.007	1.949 ± 0.002
	Subtilisin novo	1.342 ± 0.005	1.603 ± 0.002	1.873 ± 0.003
	Rhodanese	1.335 ± 0.008	1.552 ± 0.003	1.992 ± 0.004
	L-arabinose-binding protein	1.354 ± 0.008	1.563 ± 0.003	1.947 ± 0.002
	Carboxypeptidase A (bovine)	1.327 ± 0.003	1.367 ± 0.005	1.953 ± 0.001
	Glyceraldehyde-3-phosphate dehydrogenase	1.382 ± 0.005	1.369 ± 0.006	1.897 ± 0.002

Table 2 Fractal Dimensions of Some Proteins (continued)

Structural class	Protein	Fractal dimensions		
		D_L	D_M	D_G
	Phosphoglycerate kinase (horse)	1.325 ± 0.004	1.366 ± 0.007	1.917 ± 0.003
	Pyruvate kinase (cat)	1.344 ± 0.005	1.370 ± 0.005	1.905 ± 0.001
	Hexokinase A	1.321 ± 0.007	1.348 ± 0.007	1.936 ± 0.003
	d-Glucose-6-phosphate isomerase	1.332 ± 0.007	1.504 ± 0.008	1.968 ± 0.002

Table 3 The Mean Values of Fractal Dimensions for Four Structural Classes of 39 Proteins

	α	β	$(\alpha + \beta)$	α/β	Mean values of 39 proteins
Mean values of D_L	1.388 ± 0.004	1.311 ± 0.005	1.329 ± 0.006	1.343 ± 0.005	1.343 ± 0.005
Mean values of D_M	1.478 ± 0.006	1.418 ± 0.005	1.442 ± 0.005	1.453 ± 0.005	1.448 ± 0.005
Mean values of D_G	1.960 ± 0.005	1.914 ± 0.005	1.940 ± 0.003	1.945 ± 0.002	1.940 ± 0.004

In the last column of Table 3, the mean values of fractal dimensions D_L, D_M, and D_G for 39 proteins are 1.343, 1.448, and 1.940, respectively, where D_L = 1.33 is very close to the theoretical value d_f = 4/3 of the spectral (fracton) dimension proposed by Alexander and Orbach. This problem is worth pointing out and studying because it indicates that the local fractal dimension of short-range interaction in the protein is equal to the spectral dimension; hence, the D_L reflects the internal vibration information of a protein.

Table 2 shows that the D_L is close to the value of the D(S) of Isogai et al.,[30] D_M is close to the value of global dimension D_2 given by Wang et al.,[31] and D_G is very close to the D(L) of Isogai et al. This fact implies that the present values are reasonable and reliable, because the serious fluctuation of the logL(R) curve is partly eliminated by the present method.

Finally, we are inclined to rule out the possibility of the correlation between the fractal dimesion and the chain length. By plotting the scatter diagrams of the 39 proteins, the distinct negative correlation between the fractal dimension and the chain length, i.e., the number of amino acid residues in the protein has been found. From Table 2 one can say at least that the D_M value of the fractal dimension of a certain chain length is limited by a function which decreases with the increase of the chain length, even if it is hard to say that the fractal dimensions D_L and D_G decrease with the increase of chain length, because we have found the relationship between D_M and entropy S(N), which will be reported in Section 4.5.3.

The new method for calculating the fractal dimension of the protein molecule reported in this section is a useful tool for description of the conformational property of a protein chain. Availability and also limitations of the theory are made clear and the analysis is carried out focusing on the relation between the fractal dimensions and the structural classes. The correlation of the fractal dimension and the chain length is also examined, and it is shown that D_M is negatively dependent on the chain length. The mean value of fractal dimension of protein increases in the order of $D_L < D_M < D_G$ for the short-, middle-, and long-range construction, respectively, and the order of the mean fractal dimensions of 39 proteins are $\alpha > \alpha/\beta > (\alpha + \beta) > \beta$.

Certainly, this result is given validity based upon the statistics and makes good sense. In a word, the previous results suggest that the fractal dimension may be a good candidate for determining a structural quantity describing the protein chain.

4.5.2 *The dimensional calculation of protein chains*

The catalysis of enzymes has been one of the most fascinating phenomena on which studies have been focused over the years, and various concepts and techniques have been developed in this field. As is well known, enzymes are, generally, protein molecules, and their catalysis depends in part on protein conformations. Stapleton and co-workers[32] introduced a fractal model to characterize the anomalous temperature dependence of the Raman electron spin relaxation rates in proteins containing iron. The model explains the observed T^n temperature dependence ($5 \leq n \leq 7$) of the Raman spin-lattice relaxation.[34] The exponent n equals $(3 + 2d_f)$. The fractal description of the protein has developed rapidly and is widespread;[22,35,36] however, recently Yang[37] and Krumhans[38] pointed out that Stapleton's method is not satisfactory since the fracton dimension (d_f) is, in general, different from the fractal dimension D_f and the Euclidean space dimension d.

Here, we present a simple method to calculate the fractal dimensions of protein chains.[39] Then, some fractal features of proteins and their implications as well as the relation between fractal demension D_f and conformational entropy are developed.[27] We hope to gain an insight into the gross features of the native state of a protein. At the same time, we shall introduce the work of other authors regarding the fractal dimensions of protein chain conformation and will compare their results with ours.

The protein molecules are long-chain copolymers, although no bifurcation or branch is involved, and are usually folded through cross-linking as a consequence of the interaction of the contiguous amino acid residues by hydrogen bonding, Van der Waals force, etc. Protein strands are neither regularly recurrent nor strictly random fabrications of the copolymers, yet they have statistical self-similarities and can, therefore, be characterized by an average fractal dimension.

A protein molecular chain may be regarded as a space curve in three dimensions and a planar curve in two-dimensional space. Since it is a fractal object, the conformation may be characterized by fractal dimension D_f. We will first look into the fractal properties of a protein chain in a plane and then extend the results to three-dimensional space based on the principles of fractal geometry. According to Mandelbrot's theory,[1] the general form of a fractal dimension of a planar curve is

$$(\text{length})^{1/D_f} = k(\text{area})^{1/2} \qquad (36)$$

where "length" signifies the total length of the curve, "area" is the maximum potential area the curve fills, and "k" is a constant.

To use fractals practically, three decisions must be made. First, the appropriate size and shape of the limiting planar area must be determined. In the present case, the limiting area should be that which is filled by a self-avoiding random walker. A random walk chain in a plane tends to fill a circular area, so we can choose a circle as the appropriate profile of the area. Second, the appropriate units of measurement must be chosen and made explicit because estimates of fractal dimensions vary with the scale of measurement. We should choose the average step size as the appropriate unit. Third, the constant k is so chosen to ensure that the right-hand side of Equation (36) yields a true one-dimensional characteristic of the area. This straight-line characteristic can be the "linear size" or "linear scale" of the area. We can choose the diameter of the circle as the straight-line characteristic of the chosen area profile. Such choices should lead to the following general equation

$$(L/b)^{1/D_f} = (k/b) A^{1/2} \qquad (37)$$

where L is the total length of the curve, b is the average step length (b = L/N, N being the total number of steps). $k = \pi^{-1/2}$, and A is the area of the circle potentially filled by a self-avoiding random walker. Thus, the fractal dimensiom D_f is given by

$$D_f = \ln(L/b)/\ln[(k/b)A^{1/2}] = \ln(N)/\ln(Nd/L) \qquad (38)$$

For a protein molecular chain, N is the number of amino acid residues on the chain, d is the diameter of the protein, and L is the chain length. Then, L = N · b, where b is the average bond length of C–C, C–O, and C–N bonds and its value is 1.48 Å.

To test the calculated results, computer simulations were carried out using the Monte Carlo method, which is a computational technique in which various states of a system are generated with random numbers and weighted with appropriate probabilities. As models, Monte Carlo simulations are useful in the analysis of protein chain conformations. For our model, we considered a self-avoiding random walk model with massless bonds[20] and used the s-p enrichment technique[97] on an IBM 3081 computer. Further details of this method may be found in References 20, 39, and 41. We utilized the Monte Carlo method to compute the fractal dimension of a protein chain. The number of monomers N(R) is counted as a function of the radial distance (R) from an arbitrary origin. We then fit N(R) to $\overline{Ro}$;[42,43] such fits need to be done at several places within the structure. The average fractal dimension is obtained as the best fit of N(R) ~ $\overline{Ro}$ by a least-squares linear fit of ln(N) to ln(R). In the above calculations, the key procedure is to assess the appropriate planar diameter (R) of the protein chain. The simplest way to estimate this value is to find the largest distance between two points on the curve.

Based on the data determined by X-ray crystallography[44] from the literature, the fractal dimensions of some protein molecular chains are calculated by Equation (38) and the results are listed in Table 4. It is shown that D_f is the reflection of the profile of the protein molecule. Generally, a real protein molecule is an ellipsoid with three diameters, a, b, c, and volume V = a × b × c. The D_f values are calculated by the average diameter d and are in agreement with the results of computer simulations. The deviation of our results from those of the Stapleton group[34] is understanable since the latter is the fracton dimension of the backbone of protein.

The fracton dimension d_f was originally introduced through consideration of the scaling properties of both the volume and the connectivity in calculating the density of states on a fractal. Alexander and Orbach[7] pointed out that the fracton dimension of percolation in any dimension d, 1 < d ≤ 6, seems to be close to 4/3. For a linear chain, it is $d_f = 1$, no matter what its fractal dimension is. Thus, if only the backbone of the protein is taken into account, it results in $d_f = 1$.[20]

Stapleton and co-workers[21,23,34] have found that in the temperature range between 4 and 20 K the electron-spin relaxation rate (1/T_1) of low-spin ferric iron in a number of heme and iron-sulphur proteins is dominated by a two-phonon (Raman) process, of which the temperature dependence is given by

$$1/T^1 \sim T^{3+2d} f(T/\theta d_f) \qquad (39)$$

Table 4 Fractal Dimensions for Some Selected Proteins

Protein	Parameters	D_f (calculation)	D (simulation)	d_f
Lysozyme	N = 129, V = 45 × 30 × 30 (Å³)			
	d_1 = a = 45 (Å)	1.432		
	d_2 = b = c = 30 (Å)	1.615		
	d = 35 (Å)	1.536	1.54 ± 0.02	1.76
Carboxypeptidase A (Zn^{2+})	N = 307, V = 50 × 42 × 38			
	d_1 = a = 50	1.627		
	d_2 = c = 38	1.765	1.68 ± 0.02	1.56
	d = 43	1.696		
Chymotrypsin (α)	N = 245, V = 51 × 40 × 40			
	d_1 = a = 51	1.416		
	d_2 = b = c = 40	1.668		
	d = 44	1.625	1.63 ± 0.003	1.36
Myoglobin	N = 153, V = 43 × 35 × 23			
	d_1 = a = 43	1.493		
	d_2 = c = 23	1.834		
	d = 34	1.605	1.62 ± 0.03	1.66
Hemoglobin (β)	N = 146, d ≈ 55 (Å)	1.378	1.40 ± 0.03	1.64

where θ is the Debye temperature, f is a smooth analytic function of T/θ, and T is the absolute temperature. The experiment indicates that the low-temperature (4 ~ 20 K) behavior of $1/T_1$ is best described by a noninteger power law of the form

$$1/T^1 \sim T^{3+2d_f} \sim T^m \quad (40)$$

with n ≈ 6.3 for hemoproteins[23] and n ≈ 5.67 for ferredoxin.[34] Stapleton et al.[23] obtained for different proteins values of d_f between 1 and 2 using Equation (40). In general, we have $1 < D_f \leq 2$ for a protein in a plane. For a protein in three-dimensional space, the fractal dimension is $D_{ft} = (D_f + \Delta D_f)$, where ΔD_f, the increment of fractal dimension, can be calculated by the transformation of the projection, and the value of $\Delta D_f \approx 1$ for the self-affine structured.[45] Therefore, the D_{ft} values of the protein molecular chains with three-dimensional structure may be estimated through self-affine fractals.

The fractal dimensions are useful in the interpretation of certain thermodynamic properties,[46] reaction kinetics,[45] and catalysis of the protein molecules, particularly enzymes. In Section 4.4, we discussed the applications of fractal dimensions to the Hill coefficients of allosteric enzymes, and we shall further discuss this subject in Section 4.5.3.

4.5.3 *Fractal dimensions and conformational entropy of protein chain*

The nature of the surface and chains of proteins has been extensively studied from the fractal perspective by computer simulation.[2,21,35,39] Although such results can only be available in an average or statistical sense for proteins, the knowledge that emerges from a fractal treatment is valid and can serve as a stimulus for further investigation.[45]

The primary structure of the protein implies the sequence of amino acids, which dictates the native anatomy of protein. The native state of proteins has a hierarchical structure with a rather complex pattern, which is beyond the reach of traditional geometries (Euclidean or differential geometry), the application of which is restricted to regular forms, whereas the structure and conformation of proteins are of extremely irregular form. Rackovsky et al.[47] have demonstrated that various types of ordered backbone structures are well characterized in terms of differential geometric repesentation. They constructed a local frame on the protein chain and defined a curvature and torsion at any point. Therefore, bends have been classified in a very natural way in their picture; which provided a useful and objective method for comparing protein conformations. The availability of this method, however, is strictly confined to smooth and regular structures or to very short local structures of like bends. More generally, a protein chain is a fractal curve. Within the usual limitation of appropriate length scales, certain properties of proteins can only be described adequately in terms of the fractal geometry proposed by Mandelbrot.[1] With the inspiration of Mandelbrot's works, Isogai and Itoh[30] have calculated the fractal dimensions of proteins, resulting in a diagram with length of a protein molecule (L) as the function of fineness of scale (m) that has two ranges: $m \leq 10$ and $m > 10$, since the logL–logm diagram of proteins within each range can be regarded as linear. The fractal dimensions within the range $m \leq 10$ and $m > 10$ are named D_s and D_L, and their mean values for 43 proteins are 1.3397 and 1.9536, respectively. However, the D_L values of various proteins are distributed within the range of 1.5911 to 2.3445. What kind of conformational character causes such a wide range distribution of the fractal dimension? The structure of protein with $D_L > 2$ is an interesting and open question.[30]

4.5.3.1 *The fractal dimensions of protein chains*

Now, we restudy the problem above, because we find that there remain some drawbacks to the work of Isogai and Itoh,[30] and the local fractal dimension D(N) concept proposed by Havlin and Ben-Avraham[32] is of advantage in characterizing the protein molecular chain. The D(N) is defined as

$$D_{N_0} = \ln[(N+1)/N] \Big/ \ln\left[\langle R_1^2 \rangle_{N_0} \Big/ \langle R_N^2 \rangle_{N_0}\right]^{1/2} \tag{41}$$

where $\langle R_N^2 \rangle_{N_0}$ is the mean-square separation of the end points of a segment containing N residues in a protein consisting of N_0 residues. It may be given by

$$\langle R_N^2 \rangle_{N_0} = 1/(N_0 - N - 1) \sum_{i=1}^{N_0 - N - 1} \langle R_{i,i+N}^2 \rangle_{N_0} \tag{42}$$

where $\langle R_{i,i+N}^2 \rangle_{N_0}$ is the through-space distance between residue i and (i + N) separated by N sequential α-carbon segments. It could be shown that if $D_{N_0}(N) = D$ is a quantity independent of N, we then have

$$\left[\left(\langle R_N^2 \rangle_{N_0}\right)^{1/2}\right]^D = AN \quad (1 \leq N \leq N_0) \tag{43}$$

where A is a constant. The slope of the curve of the log $[(\langle R_N^2 \rangle_{N_0})^{1/2}]$–logN plot is the local fractal dimension D. The D in turn is a measure of how winding the protein chain is in a certain scale N. The existence of D is supported by the Monte Carlo simulations.[23] To obtain a stable value of fractal dimension D_f, we calculate the mean value of D, namely

$$D_f = (2/N_0) \sum_{N=1}^{N_0/2} D_{N_0}(N) \quad \text{(even numbers)} \tag{44a}$$

$$D_f = [2/(N_0+1)] \sum_{N=1}^{(N_0+1)/2} D_{N_0}(N) \quad \text{(odd numbers)} \tag{44b}$$

Using Equations (43), (44a), and (44b), we have calculated the fractal dimensions D_f of some protein chains selected in the range of $N \leq N_0/2$. The results are listed in Table 5.

Table 5 Fractal Dimensions and Conformational Entropies of Some Proteins

Protein	NO	D_s^{30}	D_c^{22}	D_f	S(ML)	S(RG)
Ferredoxin p.aerogenes	54	1.3294	1.31	1.374	82.40	84.18
Trypsin inhibitor pancrease	58	1.3070	1.23	1.489	93.20	90.38
Cytochrome B_5	85	1.3464	1.46	1.555	134.23	132.17
Cytochrome C (albacore)	103	1.4154	1.57 ~ 1.60	1.567	157.78	160.00
Thioredoxin	108	1.3590	1.41	1.530	169.32	167.73
Hemerythrin	113	1.3800	1.41 ~ 1.43	1.485	177.14	175.42
Lysozyme (hen)	129	1.4160	1.73	1.502	198.78	199.98
Flavodoxin	138	1.3148	1.38	1.457	215.12	214.10
Hemoglobin	141	1.4015	1.50	1.480	224.89	218.68
Superoxide dismutase	151	1.2960	1.32	1.456	237.51	234.15
Myoglobin (sperm whale)	153	1.4227	1.49	1.462	237.85	237.22
Agglutinin	164	1.4777	1.82	1.503	249.80	254.57
Adenylate kinase	194	1.3600	1.49	1.401	228.57	300.97
Immunoglobulin	208	1.2568	1.19	1.373	323.40	322.37
Papain	212	1.3208	1.66	1.442	328.21	328.53
Trypsin	223	1.2968	1.48	1.380	345.35	345.78
Prealbumin	228	1.2675	1.25	1.360	350.20	353.19
Concanavalin A	237	1.2591	1.28	1.355	369.10	366.99
Elastase	240	1.2859	1.43	1.377	377.56	371.63
Carbonic anhydrase B	254	1.3059	1.52	1.405	389.74	393.25
Carboxypeptidase A	307	1.3447	1.53 ~ 1.59	1.367	481.23	475.12
Thermolysin	316	1.3899	1.63	1.391	489.81	489.02
Glyceraldehyde 3P	333	1.3421	1.49	1.369	522.10	515.27
Phosphoglycerate kinase	408	1.3342	1.69	1.366	624.20	631.11
Hexokinase A	457	1.3505	1.81	1.348	689.73	706.78

4.5.3.2 Calculation of conformational entropy

In this section, the calculation of the conformational entropy $S(N_0)$ of the protein chain by the random self-avoiding walks (RSAW) model with the aid of Monte Carlo simulation[41] is described. The entropy S(n) of the chain with n residues is given by

$$S(n) = k \ln \Omega(n) \qquad (45)$$

where $\Omega(n)$ is the number of continuous RSAW ways up to n steps and k is the Boltzmann constant. Let k = 1 for short, and k be the measuring unit of S(n). If the number of RSAW ways from residue ith to (i + 1)th among the $\Omega(n)$ ways is $\omega_i(n)$, and if in the meantime we let i travel through each of the $\Omega(n)$ ways, we then have

$$\Omega(n+1) = \sum_{i=1}^{\Omega(n)} \omega_i(n) \qquad (46)$$

and the mean value of $\omega_i(n)$ is as follows:

$$\overline{\omega}_i(n) = \Omega(n+1)/\Omega(n) \qquad (47)$$

When n is very large, $\Omega(n)$ and $\Omega(n + 1)$ likewise will be very large, and it is thus difficult to calculate the denominator and numerator in the right side of Equation (47) by means of counting directly. We suggest therefore a new approximate method dubbed "the conformational counting method" to estimate the conformational entropy of a protein chain. Let m be the ways arbitrarily selected from $\Omega(n)$ RSAW ways by the Monte Carlo method. Counting for $\omega(n)$ from m ways,

$$\overline{\omega}(n) = \sum_{i=1}^{m} \omega_i(n)/m \qquad (48)$$

Applying the Rosenbluth-Rosenbluth weighting factor $W_i(n)$, Equation (48) can be written as

$$\overline{\omega}(n) = \left[\sum_{i=1}^{m} W_i(n)\omega_i(n)\right] / \left[\sum_{i=1}^{m} W_i(n)\right] \qquad (49)$$

and the entropy S(n) is given by

$$S(n) = \sum_{j=1}^{n-1} \ln \overline{\omega}(j) \qquad (50)$$

where $\overline{\omega}(j)$ is the mean value of the ways $\omega(j)$ of each RSAW of ith chain. We can obtain all $\omega(j)$ ($j \leq n$) once after finishing every one RSAW of sample chain. The method employed is the s-p enrichment Monte Carlo technique on a computer.

To test the results of simulations, we have computed conformational entropy using the formula given by LeGuillon et al.[48] with renormalization group (RG), that is

$$S(n)/k = n \ln \mu + (\gamma - 1)\ln n + \ln C_0 \qquad (51)$$

where k is the Boltzmann constant (let k = 1), with the values of $\mu = 4.6838$, $\gamma = 7/6$, and $C_0 = 1.17$[49,50] We then have

$$S(N_0) = 1.544 N_0 + 0.1667 \ln N_0 + 0.1570 \qquad (52)$$

where N_0 may be regarded as the number of residues. The values S(RG) calculated by the RG method and the results S(ML) obtained by Monte Carlo simulation are all listed in Table 5.

4.5.3.3 The relationship between fractal dimension and conformational entropy

We have for the first time studied the relation between the local fractal dimension D(N) and the number of the residues of a protein chain (N_0), as well as the length of scale N in two-dimensional space by the Monte Carlo method. By choosing the pair of parameters, S = 20 and P = 10, the equation $Pe^{-\lambda s} = 1$ can be satisfied, where λ is the attrition constant. The results are shown in Figure 2. It is evident that only when $N \leq N_0/2$, can the fractal dimension D(N) preserve a constant in equilibrium. Whereas when $N > N_0/2$, the D(N) displays fluctuation. Moreover, the D(N) tends to become the equilibrium value D = 1.36 as the $N_0 \to \infty$.

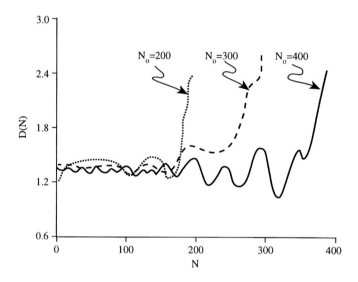

Figure 2 Dependence of D(N) for two-dimension RSAW on N, with N_0 = 200, 300, and 400.

The values of the local fractal dimensions for 25 proteins calculated by Equations (43), (44a), and (44b) and the chain fractal dimension D_c reported by Colvin and Stapleton[22] are all listed in Table 5. The values are confined within the range of $1 \leq D \leq 2$, reflecting the local conformation of the protein chain. The qualitative consistency of the results obtained by three methods implies that the structures of protein chains within the limitation of $N \leq N_0/2$ are fractals, i.e., statistically self-similar objects. This is true as far as the short-range interactions among residues through hydrogen bonds are concerned. However, in regard to long-range interactions, the properties of proteins can never be treated as fractals since there is no self-similarity. This fact indicates the fractal property of a protein is limited to a certain range of scale. From the viewpoint of molecular structure, the existence of attractive forces between the elements (atoms, segments, subunits, domain, etc.) in the short-range causes the hierarchic structures of proteins, e.g., the α-helix, β-pleated sheet, β-meanders, and consequently multilevel structures to come into being. These complex factors, however, bring about the statistical self-similarities of proteins provided that the recurrent or repetitive structural unit is included in proteins. As is well known, the fractal dimension of the trajectory of a particle under Brownian motion equals 2,[1] which represents an ideal random walk without interaction or restriction. The fractal dimensions D_L given by Isogai and Itoh[30] are in the range 1.5911 to 2.3445, and the mean value is 1.9536, being very close to 2. This indicates that the interaction force between the residues or atoms in a protein vanishes as the distance of separation increases. Therefore, the properties of a protein are mainly dictated by the sequence of amino acids.

It must be pointed out, however, that the value of fractal dimensions given by Isogai and Itoh[30] greater than 2 might be questionable, since the maximum fractal dimension of the fractal curve (e.g., protein chain, trail of RSAW, etc.) in two-dimensional space could not be larger than the dimension of the embedding Euclidean space (d = 2). The discrepancy may come from the nonlinearity and fluctuation of the relation logL–logm in the range of m > 10.

The differences among three results from Isogai and Itoh,[30] Colvin and Stapleton,[22] and our method may be caused by the alternative definitions of these three fractal dimensions. In fact, the values calculated by Isogai and Itoh are the divider dimensions which reflect the global profiles of the protein zigzag chains, and those obtained by the Colvin-Stapleton method are the chain fractal dimension D_c, which can be rationalized as the scaling exponents of the contour length with respect to the end-to-end length of a protein chain. The D_c is shown to be related to the end-to-end exponent v by $D_c = 1/v$, but the fact that the exponent v and $1/D_c$ are equal still remains ambiguous. Moreover, the v does not represent an internal self-similarity of the protein or polymer chain, and therefore it does not signify fractal dimension. Since our approach is based on a simple self-similarity assumption of a protein chain, there is, therefore, a clear physical meaning to visualize the fractal dimension of protein conformation.

The values of conformational entropies calculated by the Monte Carlo method are in good agreement with those of the renormalization group, indicating that the conformational counting method is satisfactory for computing the conformational entropy of a protein chain. However, a real protein chain does not perfectly resemble the random self-avoiding walk chain because the motion of a protein chain is not entirely free but is restricted by many factors. Hence, the conformational number and entropy are smaller than that of an ideal RSAW chain. Nevertheless, the conformational entropy here is of important theoretical significance for protein conformation.

It can be seen that the fractal dimension D_f tends to decrease as the conformational entropy $S(N_0)$ increases. This phenonenon may be accounted for by the difference in specific property of each protein. In other words, the D_f reflects the individual features of each protein molecule, and the $S(N_0)$ is only related to the number of residues. Using linear regression, we have found the following formula to approximate such a relationship

$$D_f = 1.532 - 3.000 \times 10^{-4} S(N_0) \quad (N_0 \geq 85) \tag{53}$$

where $S(N_0)$ is the entropy value calculated by the Monte Carlo method, namely, S(ML).

The theoretical basis of Equation (53) may be multifractal.[46] If we cover the support of the measure with boxes of size l and define P_i as the probability in the ith box, the dimension D_q is defined

$$D_q = \frac{1}{(q-1)} \lim_{l \to 0} \frac{\log X(q)}{\log l} \quad (54)$$

where $X(q) = \Sigma P_i^q$, $D_q = \tau(q)/(1-q)$, and $\tau(q)$ is the mass exponent.[46] When $q = 1$, the information dimension D_1 is given by

$$D_1 = \lim_{l \to 0} \frac{\Sigma P_i \ln P_i}{\ln l} \quad (55)$$

The dimension D_1 describes the scaling behavior of the partition entropy of the measure on the multifractal set.[46] The entropy $S(l)$ is defined as

$$S(l) = \sum_i P_i \ln P_i \sim D_i \ln l \quad (56)$$

Obviously, when the length of scale l stays constant, the dimension D_1 is directly proportional to $-S(N_0)$. This just happens to be similar to Equation (53). It would seem that Figure 3 displays the multifractal properties of protein structure. With the treatment of entropy in relation to a fractal as a breakthrough, we are in a position to predict that a number of thermodynamic properties of proteins may be found to relate to multifractals.

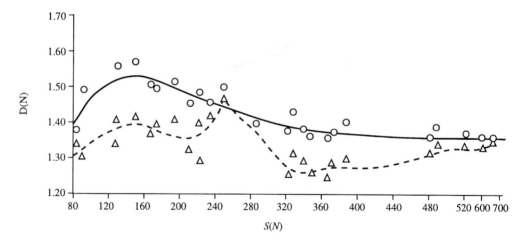

Figure 3 Dependence of the fractal dimension on the conformational entropy. △ = data from Isogai and Itoh; ○ = data from this section.

4.6 The correlative dimension for globular proteins

The correlative dimension is one of the important parameters describing the dynamical character of a system. Grassberger and Procaccia[51-52] proposed algorithms to compute the correlative dimension and entropy from a chaotic time series. They are called G-P algorithms. So far, the algorithms themselves have been improved constantly and some progress has been made.[53] At the same time, these algorithms have been widely used in many scientific fields.[45] In this section, we present G-P algorithms of correlative dimension and Kolmogorov entropy, as well as their applications in molecular biology.

4.6.1 G-P algorithms of correlative dimension and Kolmogorov entropy[54]

Let $\{x_t: t = 1, \ldots, N\}$ denote a given time series and influential factors of the system be less than n. The $\{x_t\}$ can be embedded into n-dimensional space and get a point set $J(n)$. Let us define $X_t(n,t)$ to be the vector of state variable at period t,

$$X_t(n,t) = \left(x_t, x_{t+\tau}, \ldots x_{t+(n-1)\tau}\right) \quad t = 1, \ldots, N_n \tag{57}$$

where τ is time delay and set $\tau = 1$, then $N_n = N - n$. The space of those n-history $X_t(n)$ for a fixed choice of n is called *phase space*, and the selected value of n is called the *embedding dimension*. As defined in that manner, phase space, the state vector, is defined to contain all information that is relevant to the behavior of the system during period t + 1. It therefore follows that there must exist a function f (called the *deterministic dynamical system*) such that $X_{t+1}(n) = f(x_t(n))$. We need not assume that all or any of the state variables are actually observed. We also certainly need not assume that we know f. However, we can compute the dynamical behavior of f from x_t ∀ $x_t(n) \in J(n)$, computing the distance from $X_t(n)$ to the other N – 1 points:

$$r_{ij} = \sum_{l=0}^{n-1} \left|x_{i+l} - x_{j+l}\right| \quad i,j = 1, \ldots, N_n \tag{58}$$

We can give the following recursion formula

$$r_{i+1,j+1} = r_{ij} - \left|x_i - x_j\right| + \left|x_{i+n} - x_{j+n}\right|, \quad i,j = 1, \ldots, N_n - 1 \tag{59}$$

For all $X_i(i = 1, \ldots, N_n)$, the above procedure is repeated, and one can get the correlative integral function

$$C_n(r) = 2\sum_{i,j=1}^{N_n} H(r - r_{ij}) / N_n(N_n - 1) \tag{60}$$

where r is a scale to explore the dynamic of the system, and H is a Heaviside function. According to fractal theory, at sufficiently small r the correlative integral satisfies

$$\ln C_n(r) = \ln c - D(n)\ln r \tag{61}$$

where c is a constant. In practice, with a well-behaved chaotic model, the value of $D(n)$ will "saturate" (i.e., attain its limit) at some finite level of n. This n is called the *saturation embedding dimension*, denoted by n_0. The correlative dimension of the system is $D(n_0)$ is

$$D_2 = \lim_{n \to \infty} D(n) = D(n_0) \tag{62}$$

In computation, we can start from n = 2, deciding r and $C_n(r)$ according to r_{ij}. By a broken-line regressive algorithm, we compute the scaling range and fit $(\ln r, \ln C_n(r))$.

Changing n into n + 1 and repeating the above procedure when D(n) is saturated with n gives $D(n) = D_2$.

D_2 can give the dynamical information of the system. When $D_2 = 1$, the system shows periodic oscillation; when $D_2 = 2$ the system shows quasiperiodic oscillation with two irreducible frequencies; when $D_2 \geq 3$ or is not integer, the oscillation is chaos.

The second order Renyi entropy K_2 is a lower bound to the Kolmogorov K-entropy, i.e., $K = K_1 \geq K_2$ usually, K_2 is an approximation of K, and K_2 can be extracted from experimental data. According to fractal and information theories

$$\ln C_n(r) \sim D_2 \ln r - (n+1)K_2 \qquad (63)$$

For sufficiently small r, when D_2 is saturated with n, we have

$$K_{2,n}(r) = \frac{1}{2}\left[\ln C_n(r)/C_{n+1}(r)\right] \qquad (64)$$

Giving a value of r in scaling range for n = 2, 3, ... according to Equation (64), we get a value of $K_{2,n}(r)$. As it is saturated with n, the value is denoted by $K_2(r)$. Decreasing r in scaling range and repeating the same method, when $K_2(r)$ is saturated with r, the value denoted by K_2, it is an approximation of K; $0 < K_2 < \infty$ provides a sufficient condition for chaos.

4.6.2 The correlative dimension for globular proteins[55]

The tertiary structure of a protein is determined generally by the primary structure of the amino acid sequence. Therefore, if one can find a structural parameter which reflects the tertiary structure, it may be quite important for the understanding of the protein structure. Nishikawa and Ooi[56] proposed that the structural quantity be expressed by N_{14}, which represents the number of all C_α within a sphere of 14 Å-radius centered on each C_α. This quantity can reflect information of an amino acid sequence and relates closely to the distance ρ of each C_α from the center of the mass of the protein molecule. Now we shall present some results of N_{14} and the other parameters and discuss the conjecture of Kubota et al.[55]

First, it should be pointed out that the most fundamental criterion by which a parameter can be regarded as a structural quantity is that the sequence of the quantity must show its deterministic origin.

In order to apply the G-P algorithm to a sequence of parameter value, the residue number within a protein molecule is regarded as a time series, e.g., Nishikawa and Ooi's parameter N_{14} values for alcohol dehydrogenase obtained by experiments. The coordinate values of C_α of proteins were taken from the Bernstein and co-worker's[57] protein database. The typical profile of quantity N_{14}-residue numbers for the globular protein alcohol dehydrogenase looks like a sequence of random numbers belying determinism. The data are processed by the G-P algorithm introduced in Section 4.6.1. When the embedding dimension is n, the correlation integral $C_n(r)$ was computed by Equation (60). The exponent D(n) can be calculated by least squares fit from the $\ln C_n(r) - \ln r$ plot. For n = 1, 2, ..., 30, the procedures were repeated. Kubota et al.[55] determined that for $n \geq 20$ the value of D(n) becomes constant and the value happened to be 5.5. Therefore, $D_2 = 5.5$ and the embedding dimension $d \approx 20$. The result indicates that the sequence of N_{14} values observed by experiments is generated by a certain deterministic mechanism, although outwardly it appears to be stochastic. Therefore, one can say that the protein system has deterministic-chaos behaviors.

d = 20 implies that if one could find the mechanism which can be deduced by a series of dynamical equations for protein systems, then their degrees of freedom may be 20, and the fractal dimension of the strange attractor for these equations may be 5.5.

It is worth pointing out that the experimentally observed values of N_{14} do reflect determinism qualitatively. From the quantitative point of view, $D_2 = 5.5$ and $d \approx 20$ cannot be related directly to physicochemical forces responsible for the tertiary structure. Since N_{14} fulfills the criterion of structural quantity, it is a good candidate for the structural quantity. If we choose an empirical parameter which cannot be measured directly for each residue of a protein molecule, then to study the chaotic behavior, e.g., for the values of the polarity, what is the result? In order to discuss this case, the value of the polarity for each amino acid was taken from Grantham's paper.[58] It is the same as the case of N_{14}; in the profile of the polarity-residue number, there is nothing but a sequence of random numbers. Processing these data using the G-P algorithm, the method is the same as N_{14}. In this case, the value of D(n) increasing as n increases, no flattening tendency is seen and values of D_2 and d cannot be determined, i.e., the deterministic origin cannot be detected. Therefore, the polarity of alcohol dehydrogenase cannot be a structural quantity. Of course, this does not mean that the empirical parameter has no meaning for the tertiary structure, but the parameter obtained from the primary structure does not directly reflect tertiary structural information. When other empirical parameters, such as the hydropathy index and the propensities to form α-helix and β-sheet, etc., are chosen to be analyzed by the G-P algorithm, the result is the same for polarity: D(n) divergence at a constant rate.

The results computed by the G-P algorithm are independent of the folding type. Kubota et al. chose citrate synthase (a typical α-type protein); the results for N_{14} are $D_2 = 5.4$, $d \approx 20$. For other empirical parameters, there was D(n) divergence at a constant rate. Porcine pancreatic elastase (a typical β-type protein) was selected to give $D_2 = 5.0$ and $d \approx 20$ for N_{14} and divergence for other parameters. Phosphoglycerate kinase (α/β-type protein) gave $D_2 = 7.0$ and $d \approx 20$ for N_{14} and D(n) divergence for other parameters. Hemagglutinin HA1-chain (α + β type) gave $D_2 = 5.6$ and $d \approx 20$ for N_{14} and divergence for other parameters. It is clear that deterministic-chaos features can be detected by the finite values of D_2 from N_{14} for these proteins and is independent of their folded types.

Although more conclusions could not be deduced logically from D_2 and d, the results obtained have important meaning. Kubota et al. think that $d \approx 20$ might reflect the number of the amino acid varieties in protein molecules. On the other hand, the values of D_2 distribute within the range of 3 to 7, which indicates that the tertiary structure of a protein molecule may be governed by about six virtual factors. Perhaps these factors can be selected by some method from all the frequently used parameters.

4.6.3 Estimation of the Kolmogorov entropy on EEG

Recently, Dünki[59] reported the Kolmogorov entropy from electroencephalogram (EEG) and EEG-like signals. The algorithm computing K_2 entropy has been presented in Section 4.6.1. Dunki's time series was obtained from a fetal sheep before the onset of myelinization of the brain. The correlation integral $C_n(r)$ is computed by Equation (60) for different embedding dimensions n (n = 20, ..., 30). The plot of $\ln C_n(r) - \ln(r)$ shows a parallel region. Hence, taking the value towards which these plateaus converge, Dünki found the K_2 entropy to be 2.69 bit/s. Dünki thought this is a highly misleading value; therefore, several limiting factors were studied:[59]

1. The signal is stationary only over limited periods which limits the number of useful data points.
2. A high amount of brain-generated noise could be overlaid onto the signal.
3. There might be a danger of misinterpreting a series of subsequent states as belonging to the same attractor.

For a low-dimension attractor, Dünki found that item 1 is not severe if the time series contains no further disturbance. He thought that the 6000 vectors defining the EEG attractor would be sufficient. In order to find the influence of noise, noise was added to the sine wave and the logistic map until they produced plots similar to those obtained for the fetal sheep EEG. The results indicate two opposite effects of the noise — an expanding one but also a contracting one. The former can be expected for a noisy signal: the latter seems to be rather ironic. Therefore, limitations 2 and 3 may be serious. The K_2 entropy can be estimated by the largest Lyapunov exponent. Experimentally, however, for signals of low signal-to-noise ratio, a way to produce unambiguous results has not been found. This way may be obtained with the algorithm proposed by Wolf et al.[60]

The influence of state changes has also been studied by Dünki applying a simulation process. A positive K_2 entropy and a value for the largest Lyapunov exponent λ were found, and λ is in reasonable agreement with the K_2 entropy (0.35 bit/s ≤ λ ≤ 0.62 bit/s). This comfirms the illusion of a chaotic process. However, for the case of the fetal sheep EEG attractor, it means that the positive value of the K_2 entropy is not a criterion to decide whether the underlying dynamics are chaotic or not.

4.7 Chaos — theoretical analysis for biochemical reactions

In the preceding section, some concepts relating to chaos, such as correlative dimension attractor, K_2 entropy, etc., have been discussed. The existence of chaotic behavior has been shown in a variety of biological phenomena, including biological clocks, cardiac cells, neurodynamics, and biochemical reactions. Chaotic analysis for biochemical reactions will be presented in this section.

4.7.1 Metabolite regulation and biochemical attractors[61]

In biochemistry, oscillating dynamics play a very important role in biological functions, outer- and intracellular information transmission, and cellular differentiation. Some biochemical systems, for example, the peroxidase reaction and glycolysis, display oscillations which are ever-changing and unpredictable. These oscillations have normally chaotic features. Chaos does not mean disorder; it more resembles an order having no periods.[45] Mechanisms producing chaos are deterministic; thus, more strictly speaking, *chaos* means *deterministic chaos*. In the traditional point of view, biochemical reaction dynamics often can be described as steady state. The mathematical method, such as continuous and differentiable functions or linear differential equations, are often used. Therefore, some variables and parameters relating to system features, such as flow rate, concentration, and so on, can be denoted by partial derivatives. This formalism, of course, can be applied to a large class of simple phenomena; however, for complex problems, they may be unsuitable. Biochemical reaction sequences are nonlinear, generally occurring at the so-called bifurcation points. At these points, a small change of a subsystem causes the system to jump into a new state from the original state. The simplest case is the transition from one steady state into another. Generally, the state transitions of a system may be of the following three types: (1) transition from a steady state into a periodic oscillation;

(2) transition from a steady or periodic to quasiperiodic oscillation, which can be decomposed into a sum of periodic oscillations within commensurate frequencies; (3) transition from another case to a chaotic oscillation. If the enzyme is activated by its product under suitable turnover conditions, the system jumps into a periodic oscillation from a steady state. If two product-activated enzymes are coupled in series, or if a product-activated enzyme has a periodic source of substrate, quasiperiodic and chaotic oscillations will occur.[61]

The phase-space representation is a useful method to understand the types of oscillations and the transitions between them. The phase variables are time dependent and can characterize the state of the system, e.g., metabolite concentrations. In contrast, the variables which remain constant in time and are set from outside are called *control parameters*. In the case of periodic oscillations, the changing of metabolite concentrations with time is described in phase by a closed orbit as not intersecting. This orbit is called a *limit ring*. For quasi-oscillation, the orbit of the metabolite fills up a surface which is called the *limit ring surface*. The most complex case is chaotic. The orbit of oscillation is neither the limit ring nor the limit ring surface but a strange attractor resembling a weird "fuzzy". Strange attractors generally have a certain fractal structure, which can be characterized by fractal dimensions. In fact, one can classify the feature of oscillation in biochemical reactions according to the value of D_2 (see Reference 54 and Section 4.6.1). A mathematical analysis gave the fractal dimension value of 2.2 for the chaotic attractor obtained from a glycolytic model.

If we can obtain a time series from a biochemical reaction, then the correlative dimension D_2 and embedding dimension d of the biochemical attractor can be computed by the G-P algorithm presented in the previous section. In this way, it has been deduced from experimental data that d = 3 for glycolysis. This indicates that three variables are enough to characterize the complex dynamical states of the glycolysis system, in spite of the much larger number of metabolites involved.

4.7.2 Nonlinear dynamics in glycolysis

It is well known that the dynamical behavior of glycolysis is oscillatory in intact cells and couples to membrane transport functions. Such coupling can be simulated by experiments in yeast extracts by a periodic input flux of glucose to oscillating glycolysis. The system can be described by the following dynamical equations:[62]

$$d[F6P]/dt = \overline{V}_{in} + A \sin \omega t - V_{PFK} \tag{65a}$$

$$d[ADT]/dt = V_{PFK} - V_{PK} \tag{65b}$$

where $\overline{V}_{in}$ represents the mean input constant, A and ω are the amplitude and frequency of periodic inputing flow, respectively; V_{PFK} and V_{PK} are reaction rates of PFK (phosphofructokinase) and PK(pyruvate kinase), respectively; F6P represents fructose 6-phosphate, and ADP, adenosine dephosphate.

In biochemical experiments, the nonlinear dynamics behavior in the glycolysis is expressed by measuring NADH (reduced coenzyme 1) fluorescence of a glycolysing yeast extract under the sinusoidal input flux of glucose. The input periodic T can deduce glycolytic doubling-period oscillation, such as periods T, 2T, 3T, and quasi-oscillation and chaotic oscillation. A number of techniques have been used to prove that the observed disordered patterns are indeed deterministic chaos, not quasiperiodic oscillations or noise.[62]

4.7.3 Switching process in metabolite regulation

In general, a special feature exists in nonlinear dynamical systems, i.e., multiple oscillatory states. The glycolytic model also predicts this feature, which corresponds to coexisting attractors in phase space for the same set of control parameters. What type of attractor will appear in the system? This attractor is the switching process among attractors, and the outcome of this process is decided by the initial conditions. This switching process has been verified by chemical experiments, e.g., the chloride-bromate-iodide oscillatory system. This metabolite regulation switching is expected to occur in biochemical and biological systems. In glycolytic systems, a chaotic and periodic attractor coexist in phase space in such a manner that each attractor can be reached starting from the other one, by addition or subtraction of PEP. In glycolytic systems, there are four switching possibilities, possibly corresponding to four concentrations. A remarkable feature of these processes is that switching occurs without any transients, since the final attractors are reached directly through the pulses of PEP.

Many other studies such as photosynthesis, mitochondrial respiration, intact cells, cellular aggregates, and chaotic oscillation in chick heart, algae, molluscan neurons, and so on have shown that chaos is a very widespread biochemical phenomenon,[61] and as such motivates us to understand the dynamics of structure and function in biochemical systems

4.8 The fractal dimension of the protein surfaces

This section presents our work on fat fractals and multifractals for protein and enzyme surfaces.[3] We will then discuss the Goetze-Brickmann theory of self-similarity of protein surfaces.

4.8.1 Fractal behavior of surfaces of enzymes and protein

The irregularity and roughness of enzyme surfaces are closely related to their macroscopic behavior, e.g., catalysis and selectivity. Thus, the characteristics of protein surfaces are of major importance in the association of different subunits, recognition, binding, and diffusion of a ligand. The powerful tools of molecular graphics have been widely employed in the study of protein surfaces at the qualitative and quantitative levels.[2,64,65] A possible quantitative tool which provides a measure of the surface corrugation and roughness is fractal analysis. Several groups initiated research in this field using different techniques to calculate the fractal dimension of a surface.[35,43,66] They examined protein surfaces at different levels of detail.

There are several definitions for protein surfaces. The surface is always measured by a probe which can be a stick for a two-dimensional cross-section or a small sphere for a three-dimensional representation of the macromolecule. The surface can be defined as the area of the probe multiplied by the number of probes required to cover the surface completely. Alternative definitions are concerned either with contact surfaces (the contact of the probe with the protein) or with the surface defined by the center of the sphere. A continuous description of the protein surface for a given probe size has been constructed by Connolly.[64] This surface is not related in a simple way to the number of monomers that are required to cover the surface by a monolayer. It is related to the surface covered by a sphere rolling over the surface. The difference between the surface area according to the contact algorithm and the center-of-probe algorithm is significant. Detailed calculations of the effect of the different algorithms on the scaling properties are not available.

Pfeifer et al.[43] calculated surface fractal dimensions D_s by examining two-dimensional cross-sections of the protein. They examined the length of the contour which closes the cross-section using different step lengths ε. The number of steps N(ε) required to close the cross-section is a function of the step length. By employing the relationships[5]

$$N(\varepsilon) \sim \varepsilon^{-D_s+1}$$

they were able to extract D_s. It is assumed that the system is homogeneous in the sense that different cross-sections should give the same results. Of course, the concept of a single surface dimension D_s is valid only if the system is indeed homogeneous. For its heterogeneous system, as will be seen later, multifractals are used to describe its features. A similar procedure was used by Farin et al.[66] in which the contour of a two-dimensional projection of a space-filling protein model was considered. They calculated the proportional relationship between the number of steps N and the step size ε in the same way that was used for the cross-section. The resulting exponent is difficult to correlate with the surface dimensionality. In practice, the differences found between the projection dimension D_p and the dimension of the cross-section are small.

The surface dimension D_s of protein also has been calculated by employing the method developed by Connolly.[64] The following relationship exists:[5]

$$S \sim \delta^{(D_s-2)/2} \tag{67}$$

where δ is the cross-section of a spherical unit, S is the surface area of the protein. For lysozyme, Farin and Avnir[67] found $D_s = 2.53$, and Lewis and Rees[35] found $D_s = 2.4$. The molecular graphics program of Connolly generates surfaces in the interior in empty spaces that are not necessarily exposed to the solvent. If these closed pockets are not ignored, a larger surface area can be obtained. This section presents the fractal analysis of the protein surface based on the concepts of fat fractals and multifractals.

4.8.2 Fat fractals of protein and fractal dimension computed by the variation method[3]

A new algorithm, which is a variation of a method proposed by Dubuc et al.,[68] is used to estimate the fractal dimension of a protein surface. It is a new definition of fractal dimension particularly suited for graphs of functions. The variation method is validated with both fractional Brownian (FB) surfaces and Takagi surfaces, two classes of mathematical objects with known fractal dimensions,[69] and is shown to give more accurate results than the classic algorithms. Note that we consider only the concept of Minkowski-Bouligand for the fractal dimension. The Hausdorff dimension does not appear to have any practical application and we do not consider it here.

This new algorithm may be applied to investigate the fractal dimension of a protein surface. If E is a bounded set of Euclidean space, then E(ε) is the set of all points at a distance less than ε from E, and the "thickened" set E(ε) is also called a *Minkowski sausage*. It is the union of all balls of radius ε centered on E. Denoting the volume by N, we get the Minkowski-Bouligand dimension:

$$D_M = \lim_{\varepsilon \to 0} \left(3 - \frac{\log N(E(\varepsilon))}{\log \varepsilon} \right) \tag{68a}$$

As N(E(ε)) is difficult to evaluate, it is usually preferable to count the number Ω_ε of cubes of side ε with disjoint interiors necessary to cover E. Replacing N(E(ε)) in Equation (68a) by $\varepsilon^3 \Omega_\varepsilon$ gives

$$D_M = \lim_{\varepsilon \to 0} \left(\frac{\log \Omega_\varepsilon}{\log(1/\varepsilon)} \right) \tag{68b}$$

This is usually called the *box-counting dimension*, although it is the same as the Minkowski-Bouligand dimension because Equations (68a) and (68b) are mathematically equivalent. An important point is that Equations (68a) and (68b) can be used for any kind of bounded set if the points of E are determined by their X, Y, Z coordinates. No particular choice of Cartesian axes is commanded. The surface S, as the graph G_f of a continuous function of two variables, is defined by

$$Z = f(x, y) \tag{69}$$

Strictly speaking, the functions f and Cf, where C is a constant, should define surfaces with exactly the same theoretical fractal dimension. Unfortunately, algorithms implementing Equations (68a) or (68b) are sensitive to a change of amplitude due to an accumulation of numerical instabilities. This may be explained as follows: the multiplication of all Z coordinates by C is an affine transformation. Hence, a Minkowski sausage for G_f is no longer a Minkowski sausage for G_{cf} after multiplication by C. We submit that methods to evaluate D_G should exhibit a form of invariance with respect to affine transformations, and the variation method maintains this property. Such methods should be particularly well adapted to the study of self-affine sets, because the data in Z are anisotropic to the data in X and Y, and they are good models of rough surfaces. Throughout this section we assume that the function f(x,y) is continuous and defined for all x, y such that $0 \le x \le 1, 0 \le y \le 1$.

The variation method may be stated as follows. The ε-oscillation of f at (x,y) is the difference between the two extreme values of f in an ε-neighborhood of (x,y):

$$V_f(x, y, \varepsilon) = \sup |f(x_1, y_1) - f(x_2, y_2)| \tag{70}$$

where the maximum is taken over all pairs (x_1, y_1) (x_2, y_2) such that $\max(|(x,y_1)|, |(x,x_2)|, |(y,y_1)|, |(y,y_2)|) \le \varepsilon$. Averaging v_f over all (x,y), $0 \le x, y \le 1$, gives the ε variation of f:

$$v_f(\varepsilon) = \int_0^1 \int_0^1 v_f(x, y, \varepsilon) \, dx \, dy \tag{71}$$

For all nonconstant functions f, the formula

$$D_G = \lim_{\varepsilon \to 0} \left(3 - \frac{\log V_f(\varepsilon)}{\log \varepsilon} \right) \tag{72}$$

chapter four: Protein Conformation and Enzymatic Kinetics   93

is valid. The results indicate that this method exhibits invariance under multiplication of all z-coordinates by a constant C. This is the reason why the variation method is more stable than the others with respect to a change of the amplitude of the function f. How to implement the variation method is seen in Reference 68. To test the reliability of the algorithm for estimating the fractal dimension of a rough surface, we apply it to surfaces with known fractal dimensions, e.g., Takagi and FB surfaces, because they are mathematical objects that span much of the range of possibilities that arise in practice, and because they can be generated with dimension as a parameter. The results indicate the values calculated are in good agreement with the theoretical ones (Table 6). Here, the D_G values are also fat fractal scaling exponents.

Table 6 Estimation of the Fractal Dimension of Takagi and FB Surfaces

Takagi surfaces			FB surfaces		
$D_{G(theor)}$	$D_{G(calc)}{}^a$	$D_{G(this)}$	$D_{G(theor)}$	$D_{G(calc)}{}^a$	$D_{G(this)}$
2.40	2.427	2.415	2.40	2.386	2.437
2.50	2.494	2.486	2.50	2.487	2.488
2.60	2.577	2.589	2.60	2.588	2.575
2.70	2.730	2.724	2.70	2.649	2.684

[a] See Reference 68.

The characterization of surface roughness is often done by measuring the altitude of the different points of the surface with respect to a reference plane by tactile or optical techniques. This altitude f is usually a random variation. Therefore, statistical methods based on the determination of the correlation function, the height distribution moments, the power spectral density, etc., which are used for characterizing random signals, have been directly applied. The noninteger fractal dimension D_G characterizes the space-filling capacity of the surface; it can then be used as a roughness index characterizing the texture of the corresponding surface. Now, we illustrate the application of the variation method to estimate D_s for protein surfaces; namely, let D_G be the surface fractal dimension.

Consider protein surfaces of a size that lies in the range $0.05 \sim 200$ Å^2, a molecular weight of $M = 50 \sim 50000$, and for which protein atomic coordinates are obtained from Reference 70. The results calculated are given in Table 7. The D_G values depict the roughness and texture features of proteins and enzymes.

4.8.3 *The multifractals and mass exponents for protein surfaces*

Proteins are fat fractals, and the corrugation and roughness of their surfaces may be described by multifractals. Now we explore the physical basis and implications of an infinite hierarchy of scaling exponents with a view to developing some understanding of the surface of fractal objects.

4.8.3.1 *Multifractals for biological macromolecules*

The definition of the surface of an irregular ramified object, such as a biological macromolecule or protein, is a question of interest that was addressed by Coniglio and Stanley.[71] They noted the distinction between the geometric surface of an object and the vanishingly small subset of that surface that can actually be reached by some probe. Consider, for example, a protein. Naively, one expects that the total surface area doubles when the mass doubles. A large portion of the surface is so well screened by the outer part of the protein that the part of the surface which is unscreened is vastly smaller than the total surface. Moreover, this unscreened surface scales entirely

Table 7 Fractal Dimensions of Protein Surfaces

Protein	D_G
Chicken lysozyme	2.491 ± 0.003
Lysozyme	2.518 ± 0.003
Mouse immunoglobulin AFab	2.473 ± 0.004
Bacterial serine protease A	2.325 ± 0.004
α-Cobtratoxin	2.378 ± 0.007
Cytochrome C_3	2.544 ± 0.006
Ribosomal protein L7/L12	2.507 ± 0.005
Retinol binding protein	2.521 ± 0.003
Prealbumin tetramer	2.617 ± 0.004
Trypsin	2.577 ± 0.004
Chymotrypsin (α)	2.624 ± 0.003
Carboxypeptidase A(Zn^{2+})	2.173 ± 0.005
Myoglobin	2.644 ± 0.003
Hemoglobin (β)	2.519 ± 0.003

differently from the total surface. When we recall that many biological macromolecules are triggered by the arrival of diffusing particles, we realize that the properties of these unscreened perimeters are responsible for controlling fundamental biological processes. To make these concepts quantitative, Coniglio and Stanley formulated a simple mean-field treatment of the unscreened perimeter of a fractal object.

They found that M_u, the number (or mass) of unscreened surface sites, scales with the molecular diameter L as

$$M_u \sim L_u^d \tag{73}$$

$$d_u = (D-1) + (d-D)/d_p \tag{74}$$

Here, d_u is the fractal dimension of the unscreened perimeter, D is the fractal dimension of the cluster or macromolecule, and d_p is the fractal dimension of the walk taken by the incoming particles (e.g., $d_p = 2$ for a diffusing particle and $d_p = 1$ for linear trajectories). Equation (74) has a simple physical interpretation. The first term in parentheses corresponds to the cookie-cutter perimeter (which is found if we simply cut the fractal) whose length scales as L^{D-1}. The second term corresponds to the fact that walks with noninfinite fractal dimensions will penetrate the fractal to a mean depth $\lambda \sim L^{(d-D)/d_p}$. It is physicaly plausible that the degree of penetration is controlled by the codimension of the fractal (d – D) and by the fractal dimension of the walk, d_p.

Equation (74) was put to a direct test by an extensive series of numerical calculations by Meakin et al.[72] To quantitatively analyze protein, the following family of moments was introduced:

$$Z_q \equiv \sum_{i=1}^{M} p_i^q$$

This formula is the same as Equation (6); however, where p_i in that equation is the probability that perimeter site i is the next to be hit, M here is the total number of sites. Based on the analysis of mean-field theory, we can obtain the results as follows:

$$Z_q = M_u^{-(q-1)} \sim L^{-(q-1)d_u} \tag{75}$$

$$p = 1/M_u \tag{76a}$$

For large values of q, $d_u = D(q)$ is q dependent and

$$Z^q = \sum_{\ln P} n(p) P^q \tag{76b}$$

where n(p) is the number of sites whose lnp values fall in the range (lnp, lnp + Δlnp). Let P_{max} be the maximum value of p, then

$$P_{max} \sim L^{-\alpha_{max}} \tag{77}$$

$$n(P_{max}) \sim L^{f_{max}} \tag{78}$$

For sufficiently large q, we find

$$(q-1)D_q = q\alpha_{max} - f_{max} \tag{79}$$

Matsushita et al.[73,74] have pointed out that the mean generalized dimension

$$D(q) = \frac{2}{3} + 1/(q+2) \tag{80}$$

for $q \geq 1$ on the two-dimensional DLA with $D_f = 5/3$. Note that Equation (79) would reduce to the mean-field result if $f_{max} = \alpha_{max} = d_u$. For general q, $P^*(q) \sim A(q)L^{-\alpha(q)}$ and $n(P^*) \sim B(q)L^{f(q)}$, where P^* is the value of P that maximizes the sum, and $n(P)p^q = \exp[\ln n(p) + q\ln(p)]$. It is the solution of the equation

$$\frac{d\ln n(p)}{d\ln p} = -q \tag{81}$$

Applying algorithms proposed by Ohta et al.[74] and Connolly,[64] assuming $P_i = M_i/M$, M_i (i = 1, 2, …) being the increasing weight at i zone and M being the total weight of enzyme, the multifractal features of lysozyme (N = 129, M = 14600, V = 45 × 30 × 30 Å³) and chymotrypsin (N = 245, M = 25000, V = 51 × 40 × 40 Å³) have been studied by the simulation method. The method employed is the s-p enrichment Monte Carlo technique on a computer, and the f-α spectrum is given in Figure 4.

4.8.3.2 Mass exponents for macromolecules

In order to study the fat fractal for macromolecules, we have discussed the surface mass exponents of protein and enzyme by simulation. From the contact probability distribution, an infinite family of surface size measurements can be obtained using the definition of surface size given earlier.

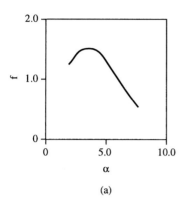 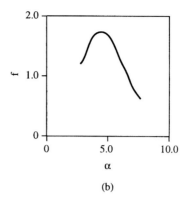

Figure 4 The f-α spectrum for enzymes obtained by simulation (Δq = 0.02). (a) Lysozyme. (b) Chymotrypsin.

$$\mu_j = \left[1/\sum_j P_i^{j+1} \right]^{1/j} \quad (82)$$

where P_i is the contact probability for particle i in the cluster (or unoccupied surface site).

$$\mu_j \sim M^{1/j} \quad (83)$$

$$D(q)/(q = j+1) = Dr_j(q = j+1) \quad (84)$$

where r_j is the surface mass exponent. Using Monte Carlo simulation, Meakin et al.[25] found $r_1 = 0.882$, $r_2 = 0.727$, $r_3 = 0.651$, $r_4 = 0.611$, $r_5 = 0.587$, $r_6 = 0.570$, $r_7 = 0.559$, and $r_8 = 0.550$ using 10^6 particle trajectories for 15 cluster (star) sizes in the range M = 601 ~ 6001 or L = 50 ~ 500 diameters. According to Equations (80) and (84), the relationship

$$r = \frac{D(q)}{D} = \frac{1}{D}\left(\frac{2}{3} + \frac{1}{q+2}\right) = \frac{1}{D}\left(\frac{2}{3} + \frac{1}{j+3}\right) \quad (85)$$

is obtained. The results calculated are given in Table 8. Assume that the ratio $\mu_j^j / \mu_{j+1}^{j+1}$ is related to M by the power-law expression

$$\mu_j^j / \mu_{j+1}^{j+1} \sim M^{-r_j''} \quad (86)$$

The exponent r_j'' is obtained by least-squares fitting of straight lines to the dependence of $\ln(\mu_j^j / \mu_{j+1}^{j+1})$ on $\ln M$. The results of numerical investigation imply that the exponent r_j is very close to the asymptotic value $r_\infty'' = r_\infty$. The r_∞ is obtained by the formula $p_{max} \sim M_\infty^{-r}$, and p_{max} is the maximum contact probability. From simulations carried out using 26 cluster masses of between 10 and 50,000 particles and 250,000 random-walk trajectories for each cluster, the results are listed in Table 8. For protein and enzyme, M is the molecular weight, and $P_j = N_j/N$, N_j (j = 1, 2, ..., N) is the increasing number of amino acid residues at the ith stage, and N is the total number

Table 8 The Surface Mass Exponents Obtained by Simulation

	Semilattice model[a]	Matushita model		Screened-growth model[b] (D = 5/3)
j	r_j	r_j''	r_j	r_j''
1	0.523 ± 0.005	0.463 ± 0.003	0.450	0.539 ± 0.007
2	0.493 ± 0.005	0.436 ± 0.005	0.440	0.482 ± 0.009
3	0.474 ± 0.004	0.420 ± 0.006	0.433	0.450 ± 0.009
4	0.461 ± 0.005	0.409 ± 0.007	0.428	0.429 ± 0.010
5	0.450 ± 0.004	0.402 ± 0.006	0.425	0.416 ± 0.010
6	0.442 ± 0.004	0.394 ± 0.007	0.422	0.406 ± 0.010
7	0.436 ± 0.004	0.379 ± 0.006	0.420	0.397 ± 0.010
8	0.431 ± 0.004		0.418	0.393 ± 0.010
⋮	⋮		⋮	⋮
∞	0.390 ± 0.004		0.400	0.347 ± 0.009

[a] See Reference 72.
[b] See Reference 73.

of amino acid residues. Based on the above definition, we have calculated the surface mass exponent for some proteins and enzymes. The values are given in Table 8. In fact, the mass exponents are the fat fractal exponents as well.

From the values in Table 7, one knows that the results calculated by the variation method are different from the other methods.[65-67] The origin of the difference lies in the fact that the covering surface and cross-section methods are not all applicable to the self-affine surfaces, whereas only the variation method is applicable. Thus, the fractal dimensions of protein surfaces reported in this section are probably reliable and true. On the other hand, since the surface of protein is rough and corrugated, the multifractal theory is a useful tool for the study of enzyme catalysis and selectivity.

Note that the situation examined here is limited to two-dimensional space. Thus, the value of the multifractal spectrum $f(\alpha)$ is smaller than 2, and its maximum value $f_{max}(\alpha) = D_f$ (Hausdorff dimension). For example, the $D_f = 1.536$ for lysozyme and $D_f = 1.625$ for chymotrypsin (α).[39] The Hausdorff dimension is only valid for homogeneous systems, whereas $f(\alpha)$ provides detailed information about heterogeneous systems. For aggregate systems (e.g., proteins) in three-dimensional Euclidean space, the global fractal dimensions of surfaces are equal to $(D_f + 1)$, according to the principles and simulation of fractals. Then, lysozyme $D = 2.536$, and chymotrypsin (α) $D = 2.625$. These values are in good agreement with those reported by Farin and Avnir;[67] namely, $D_s = 2.53$ for lysozyme. In addition, they have calculated D_s for trypsin using Equation (67); the global value is 2.62 ± 0.01, and the dimension of the active site is 2.80 ± 0.04. They proposed that binding is enhanced on a more corrugated surface; however, they commented that active sites include very specific interactions, and a single parameter (such as the fractal dimension) may be insufficient to describe this complex system.

On the other hand, Lewis and Rees[35] obtained different results for lysozyme and superoxide dismutase in which the active site dimension was found to be lower than the global value. They rationalized this result following arguments similar to those of Pfeifer et al.[43] in regard to the surface diffusion of protein ligands. They suggested that the active site should not bind the ligands strongly, since the final stage of the enzymatic reaction (the release of the products) would be slow. Hence, in order to keep the catalytic efficiency, both the trapping and the desorption should be optimized. Pfeifer et al.[43] suggested that the optimum fractal dimension between efficient trapping and efficient diffusion on the protein surface would be the values 2.1 ~ 2.2.

For a larger D_s value (corrugated surface), the trapping would be more efficient but the diffusion (to the active site) would be too slow. Aqvist and Tapia[65] also calculated the local fractal dimension by assigning average D_s for each amino acid. They found (in accord with Lewis and Rees[35]) high local dimension at interfaces for the satellite tobacco virus.

For prealbumin, the variations in the surface dimension at interfaces is small. They also found low dimensions at active sites. As a matter of fact, the contradictions and controversy may be interpreted by multifractal theory. As seen in Figure 4, the fractal dimension (D_s) of protein and enzyme surfaces is the spectrum $f(\alpha)$, i.e., there are many D_s values, not just one. The D_s is a different value for the different areas on (in) the protein surface; the more corrugated the surface is, the larger the D_s value is. The D_s of an active site is larger than that of a global surface, because the active site is more corrugated than the global surface. This way, the ligands (or substrate) are effectively absorbed on the active site and changed into products. The above analysis indicates that multifractal theory is particularly applicable to the characteristics of protein and enzyme surfaces.

Based on computer simulations of the surface mass exponent (r) for proteins, we have proved that the protein is a fat fractal because its surface is a fractal but of nonzero volume. In fact, the r is also another kind of fat fractal scaling exponent, and it describes the relationship between probability measure and molecular weight for protein and enzyme. The results in Table 9 indicate that the values of the surface mass exponents lie in the range 0.485 ~ 0.392, and this is consistent with the Matsushita model.[73] The results in Table 8 show the Matsushita model is in good agreement with the semilattice and screened-growth model. The exponent r''_j is close to the asymptotic value $r''_\infty = r_\infty$. For DLA, r_∞ should have a value of $(D_f - D)/D_f$ where D_f is the fractal dimension for DLA, its value is about 1.7, and $r_\infty = 0.7/1.7 \approx 0.412$. For the semilattice model, the r_∞ obtained by simulation has a value of $0.390 + 0.004$ in good agreement with the results obtained by the Matsushita model ($r_\infty = 0.400$) and by the ratio $(\mu_j)^j/(\mu_{j+1})^{j+1}$ method ($r_\infty = 0.40$). This fact indicates that the protein and enzyme aggregate systems are complex. In other words, the protein and enzyme are fat fractals and their surfaces may be studied by the variation method and multifractal theory.

Table 9 The Surface Mass Exponents of Protein and Enzyme Obtained from Estimates of the Aggregation Model

Protein	r_1	r_2	r_3	r_4	r_5
Lysozyme	0.482 ± 0.004	0.473 ± 0.004	0.461 ± 0.005	0.455 ± 0.004	0.433 ± 0.004
Trypsin	0.475 ± 0.005	0.465 ± 0.004	0.451 ± 0.004	0.434 ± 0.005	0.427 ± 0.004
Chymotrypsin (α)	0.485 ± 0.004	0.463 ± 0.005	0.449 ± 0.004	0.422 ± 0.005	0.418 ± 0.005
Carboxypeptidase A (Zn^{2+})	0.455 ± 0.005	0.437 ± 0.005	0.425 ± 0.004	0.408 ± 0.005	0.387 ± 0.004
Hemoglobin	0.452 ± 0.004	0.438 ± 0.006	0.414 ± 0.007	0.402 ± 0.006	0.392 ± 0.005

Recently Geotze and Brickmann[63] proposed a method to study the self-similarity of protein surfaces. They studied 53 proteins and found that the surfaces show self-similarity within a yardstick range of $1.5\text{Å} < \varepsilon < 15\text{Å}$. They applied two different algorithms for the calculation of the fractal dimension, i.e., Connolly's contact surface (CS) method and Richards' solvent accessible surface (SAS).[63] The fractal dimensions based on the two methods are different. The CS type approach leads to relatively high values of D = 2.5 to 2.6, while the SAS approach gives fractal dimension of D ≈ 2. Geotze and Brickmann[63] mainly discuss whether self-similarity of the surfaces is a generic property of all proteins and, if this is true, whether there are systematic trends in the fractal dimension when the size of the protein is changed. The results show that the self-similar dimension (and therefore the self-similarity of a protein surface) is highly related to the calculating method; hence, the actual value of fractal dimension D obtained with one particular algorithm for the generation of the surface of a protein might have not any universal importance.[63]

4.9 Fractal reactions of enzyme surfaces[77]

In recent years, studies in catalysis and enzyme surface have verified the existence of varied and exotic surface phenomena during catalysis.[78] Phenomena that occur on the molecular level, such as surface diffusion and interaction of substrate molecules with the enzyme surfaces and each other, have been observed to produce rather interesting behavior over the relatively large length scale of the catalytic surface. Phenomena occurring over the length scale of the catalytic surface in turn affect catalyst activity and, thus, they affect the reaction kinetics on the macroscopic level. In this section, we discuss the advantages of utilizing Monte Carlo methods and fractal scaling relationships to model surface phenomena which occur during catalysis. We propose that Monte Carlo methods can be quite beneficial in understanding surface phenomena and their effects on global rate behavior in enzyme catalytic systems.

Surface representations of proteins have provided a powerful approach for characterization of the structure, folding, interaction, and properties of enzymes.[2] A fundamental feature of enzyme which has not been characterized by these representations, however, is the texture (roughness) of enzyme surfaces and its role in molecular interactions. The ensuing fractal surface dimension controls biologically relevant processes. The rate of substrate arrival (by diffusion) at an enzyme receptor depends very much on whether the diffusion space is three, two, or one dimensional. The maximum rate of a reaction will depend on the encounter probability of the components. The diffusion process can be treated in terms of Brownian motion or the random walk process. Such diffusion on complex enzymes with fractal structures may have important biological implications.

In our works, we have considered an enzymatic reaction in which the substrate molecules can attach to certain points on the enzyme surface (other than the active site), and then diffuse towards the active site. The amount of substrate that can attach to the surface will greatly depend on the fractal dimension of the surface, surface area of enzyme exposed to the medium, and the size and bulkiness of the substrate. On reaching the active site, the reaction occurs and the substrate is converted to product. Either the intrinsic rate of conversion of the substrate or the availability of it due to surface diffusion would then determine the rate of the reaction. Diffusion on fractal structures is regarded as random motion on DLA networks.[79] If movement is allowed from one site of the DLA cluster occupied by the substrate only to a nearest neighbor site of the cluster, the motion is restricted. Attachment of a substrate molecule to the enzyme surface and the reaction mechanism have been discussed in Reference 77.

To simplify the simulation fractal, the model is described by the following statements:

1. The enzyme is rough in a way that can be described by a fractal model.

2. The enzymes can form spontaneously supramolecular structures which are complex geometrical clusters. In the meantime, enzyme molecules can be bound to the polymer carriers, and because of these two interactions, the enzyme immobilization is a DLA cluster.
3. There are no interactions between individual substrate molecules.
4. The substrates can attach only to the edges of the DLA cluster to account for the fact that only certain patches on the enzyme surface are capable of interacting with the substrate.
5. Once the substrate is bound to the surface, it moves towards the active site by the shortest available distance.
6. The dissociation constant for the complex formed is very low.

The reaction mechanism is simulated by the Monte Carlo method. The enzyme agglomerate is assumed to be in the form of a DLA cluster, and each individual enzyme molecule is a percolation cluster at the threshold.[6] The approach is well suited for modelling microscopic surface phenomena associated with catalysis because these phenomena are discrete and microscopic. As a model, Monte Carlo simulation has been useful in catalysis for predicting selectivity when it is governed by the surface configuration of adsorbed species.[78] Since the Monte Carlo environment is exactly defined, it becomes possible to elucidate the effect of a particular microscopic mechanism on the macroscopic kinetic behavior of a catalytic system. The model catalytic reaction we have chosen is the bimolecular reaction such as the Langmuir-Hinshelwood reaction. The two-dimensional computer array that is an on-off lattice is used to represent the catalyst surface. The catalytic surfaces in our simulations have square coordination with periodic boundary conditions.

On the DLA cluster, the substrate particle is allowed to follow a biased random walk towards the active site by the shortest available distance. Diffusion of substrate particles bound on the enzyme surface is simulated simultaneously. If a particle tries to attach to a position where another substrate is already present, it is rejected. On reaching the active site, the complex is formed and reaction takes place with a certain high probability.

How active sites are distributed on the surfaces of enzymes is a significant problem. Diffusion-limited aggregation (DLA) has become important for describing a wealth of diverse physical, chemical, and biological phenomena. One of the central problems is to understand how such a nontrivial pattern emerges from a simple set of rules for growth. The dynamics of DLA, for a given cluster, can be described by the set of growth probabilities $\{p_i\}$, where p_i is the probability that site i will grow next. Lee and Stanley[80] proposed an analytic solution of the growth-site probabiity distributions for a family of hierarchical models for the structure of DLA clusters. These models are characterized by self-similar voids that are delineated by narrow channels. The growth-site probability distributions for all the models are shown to have the same form, $n(a,M) \sim \exp\{-(A/\ln M)[\alpha - \alpha_0(M)]^2\}$, where $n(a,M)$ is the number of growth sites with $a < -\ln p_i / \ln M < a + da$, p_i is the growth probability at site i, M is the cluster mass, $a_0(M) \equiv B \ln M$, and A, B are constants. The p_i at perimeter site i is defined as the probability that the random walker steps i for the first time. According to the universality of the growth-site probability distribution of DLA clusters, we obtain the probability distribution formula for the active sites of enzymes on their sufaces, as follows:

$$D_{AS}(\alpha, M) \sim \exp\left\{-(A/\ln M)[\alpha - \alpha_0(M)]^2\right\}$$

Here, M is the localized cluster mass of enzymes and p_i is given by $p_i = M_i/M_0$, M_i (i = 1,2,...) being the increasing weight at i zone, M_0 being the total weight of enzyme. In general, the scaling form of the generating function is only on the exponent α, not on the amplitude B. For α ≤ 3, the scaling is independent of both α and B, since the Gaussian approximation is strongly localized and can be considered to be a δ function. The maximun value of α for the Gaussian polynominal is 3.

The fractal dimension of active site distribution on the enzyme surface D_g is

$$D_g = \ln[(b-\theta)(b-\theta+1)/2 + 2\theta]/\ln b$$

where b is the linear size, namely, the radius of gyration given by the equation

$$b^2 = \sum\left[(x - x_{cm})^2 + (y - y_{cm})^2\right]/M$$

where x_{cm}, y_{cm} are the coordinates of the center of mass and θ is the number of empty layers of enzyme cluster. It is shown that the fractal dimension is uniquely determined by the two parameters b and θ.

From computer simulations, we have found that the surfaces of some enzymes have features typical of islands. The relationship between the mass M and the radius of gyration for the enzyme molecule is $M(b) \sim b^{D_f}$ with $D_f = 1.33 \sim 1.90$. The perimeters of the enzyme islands are fractal entities, scaling with the radius of gyrations as $P(b) \sim b^{D_p}$. Here, P is the number of perimeter sites of an island, and D_p is the fractal dimension of the perimeter. Based on the above definitions, we have calculated the active site probability distribution of lysozyme and the fractal dimensions of some enzymes selected from the Protein Data Bank.[70] The results are listed in Table 10. The simulations are perfomed on a VAX II computer.

Table 10 Fractal Dimensions of Some Enzyme Islands

Enzyme	D_f	D_P	D_G
Lysozyme	1.47 ± 0.03	1.54 ± 0.05	1.51 ± 0.03
Trypsin	1.56 ± 0.03	1.60 ± 0.03	1.60 ± 0.03
Chymotrypsin (α)	1.83 ± 0.04	1.68 ± 0.04	1.79 ± 0.04
Carboxypeptidase A (Zn^{2+})	1.39 ± 0.05	1.42 ± 0.05	1.41 ± 0.05

From Table 10, the D_f values are very close to those of D_g. Simulations for all kinetic regime are run on 1024 × 1024 square lattices with periodic boundary conditions. The surfaces in our simulations possess several fractal attributes, each providing further definition of the surface mechanism in the system. A question naturally arising from our observations is whether the enzyme islands would continue to grow indefinitely, resulting in "poisoning" of the surface, or whether a steady-state, reactive surface would emerge. From the simulations, we have found the enzyme islands continued to grow until poisoning of the surface occurred. Furthermore, the "poison" time is dependent on the size and properties of enzyme. This relationship is given by $T_p \sim L^z \cdot D_p^\beta$, where z = 1.96 ± 0.002, β ≈ 1.5. This scaling relation held regardless of the simulation, with differences only in the pre-exponential factor. The fractal dimension of the surface as a whole is the most useful in our studies because it has the greatest sensitivity to island growth dynamics.

Assuming different surfaces, diffusion coefficients $D \sim D^\gamma$ (γ = 0.11) in the range of 0.01 to 10 formed as a function of time is studied on the DLA lattices. This fact can be explained using the following observations during the simulation. At the beginning of the reaction, very small amounts of substrate are bound to the enzyme surface. If diffusion is fast, these substrate particles can reach the active site faster than they could have in the case of slow diffusion.

The amount of substrate attached to the surface is a function of substrate concentration. At low substrate concentrations, the amount of substrate that can bind to the surface will not be affected by the perimeter. This is because at this concentration of the substrate, the ratio of the surface area occupied by the substrate in comparison to the overall area to which the substrate can bind, is small. At intermediate substrate concentrations, the amount of substrate that can bind to the surface will now depend on the perimeter and surface area available for movement towards the active site. The perimeter of a DLA cluster is larger than that of a square lattice, and the amount of substrate that binds to the cluster is larger. It faces restrictions in its movement and so the rate starts falling. The fall in rate of reaction on the square lattice is faster as it undergoes a larger change from less surface-bound substrate to saturation of the surface and has a greater tendency to get saturated with the substrate. At a high substrate concentration, the average amount of substrate that can bind to the surface of the DLA cluster is lower, even though the perimeter of the same is larger than that of the square lattice. This is because the restriction to movement on the DLA cluster is larger due to the presence of unoccupied sites. Thus, the perimeter is not made vacant easily and further attachment is hindered. In the case of a DLA cluster, the profiles of reaction rate vs. substrate concentration depict significant substrate inhibition; this phenomenon is not observed in the case of a square lattice. The reasons for this can be attributed to the complex kinetics arising from the fractal nature of the enzyme surface.

The kinetics of enzymatic reactions have been studied exhaustively and experimentally over the last 40 years, and the literature is replete with data for a wide range of enzyme-substrate systems. The comparison of the results of the simulation with the experimental data has been discussed elsewhere.[45]

In summary, the computer simulations indicate the effects of fractal geometry on the reaction of enzyme catalysis, and the substrate inhibition in certain cases can be attributed to surface diffusion of substrate molecules towards the active site. Catalytic reaction kinetics are related to the fractal nature of the enzyme surface.

4.10 Fractal approach for protein and enzyme kinetics[6]

In this section, we will discuss investigations made on the reaction of proteins and enzymes by the statistical methods of random walks on fractal structures. The content includes fractal chemical kinetics, one-step conformational change, multistep change, the generalization of rate equations based on fractal reaction kinetics, etc. They are our new explorations.

4.10.1 Fractal chemical kinetics

We are now in a position to present a general theory for studying the chemical kinetics on a fractal structure, with particular emphasis on the relationship between the reaction order (X) and the fractal dimension for a heterogeneous reaction. For a single-reactant bimolecular reaction:

$$A + A \rightarrow \text{products} \quad (I)$$

as well as for a two-reactant bimolecular reaction:

$$A + B \rightarrow \text{products} \quad (II)$$

Second-order reaction rates for the classical or homogeneous (that is, all the concentration dependence of the reaction) can be expressed as follows:

$$\text{Rate} = -d[A]/dt = -d[B]/dt = K[A]^2 = K[A][B] \tag{87}$$

where [A] and [B] are the reactant concentrations (or density) of A and B, respectively, and K is the rate constant. Note that K is independent of time. If [A] = [B], the solution (the integrated rate equation) of Equation (87) is given by

$$[A]^{-1} - [A_0]^{-1} = Kt, \tag{88}$$

where $[A_0]$ is the initial concentration (at t = 0). However, for the heterogeneous chemical reaction, the reaction rate is given by

$$\text{Rate} = d[A]/dt = K[A]^x \quad (x > 2) \tag{89}$$

where the power (order) x may be much larger than 2. It seems necessary to present here a brief argument relating X to the fractal dimension of random walks in a fractal structure. The mean number N (the number of excited molecules) of distinct sites visited by a reactant molecule (random walker) is given by Equation (15). It has been shown that the rate for the simple binary reactions I and II is[81]

$$\text{Rate} = -d[A]/dt = K_0 dN/dt \cdot [A]^2 = K_0 t^{(d_f/2 - 1)}[A]^2 \tag{90}$$

This has been verified by extensive simulations.[82] The integrated rate equation obtained from Equation (90) is

$$[A]^{-1} - [A_0]^{-1} = K_0(2/d_f) t^{d_f/2} \tag{91}$$

where $[A_0] = [A]$ (t ≡ 0). Hence,

$$t = \left(K_0^{-1} d_f/2\right)^{2/d_f}[A]^{-2/d_f} \quad ([A_0] \rightarrow \infty) \tag{92}$$

A simple substitution of Equation (92) into Equation (90) gives

$$\text{rate} = K[A]^{(1+2/d_f)} \quad (t \rightarrow \infty) \tag{93}$$

where $K \equiv K_0^{(2/d_f)}(d_f/2)^{(1-2/d_f)}$. An elementary reaction order X is given by X = 1 + (2/d_f), thus relating X to the effective spectral dimension of the medium. Obviously, for the case of d_f = 2, the classical result X = 2 is regained. For the Sierpinski gasket, d_f = 1.36, and X = 2.47; whereas, for the percolating cluster d_f = 1.33, X = 2.50. The above new approach to low-dimensional chemical kinetics might reveal new insight into the heterogeneous reactions that are of technological and biological importance.

4.10.2 Mechanisms for the allosteric effects of proteins and enzymes

For many years it has been known that the binding of oxygen to hemoglobin follows a sigmoid curve which differs appreciably from the typical Mechaelis-Menten equation covering the same concentration range. This remarkable phenomenon has aroused much interest because it cannot be interpreted in terms of the classical theories of enzyme action. As is well known, hemoglobin is not an enzyme but performs the function of transporting oxygen. The study of the binding of oxygen to hemoglobin has contributed much to the understanding of allosteric effects and cooperativity. In 1910, Hill proposed the following equation, which is now commonly known as the Hill equation,[83,84] to account for the oxygen-binding curves that he and others had observed from hemoglobin:

$$Y = K_h[S]^h / \left(1 + K_h[S]^h\right) \tag{94}$$

where [S] is the concentration of substrate or the partial pressure of oxygen and K_h is a constant. The exponent h is now commonly known as the Hill coefficient and Y is the fractional saturation, which is defined as $Y = N(t)/N_0$, where $N(t)$ is the number of occupied binding sites and N_0, the total number of binding sites. Equation (94) is purely empirical and the theoretical background is not clearly understood. It has been thought that the Hill coefficient measures the cooperativity of the enzyme subunits or the number of substrates needed to bind thereon, i.e., in a sense it implies the order of the reaction. The h is widely used as an index of cooperativity, and the degree of cooperativity is considered to increase as h increases. At the upper limit, h is equal to the number of binding sites. If h = 1, there is no cooperativity; if h > 1, there is positive cooperativity; if h < 1, there is negative cooperativity. The property of responding with exceptional sensitivity to changes in metabolite concentrations is commonly dubbed as cooperativity. The allosteric effect implies the phenomenon of conversion of protein conformation from one state to another. Many allosteric enzymes are also cooperative and vice versa. Although allosteric effects and cooperativity are related to the change of protein conformation, this does not mean that the two concepts are interchangeable. In fact, they describe two different properties and should be clearly distinguished.

In general, the Hill coefficient h is equivalent to the varieties of the distribution of liganded species. However, as Monod et al.[33] pointed out, h is not the number of interacting sites, but an interaction coefficient. Under certain conditions, h can be interpreted as a measure of the free energy of interaction between sites. To clarify the nature of allosteric effects and the physical basis of the Hill coefficient, the well known MWC and KNF models were proposed by Monod et al.[83] and Koshland et al.,[84] respectively. The former is the so-called Monod-Wyman-Changeux concerted mechanism, which assumed that the quaternary structure of the protein is always symmetrical and the protein exists in an equilibrium of two states, T(tense) and R (relaxed); the T state has a lower affinity for ligands. The latter model is the so-called Koshland-Nemethy-Filmer sequential model, which, instead of the assumption of symmetry, assumes that the progress from T to the ligand-bound R state is a sequential process. The conformation of each subunit changes alternatively as it binds to the ligands, and there is no abrupt switch from one state to another. The MWC model uses a quaternary structural change, whereas the KNF model uses a series of tertiary structural changes. The KNF model is at the expense of simplicity, more generally, and is probably a better description of some proteins than the MWC model. In return, the explanation of phenomena is often somewhat more complicated. Recently, Schweitzer-Stenner and Dreybrodt[85] brought forth an extended MWC model, the mathematical basis of which had been formulated by Herzfeld and Stanley. Such a model was used to fit oxygen and carbon monoxide binding curves of hemoglobin trout IV measured at different pH values between 8.0 and 6.0. It has been shown that the interaction between the quaternary T → R and tertiary t → r transition is different from the α and β subunits.

The experimental value of h, often nonintegral, is rarely equal to the number of ligand-binding sites on each molecule of protein. In general, the values of h range from 1 to 3.2. Why? The nonintegral phemonena could not be interpreted by the MWC and KNF models. We wish to submit and discuss a new mechanism, fractal kinetics, for this interesting phenomenon.

Currently, several lines of evidence suggest that both the structure and dynamics of proteins and enzymes are fractal. Stapleton and co-workers[21,32,34] found that the geometry of the carbon backbone determined from X-ray diffraction data and the vibrational dynamics of proteins as measured by Raman scattering are both fractals. Measurements on 70 proteins showed that the fractal dimension determined from the structure correlates elegantly with that determined from the dynamics. Experimental evidence indicates that each substate of the protein has in itself a large number of substates, and the potential energy function is statistically self similar, having the same form on many different scales. Lewis and Rees[35] found that the surfaces of proteins are fractals. Liebovitch and co-workers[36] (see also Chapter 3) proposed a fractal model of ion channel kinetics, which is more consistent with the conformational dynamics of proteins. In addition, for the conformational motion of proteins and the substrate-enzyme reactions based on random walks in nonintegral dimensions, many outstanding theoretical and experimental studies in this respect have been done in recent years.[2,20,27,39] Dewey and Datta[86] determined the fractal dimension of membrane protein aggregates using the fluorescence energy transfer method. This technique provides a means of assessing the nature of protein-protein interactions in membranous systems. Our purpose here is to extend the kinetics of allosteric proteins and enzyme fractals and to suggest a plausible physical model for the allosteric effect. We hope that the Hill coefficient h can be calculated theoretically by using this fractal theory.

To study the allosteric effects of proteins and enzymes and the implications of the Hill coefficient, Equation (94) is rearranged to give $Y/(1-Y) = K_h[S]^h$. Assuming $R = Y/(1-Y)$, we have $R = K_h[S]^h$, where R is the total reaction rate and h denotes the reaction order. As will be shown later, the exponent h is the Hill coefficient to the fractal dimension of proteins and conformation.

In this section, we attempt to propose a fractal mechanism for the allosteric effect of a protein. Before describing the model, we first define the terminology to be used as follows

1. The excited molecule: the one with sufficient energy to lead to effective collision.
2. The random walker: chaotic motion without limitation.
3. One-step conformational change: no transient state from one state to another.
4. Multi-step conformational change: subsequent middle states from one state to another.

The model is described by the following statements:

1. The protein and enzyme are fractal objects.
2. The protein may exist in two conformational states, T and R; in the absence of ligands, the protein exixts in conformation T and R in negligible amounts.
3. The R state has a high affinity for ligands, i.e., substrate molecules S, and may be viewed as the active site.
4. Whether the transition from the T to R state is a one- or multistep conformational change depends on external conditions and the properties or variety of the attacking excited molecule.
5. The annihilation of an excited molecule means the birth of an R state and the association between the R state and substrate S with rapidity and ease.
6. The R and T states may be regarded as the "trap" site and "free" site for the substrate S, respectively; the R and T are randomly distributed on or in the surface of enzyme or protein, forming random fractal networks consisting of "trap" and "free" sites.
7. The diffusion of substrate molecules on the protein may be regarded as a random walker until it hits the active site.

The present model embodies the idea of the MWC and KNF models and fractal theory, which postulates that the binding of a substrate to an enzyrne may cause conformational changes that align the catalytic groups in their correct orientations. Using these assumptions, it is possible to describe the fractal mechanism of the allosteric effects of proteins and reveal the meaning of the Hill coefficient.

4.10.2.1 One-step conformational change

In this case we suppose that an excited substrate molecule S* is directly bound to the protein. The reaction processes are two-step elementary bimolecules, as follows:

$$E_T + S^* \underset{}{\overset{k_1}{\rightleftharpoons}} E_R \cdot S \qquad \text{(III)}$$

$$E_R \cdot S + S^* \underset{}{\overset{k_2}{\rightleftharpoons}} [S \cdot E_R \cdot S] \underset{}{\overset{k_p}{\rightleftharpoons}} P + E_T \qquad \text{(IV)}$$

where k_1, k_2, and k_p are rate constants. We must first find the survival probability p(k;t) of the excited S* and then calculate the reaction rate. We start our considerations by assuming that the R site and the S* are embedded in a fractal. For a fixed protein configuration K and a certain time t, the p(k;t) of the excited S* (assumed at the origin) is exponential:

$$p(k;t) = \prod_{j \in k} \exp\left[-t\varphi(R_j)\right] \qquad (95)$$

where $\varphi(R_j)$ denotes the transfer rate to an R site at position R_j and the product extends over all R sites of a protein. The $\varphi(R_j)$ is defined by

$$\varphi(R_j) = a_m R_j^{-n} \qquad (96)$$

where a_m and n are polar interaction constants. The quantity of experimental interest is not p(k;t) but rather its ensemble average over all possible configurations of the protein molecules distributed on the fractal P(t). That is, $P(t) = \langle p(k;t) \rangle$. If the R sites are randomly occupied by the S* with probability Ψ, we obtain from Equation (95):

$$P(t) = \prod_i \left\{ 1 - \Psi + \Psi \exp\left[-t\varphi(R_j)\right] \right\} \tag{97}$$

Here the product extends over all sites of the fractal structure with the exception of the R sites. This means that the ensemble average reproduces the details of the structure, and in the process of the direct transfer, the whole fractal structure is being sampled.

For low-density R sites, $\Psi \ll 1$, an approximation of P(t), can be derived from Equation (97). Distinct from Equation (95), this form does not depend any more on the position of the substrate molecule. In the continuous description we obtain

$$P(t) \approx \exp\left(-\Psi \int dR \rho(R) \left\{ 1 - \exp\left[t\varphi(R_j)\right] \right\} \right) \tag{98}$$

where $\rho(R)$ is the density of the R sites on the fractal structure. Because protein is a fractal object, we then have $\rho(R) = \rho_0 R^{D_f - d}$, where ρ_0 is a proportionality constant. Equation (98) for isotropic interactions is then $\varphi(R)$, or

$$P(t) \approx \exp\left(-\Psi \rho_0 dV_d \int dR\, R^{D_f - d} \left\{ 1 - \exp\left[-t\varphi(R_j)\right] \right\} \right) \tag{99}$$

where V_d is the volume of the d-dimensional unit sphere. Inserting Equation (96) into Equation (99), we can obtain

$$P(t) = \exp\left(-\Psi A t^{D_f/n}\right) \tag{100}$$

where A is independent of time. Equation (100) is an extension of a known result for Euclidean dimension d to fractal dimension D_f. Equation (100) is directly verified by the experimental work on closed-open transitions in ion channel proteins.[36] Based on the assumptions involved in the model, the rate of the conformational change is given by

$$r = -[dP(t)/P(t)]/dt = -d \ln P(t)/dt = \Psi A(D_f/n) t^{(Dt/n - 1)} \tag{101}$$

For reactions III and IV, the total rate R is given by

$$R = -d[S*]/dt = k_0 r[S*]^2 = k'[S]^h \tag{102}$$

where [S] is the concentration of the reactant S; k_0 and k' are the rate constants. Inserting Equations (100) and (101) into Equation (102) gives the integrated rate equation as follows:

$$[S]^{-1} - [S]_0^{-1} = k_0 \Psi A t^{D_f/n} \tag{103}$$

where $[S]_0 = [S]$, (t = 0). Hence,

$$t \sim [S]^{-n/D_f} \tag{104}$$

A simple substitution of Equation (104) into Equation (102) gives:

$$R \sim [S]^{(1+n/D_f)} \tag{105}$$

and comparing with $R = K_h[S]^h$, we obtain

$$h = (1 + n/D_f) \tag{106}$$

This form indicates the relation between the Hill coefficient h and the fractal dimension D_f. In fact, n = 6 for the multipolar interaction and n = 2 for the elementary reaction III according to our computer simulation results. Thus, when n = 2, then h = 1.66 ~ 3, because $1 < D_f < 3$. In general, D_f = 1.33 ~ 1.66 for a fractal structure in two-dimensional space, and h = 2.50 ~ 2.20, e.g., for the Sierpinski gasket (d = 2), D_f = 1.585, h = 2.26.

4.10.2.2 Multistep conformational change

A different mechanism for reactions III and IV is the multistep process, where the excitation of a reactant molecule migrates among the fractal sites until it encounters an active site of protein. The active sites (R sites) are randomly distributed on the protein structure with probability Ψ. The microscopic transfer rates from a site to its neighboring sites are assumed to be equal. The above processes may be regarded as random walks on fractal structure. The walker gets trapped at the first encounter of a trap.

For a particular realization of the random walk on the trap-free fractal structue. Let R_n denote the number of distinct sites visited in n steps. Here, the stochastic variable R_n depends both on the starting point on the fractal and the sequence of directions of steps. Let P_n denote the probability that trapping site has not been occupied up to the nth step in the ensemble of fractal structure with traps. The measurable survival probability of an excited substrate molecule is given by

$$P_n = \langle (1 - \Psi)^{R_n - 1} \rangle \tag{107}$$

Note that the average in Equation (107) also includes the average over starting points and may be viewed as a double average. Introducing $\lambda = \ln(1 - \Psi)$, Equation (107) allows a straightforward cumulant expansion:

$$P_n = e^\lambda \tilde{P}_n = e^\lambda \exp\left[\sum_{j=1}^{\infty} k_{j,n} (-\lambda)^j / j!\right] \tag{108}$$

where $k_{j,n}$ are the cumulants of the distribution of R_n. As an example, the first two cumulants are $k_{1,n} = \langle R_n \rangle \equiv S_n$ and $k_{2,n} = \langle R_n^2 \rangle - \langle R_n \rangle^2 \equiv \sigma_n^2$, where S_n and σ_n^2 are the mean and the variance of R_n. In general, for the first cumulant $\tilde{P}_n = \exp(-\lambda S_n)$. This equation corresponds to the first-passage time approximation in the fractal field. Introducing the second cumulant, we can obtain

$$P_n = \exp\left(-\lambda S_n + \lambda^2 \sigma_n^2/2\right). \tag{109}$$

Translating the number of steps into time,

$$S(t) = at^{d_f/2}, \quad \sigma^2(t) = bt^{d_f} \tag{110}$$

We have the following results

$$\tilde{P}(t) = \exp\left(-\lambda a t^{d_f/2}\right) \tag{111}$$

for the first-passage time limit and

$$\tilde{P}(t) = \exp\left[-\lambda a t^{d_f/2} + \left(\lambda^2/2\right) b t^{d_f}\right] \tag{112}$$

for the correction, where a and b are constants. Clearly, the survival probability is dominated by the spectral dimension d_f. For the short-time behavior, the expansions of P(t) are

$$P(t) \sim \exp\left(-c\lambda^{d_f/2}\right)(d_f < 2), \quad P(t) \sim \exp(-c'\lambda t)(d_f > 2) \tag{113}$$

where c and c' denote numerical factors. These results are valid at short times and all concentrations of traps. Taking into account the fluctuations of the trap density, it is clear that reactant molecule survival for a long time will occur only in sufficiently large trap-free regions. These regions are rare but govern the limit of large t. The result is given by

$$P(t) \approx \exp\left[-a\Psi^y t^x\right] \tag{114}$$

where $x = d_f/(d_f + 2)$ and $y = 1 - x = 2/(d_f + 2)$. Equation (114) is a generalization of the long-time survival probability Euclidean space, and it reduces to the Euclidean result by replacing d_f by d. Now, we propose a scaling law for the survival probability $P_n(t)$ in a moderate-time regime, which interpolates between the two time regimes. Using Equations (108) and (112) and an extension of Equation (114) to all Ψ, we obtain the following expression

$$P_n = e^\lambda \exp[-g(z)] \tag{115}$$

where g(z) is a universal function which reduces in limiting cases to

$$g(z) = \begin{cases} z & (z \ll 1) \\ z^{2/(d_f+2)} & (z \gg 1) \end{cases} \tag{116}$$

and $z = \lambda t^{d_f/2}$. Let us return to the calculation of the Hill coefficient. The relationships between spectral dimension and Hill coefficient are derived from Equation (116) by a manner similar to that previously employed in deriving Equations (101) and (102), and the results are

$$h_a = 1 + 2/d_f \quad (z \ll 1) \tag{117a}$$

$$h_a = 2 + 2/d_f \quad (z \gg 1) \tag{117b}$$

Based on the Alexander-Orbach conjecture $d_f \approx 4/3$, we obtain $h_a = 2.50$ and $h_b = 3.50$, which is probably the upper limit of the Hill coefficient.

4.10.2.3 Generalization of the rate equation based on the fractal reaction kinetics

We are now in a position to suggest a general form of the rate equation as follows:

$$d[x^*]/dt = kr^n[x^*]^\theta = k_0[x]^h \tag{118}$$

where $[x^*]$ is the concentration of excited reactant molecules; k and k_0 are rate constants; and η, θ are the scaling exponents whose definitions and meanings can be seen from Equation (102). Applying Equations (100) and (116), the relationships between the Hill coefficients and other scaling exponents are derived from Equation (118) and $R = K_h[S]^h$ as follows

$$h' = \theta + \eta(1-\theta)(D_f - n)/[n + \eta(D_f - n)] \tag{119}$$

$$h_b = \theta + 2\eta(\theta - 1)/(d_f + 2 - 2\eta). \tag{120}$$

These formulae are, in essence, generalizations of Equations (106) and (117b). Assuming $\eta = 1$ and $\theta = 2$, the same results as in the previous sections are obtained from Equations (119) and (120).

The occurrences of any bimolecular chemical reactions all require contact and collision among reactant molecules. The allosteric phenomena are induced by attack of the excited substrate molecules. From the above studies, we have seen that non-integral characters of the Hill coefficients are direct reflections of the fractal properties of proteins. Thus, the h value is a kind of fractal dimension, not the order of reaction. This is an interesting discovery in our research. If we consider the protein chain as Brownian motion, the d_w in Equation (16) is equal to 2, and $D_f = d_f$; therefore, Equations (106) and (117a) are equivalent. The exponents n and η are related to the properties of protein conformations. In particular, the θ and n are measures of interactions among the protein subunits and, therefore, are also indices of cooperativity.

Using Equation (106), (n = 2), or Equation (117a), we calculated the values of the Hill coefficients according to the values of fractal dimensions. The results are listed in Table 11. The calculated value of h is ~2.5. because $D_f \approx 1.35$ and is close to the experimental values. However, the theoretical h values of the first two proteins in Table 11, compared with the experimental results, are different from the latter in numerical values. This indicates that Equation (117a) is not very suitable for them. If applying Equation (117b) and taking $D_f = 2.15, 2.09$, then h = 2.93 or 2.95 and is in good agreement with the experimental results. It is shown that there are long-range interactions among subunits of the two proteins. For glyceraldehyde-3-phosphate dehydrogenase, the theoretical average value in Table 11 equals 2.28 and is in good agreement with the 2.3 experimental value. This indicates that Equation (117a) is suitable for the protein and that there are short-range interactions among subunits of the protein.

Table 11 Allosteric Constants for Some Proteins

Protein	Ligand	Number of binding sites	h (exp)	d_f (30)	h (cal)	h (simulation)
Hemoglobin	O_2	4	2.8	1.40 ~ 2.15	2.43 ~ 1.93	2.93 ± 0.09
Pyruvate kinase	Phosphenol pyruvate	4	2.8	1.34 ~ 2.09	2.49 ~ 1.97	2.87 ± 0.07
Glyceraldehyde-3-phosphate dehydrogenase	NAD^+	4	2.3	1.34 ~ 1.87	2.49 ~ 2.07	2.49 ± 0.05

Let us now return to Equations (119) and (120). When $\theta = 1$, then h' and $h'_b = 1$. There is no cooperativity. The enzyme molecule has no change of conformation. When $\eta = 0$, there is no cooperativity and change of conformation. The values of all Hill coefficients equal θ, and this is a case of classic chemical reaction. If $\theta < 1$, then $h < 1$. This indicates negative cooperativity. For example, taking $\theta < 0.5$, and $\eta = -1$, then $h' = 0.68$ (n = 2), and $h'_b = 0.35$.

It is known that the Hill coefficient is not a constant for a given protein but depends on conditions. This fact indicates that the change of protein conformation in a one- or multistep process is dominated by the substrate molecule and circumstance because the fractal and spectral dimensions are related to the changes of protein conformations. The protein is a biological macromolecule consisting of amino acid residues whose branches can form fractals; the multi- and hierarchical structures of proteins are the basis for the statistical self-similarity of proteins and enzymes. Furthermore, the surface of enzymes is multifractal because of its rough and coarse features.[30] The substrate molecules randomly walk over the fractal networks of amino acid residues until they hit the active sites, or, R states of protein. The substrate molecule could be bound to these residues by means of the interaction of hydrogen bonds and Van der Waals force.[45]

If the substrate-enzyme reaction is diffusion limited, the Hill coefficient can also arise from diffusion to the protein and on the protein. In this case, the rate constant K is linearly proportional to the diffusion constant D for homogeneous reaction in three-dimensional systems.[87] Both K and D are time independent; however, this is not true for lower dimensions, because the K is related to the fractal dimension. Kopelman[88] has discussed this point in detail.

In summary, the allosteric effects are of relevance to changes of protein conformation, and the Hill coefficient characterizes the allosteric effects and cooperativity, which reflect the fractal properties of proteins and enzymes, not the number of binding sites on each molecule of protein.

4.11 Fractal art of enzyme model design[87]

In general, an enzyme is a protein. The enzyme model implies a small organic molecule or molecular cluster which can imitate the specificity and function of a enzyme. Enzymes are biocatalysts; however, the fractal model previously mentioned cannot explain the catalysis and conformation of enzyme. As a matter of fact, the fractal dimension D_f is related to the conformation and shape of protein chain,[39] or properties of the protein. The function of the enzyme is determined by its three-dimensional structure, which is influenced by the primary structure consisting of amino acid residues. In a word, the fractal dimension provides much information related to the structure and conformation of an enzyme.[65]

4.11.1 Fractal principle of enzyme model design

A good enzyme model should contain the main features of the enzyme. In other words, it should be a generator. It not only has a similar shape and function of the global, but it also can evolute the global. We have summarized some fundamental laws and used them to study enzyme models. We find that the following main fractal principles are very useful in enzyme model design.[89]

> Theorem 1 (Multiplication Principle): Let the fractal set S be the product of two subsets S_1 and S_2 having fractal dimension D_1 and D_2, respectively, then the fractal dimension of S equals the sum of D_1 and D_2.
>
> Theorem 2 (Union Principle): Let the fractal set S be the sum of two subsets S_1 and S_2 having fractal dimension D_1 and D_2, respectively; moreover, $D_1 > D_2$. Then the fractal dimension of S is situated between D_1 and D_2 and the maximum value equals D_1.
>
> Theorem 3 (Match Principle): If the subsets S_1 and S_2 can be effectively combined in fractal set S, the fractal (or local fractal) dimension of S_1 must be equivalent or close to that of S_2.
>
> Theorem 4 (Inclusion Principle): If the fractal set M is to include the subset S, the fractal dimension of S is not greater than that of M.
>
> Theorem 5 (Construction Principle): The enzyme model is the generator of an enzyme, which resembles the behavior of the enzyme, and its fractal dimension is equivalent or close to that of the active site in the enzyme.

4.11.2 Fractals of hybrid orbitals

The primary structure of enzymes resembles the Koch curves, the shape of which is determined by the generator,[1] namely, the bond angle of the atomic orbital. Thus, the relationship between the fractal dimension and the hybridized state of atomic orbitals should be established first. The angle may be extracted from the protein chain and Koch curve. Assuming ao = bo ∠ aob = θ, then the number of intervals N = 2, the similarity ratio $\gamma = [2(1 - \cos\theta)]^{1/2}$, and the fractal dimension is given by

$$D_f = \ln N / \ln(1/r) = \ln 2 / \ln 2[1 - \cos\theta]^{1/2} \tag{121}$$

For a given molecular chain, based upon the principles of quantum chemistry such as the orthogonality of the hybrid molecular orbital, the bond angle θ_{ij} between ψ_i and ψ_j orbital is given by

$$\cos\theta_{ij} = \left[a_i a_j / (1 - a_i)(1 - a_j) \right]^{1/2}$$

where a_i and a_j denote the fraction of s orbital in the hybrid orbital ψ_i and ψ_j, respectively. For the equivalent hybrid orbital ($a_i = a_j = a$),

$$D_f = 2\ln 2 / \ln[2(1 + a/(1-a))] \qquad (122)$$

Obviously, D_f depends on the bonded state of atomic orbitals. Using Equation (122), the calculated values of D_f can be obtained as listed in Table 12. In addition, the bonding capacity of hybrid orbital, f is also related to the fractal dimension. That is, $f = [1 - 2\exp(2\ln2/D_f)]^{1/2} + [6/\exp(2\ln2/D_f)]^{1/2}$. The relation is a convex curve. According to the $df/dD_f = 0$, we have $f_{max} = 2.00$ only when $D_f = 1.404$.

Table 12 Dependence of Fractal Dimension on the sp^n Hybridized State of Atomic Orbital

Hybrid type	Bond angle θ	α	D_f	f
sp	180°	1/2	1.00	1.933
sp^2	120°	1/3	1.262	1.991
sp^3	109.5°	1/4	1.404	2.000
p	90°	0	2.00	1.732

For some protein molecules containing transition metal atoms, such as hemoglobin, myoglobin, cytochrome, and so on, the d orbital of the metal atoms can participate in the bonding action. Based on the orthogonality of the hybrid orbital, the bond angle θ for the equivalent d-s-p hybridization obeys the following formula:

$$\alpha + \beta \cos\theta + \gamma(3/2 \cos^2\theta - 1/2) = 0 \qquad (123)$$

where α, β, and γ are the fractions of s, p, and d orbitals in the hybrid orbital, respectively, and ($\alpha + \beta + \gamma$) = 1. For example, we obtain $\theta = 90°$, $\theta = 180°$ using Equation (123) for the d^{2}sp^3 hybridization, and $D_f = 2.00$ and $D_f = 1.00$. The fractal dimension for the dsp^2 hybridization is the same as those of the d^{2}sp^3 type. This implies that these structures are of the Peano type and are a kind of fat fractal.[4] It is important to note that fat fractals are, in general, not self-similar objects; they have more in common with the Peano curve since these structures are typically made of parts with dimensions smaller than the embedding dimension (d). The whole object has a finite measure in d.[3] There are many physical systems in which fat fractals are expected to occur. In addition, ballistic aggregation clusters, bronchioles in the lung, or coral colonies are most likely to have the structure of fat fractals.

The fundamental structure skeleton (ring, segment, and substituent) may be extracted from various organic compounds by statistical classifications.[27,39] Furthermore, they may be classified into two categories: molecular fragments (motherbackbone) and substituents, and are regarded as generators. Their fractal dimensions are easily calculated by using the theorems. The molecular fragments are motherbackbones, on which the various substituents are combined, and produce the large number of imitative enzyme models which are functional organic compounds with catalytic activity resembling the bioenzyme. The models are organic compounds synthesized and modified by chemical methods such as cyclodextrin (CD) and metalloporphyrin. In designing and selecting models, the requirements of Theorems 4 and 5 must be considered. To illustrate the importance of fractal theory, let us examine the features and properties for some selected models of enzymes.

4.11.2.1 Stereospecificity of enzyme catalysis

It was pointed out some 50 or 60 years ago that the recognition of a chiral carbon by an enzyme implies that at least three of the groups surrounding the chiral carat carbon must be bound to the enzyme. This is called the *three-point attachment theory*. This theory reflects the fractal character of enzyme-complexing substrates because the three points form a planar triangle which is the generator for the Koch snowflake curve, and its fractal dimension lies in the range $1.262 \leq D_f \leq 1.404$, based on Equation (121). According to Theorem 1, the fractal dimension of the attachment surface part is given by $D_s = (D_f + 1) = 2.262 \sim 2.404$. The results are in very good agreement with those measured by Lewis and Rees.[35] This fact indicates that the fractal feature of the three-point attachment theory is the reason why it still remains an excellent model of enzyme catalysis.

4.11.2.2 Cyclodextrin

Cyclodextrin is a naturally occurring, doughnut-shaped macrocyclic glucose polymer containing a minimum of six-D (+)-glucopyranose units attached by α-(1,4)-linkages. The internal diameters of α-, β-, and γ-cyclodextrins estimated from X-ray analysis are approximately 4.5, 7.0, and 8.5 Å, respectively. The interior of the doughnut lined with CH groups provides a relatively hydrophobic environment. Cyclodextrin is a molecular fragment; its cavity can include the substrate molecules with a geometrical fit. If attached by the active substituent, it shows catalytic activity. β-CD has been extensively studied as an enzyme model. The perimeter fractal dimension of the β-CD cavity is 1.333, and the surface fractal dimension of the cavity equals 2.333; thus, only the substrate molecules with fractal dimensions smaller than 1.333 can penetrate into the cavity of β-CD, based on Theorem 5. The experiments show that the substrates catalyzed by β-CD have, in general, benzene or naphthalene groups and a fractal dimension of 1.262 and the ability to enter entirely or partly into the cavity of β-CD. For example, for the reaction of 2,4,6-trimethylphenol and bromopropene in the presence of β-CD, the para-product yields are close to 100%, while in the absence of β-CD, the yields are 33%. The propenyl group attacks specifically the 4-site, not the 2- or 6-sites, because of the limitation of cavity volume and the action of the hydrophobic field. Obviously, the behavior of β-CD complexing substrates is entirely analogous to those of the enzyme systems.

Many models of imitative enzymes have been investigated: calixarene, macrocyclic ligand (crown ether and shiff base, etc.), micelles, polyelectrolytes, and so on. The common feature of these models is that they can form circular microenvironments during the reaction process. The fractal theory for the imitative enzymes requires that a good model must have, or must be able to form, the circular structure of active sites and must have an active group (or metal ions). When complexed with substrates, it must obey Theorems 4 and 5. However, these are only basic rules; the actual situations are quite complicated. The fractal theory of enzymes and imitative enzyme is a very attractive and interesting field for further studies.

4.11.3 Structural parameters and enzyme model catalytic properties

The study of function of imitating natural enzymes is an advanced subject in today's chemical science. Enzyme models are smaller organic molecules which can reflect the main function of a protein enzyme. We have studied many kinds of enzyme models; metalloporphyrins are one of these models. We find that the catalytic activity will far exceed the corresponding metalloporphyrin if we introduce a steriod structural unity which has a rigidity plane dredging water, and the deduced period of reaction will clearly decrease. For metal ions, the catalytic activity sequences Co(II) > Mn(II) > Fe(II) ≫ Zn(II) are shown in steroid-metalloporphyrins. Here, we shall explore the differences of metal ion activity from the viewpoint of structural parameters (fractal dimension, topological index, information parameter).

4.11.3.1 The fractal dimension of steroid-metalloporphyrin

There have been some studies of catalytic activity and selectivity utilizing fractal dimensions, such as that of Farin et al.[78] However, enzyme models studied by fractal dimension have not been published. Here, metal complex compounds (enzyme model) is regarded as a generator and the Hausdorff dimension is calculated. Therefore, we have explained the catalytic activity sequences combined with the fractal theorem of enzyme models.[89]

The fractal dimension is calculated by formula $\sum r_j^D = 1$, and self-similar rate $r_i = l_i/d$, where l_i is bond length, d is the diameter of complex compound and N, the total number of bonds. The basic data used in the calculation are listed in Tables 13 and 14, and the results are listed in Table 15.

Table 13 Radii of Four Metal Atoms (Å)

Metal atom	r_m	r_c	r_i
Co	1.25	1.16	0.78 (+2)
Mn	1.29	1.17	0.80 (+2)
Fe	1.26	1.16	0.76 (+2)
Zn	1.33	1.25	0.74 (+2)

Note: r_m is the metallic radius, r_c is covalent radius, and r_i is the ionic radius.

Table 14 Bond Length Data

	Bond type					
	C–O	C=O	C=C	N–C	N=C	N⋯C
Bond length (Å)	1.43	1.22	1.34	1.47	1.27	1.39

Table 15 Fractal Dimensions for the Steroid-Metalloporphyrins

	Fractal dimension		
	Ethinyl estradiol–porphyrin	Estrone–porphyrin	Estradiol–porphyrin
Co	1.252	1.357	1.327
Mn	1.367	1.378	1.349
Fe	1.374	1.385	1.366
Zn	1.816	1.768	1.673

From Table 15, we know that the fractal dimension sequences are Co < Mn ≪ Zn for all three kinds of metal complex compound. If we regard isopropyl benzene as a generator also, the fractal dimensions are distributed in the range of 1.262 to 1.333. Because the fractal dimensions of iron, cobalt, and manganese complex compounds are almost identical to that of substrate or greater than that of substrate (which agrees with the match principle and inclusion principle), they have better catalytic activity. But the fractal dimension of the zinc complex compound has a greater difference from that of substrate and can form a midstate effectively;

therefore, the catalytic activity is small. Among the steroids, there is no contrasting relationship. Since crossing a metal element easily can change its valent state, which influences the radius and the fractal dimension of complex compounds, the phenomena of activity crisscrossing and inverting may occur in the catalytic reaction. These phenomena have been observed in the experiments.

We also should pay attention to the following problem. When we investigate the catalytic action using fractal theory, we must correlate it with the substrate; otherwise, there is no meaning and the above conclusion may be useless.

4.11.3.2 Application of the topological indices in enzyme models

The molecular topological index is an index which digitizes the molecular structure. It is a topological invariant quantity of the molecular diagram. The molecular topological index theory has become an effective method to study structure and physical nature as well as chemical and biological activity. Since Winer proposed the first topological index, 40 different kinds of topological indices have been reported in the literature. Topological indices can be used to characterize the size, shape, degree of side-chain, and other physical characteristics of a molecule. The previous topological indices are mainly used to study hydrocarbon. In order to extend the application range of topological indices, Kier and Hall[91] proposed the valent connectivity index, Lall and Srivastava[92] proposed the weight indices of edges, and Barysz[93] proposed the index of atomic electric charge density. These indices can be used for the heteroatomic system to a certain extent. In order to study the topological properties of the compound containing the metal iron, Zhang et al.[94] proposed the concept of the topological index of bonding parameter, the general formula of which can be denoted by

$$H_t = \left(\sum_i e^{-(x_i - y_i)} (p_i q_i)^{-1/2} \right)^2 \quad (x_i > y_i) \tag{124}$$

where p_i, q_i represent the number of degrees of atoms at two endpoints on the ith edge in the molecular diagram; x_i, y_i correspond to the bonding parameter of atoms at two endpoints; and t represents the kind of bonding parameter at the vertex point. Four classes have been classified at present. Here, we consider t = 3. In this case, x_i and y_i represent the electron-charge/radius ratios of atom of element $(x_i > y_i) Z/r_c$, where Z is the number of electrovalence of the atom, and r_c is the covalent radius.

The topological index of bonding parameter can be expressed as

$$H_{tn} = \left[\sum_i \left(1 + \Delta_i + \Delta_i^2/2! + \Delta_i^3/3! + \ldots\right)(p_i q_i)^{-1/2} \right]^2 \tag{125}$$

where $\Delta_i = x_i - y_i$, and n represents the term order of the spreading formula. When n = 0, $H_{t0} = [\sum_i (p_i q_i)^{-1/2}]^2$, and it is the square of the Randic molecular connectivity index X.

In enzyme models, we study only the topological indices of bonding parameter in the active center position; i.e., we study the M – N_4 ring of metal steroid only and calculate H_{31} and $H_{3\infty}$. The data used in the calculations are listed in Table 16.

For the M–N_4 ring of metal steroid, $p_i = 3$, $q_i = 4$, i = 4. Using Equation (125) and the data of Table 16, the topological indices H_{31} and $H_{3\infty}$ of four kinds of metal steriods are calculated and listed in Table 17. From Table 17 we can see that the topological index of the bonding parameter of the zinc complex compound is the least of all

Table 16 The Electrovalence/Radius Ratios of Elements

Parameters	Elements				
	Co	Mn	Fe	Zn	N
Z	2	2	2	2	−3
r_c (nm)	0.116	0.117	0.116	0.125	0.074
Z/r_c	17.24	17.09	17.24	16	40.54
Δ_3	23.30	23.45	23.30	23.54	0

indices; therefore, the catalytic and oxidating activities are the lowest. Because H_t is related to the polarization ability of centric metal atoms, the greater the H_t value is, the stronger the polarization ability will be, and the easier it is to complex O_2. Conversely, the less the H_t value is, the weaker the polarization ability will be and the more difficult it is to complex O_2. Clearly, one can explain the catalytic property coinciding with experiments from the viewpoint of topological indices. However, the difference among the indices of Co, Mn, and Fe is difficult to distinguish and the reason is not clear. Maybe the catalytic activity itself is not clear, the valent state changes easily, or the sequence of order is not clear. From experiments, we know that the activity of Mn is less than that of Fe in the first period of reaction, but the case is reversed in the later period. This shows that the slight difference of the activity of metal ions cannot be determined from the topological indices of bonding parameters only.

Table 17 The Topological Indices of Bonding Parameters

	Co–N$_4$	Mn–N$_4$	Fe–N$_4$	Zn–N$_4$
$H_{31} \times 10^3$	2.258	2.231	2.258	2.044
ΔH_{31}	0.060	0.033	0.060	−0.15
$H_{3\infty} \times 10^{21}$	7.706	5.709	7.706	0.6453

4.11.3.3 Application of the structural information index

Information is an important concept in science of the 20th century. At first, information is regarded as a measurement that decreases indefinitely. For chemical structures, the information generalized by Ashby et al.[95] is very convenient and suitable. In this point of view, information may be regarded as a measurement eliminating homogeneity and isotropism in the system or a measurement of giving manifold properties of the system, or, in other words, "a nonhomogeneous measurement of material and energy distribution in time and space." Here we use the definition given by Mowshovitz.[96]

Suppose each possible structure consists of definite N elements which can be divided into equivalent classes 1, 2, ..., k by some selection and contain N_1, N_2, ..., N_k elements, respectively, then the distributions are $P = N_i/N$, i = 1, ..., k, and each element of the structure given by the division has the mean structural information

$$\bar{I} = -\sum_{i=1}^{k} P_i \log_2 P_i \tag{126}$$

then the unit is bit. The structural total information $I = N\bar{I}$. For metal ions of metal steroid, for convenience, we study only the electronic distribution on the third orbit and divide them into single electron and couple electrons, i.e., k = 2. By Equation (126), we have

$$I = N\bar{I} = -3.322 \log_{10}\left(Z_i Z_j / Z^2\right) \tag{127}$$

where Z_i is a single electronic number, Z_j is a douple electronic number, and Z is the total number of electronic informaton parameters. The calculated results are listed in Table 18.

Table 18 The Electronic Information Parameters for Metal Ions

	Co^{2+}	Mn^{2+}	Fe^{2+}	Zn^{2+}
Configurational information (bits)	2.030 3.029	2.059 1.644	2.170 2.170	
Total information (bits)	5.059	4.703	4.340	0

According to the data of Table 18, the catalytic activity sequence of metal ions has been illustrated successfully and the activities of Co, Mn, and Fe in two-order states also have been distinguished. If we regard the electronic information parameters as the manifold measurment of atomic systems, we can consider that the more distributive the pattern of a metal atom, the better the catalytic activity will be. This provides a theoretical reference for choosing a centric metal element which usually has a different valent state. There are many electronic distributive patterns; therefore, the centric element can be chosen as a candidate for the catalyst group. In general, the system with several kinds of construction and conformations usually has greater information parameters and therefore may make a good enzyme model. Perhaps a system too rigid and simple cannot become an effective enzyme model. A biological enzyme usually has a long chain and the conformation is ever changing, the information parameters are very large, and the catalysis is very effective. When we design enzyme models, we must consider manifold and variable characteristics.

Fractal dimension, topological index, and information parameters have important applications in imitative enzyme studies. We find that fractal dimensions are related to the information parameter. There is a linear relationship between information parameters and fractal dimensions of Co, Mn, Fe, and Zn, four kinds of metal complex compounds. The fractal dimension also reflects the molecular information. In fact, the electronic information parameter only reflects the electronic distribution in an atom, and the global geometric structure of the model material is not considered, although this structure is very important. Hence, the electronic information parameter is a macroparameter, and there are many difficulties in predicting the global macroproperties for a model material. A fractal dimension is a more comprehensive global information parameter, as it is calculated from the bonding length parameter of a molecule, and bonding length itself contains the information of the atomic (ion) radius, valent electronic number, etc. Besides this, the geometric structure is an important factor which is part of the fractal dimension. Therefore, it is well worth paying attention to the fractal dimensions as parameters characterizing enzyme models.

4.12 Discussion and conclusion

We have discussed several applications of fractal and chaotic analysis to protein conformation and enzymatic kinetics. The power of this approach is in its simplicity and generality.

Fractal dimension can be useful in analyzing the fractal aspects of polymers. The universal aspects of protein structure have been discussed by scaling laws related to the radius of gyration, volume, and surface area. The fracton dimension d_f can be used to describe the atomic vibration and macrodynamical behavior of proteins. For fractal analysis of protein chain conformation, we have proposed a new method to calculate the fractal dimension of protein molecular chains, and the results suggest that the fractal dimension may be used to characterize the tertiary structure of proteins and enzymes. As a structural quantity, it reflects information of the morphology of the protein chain based on the amino acid sequence. For each chain, the fractal dimension is based on the number of the amino acid residues, and the total length and the planar diameter of the chain are related to the shape of the protein chain. These fractal dimensional values are different from those of Stapleton's group, but are in agreement with the Monte Carlo simulation. Further investigation shows that there is a linear relationship between the fractal dimension and conformational entropy. The two parameters all reflect the conformational properties of a protein molecular chain. The results suggest that the thermodynamical behaviors of protein may be related to multifractals. The correlative dimension D_2 and and Kolmogorov entropy K_2 are impotant parameters to describe the dynamical character of proteins and enzymes. By these parameters, the characters of globular protein, the EEG of fetal sheep, and biochemical reactions have been analyzed. Although further conclusions have not been deduced logically from D_2 and K_2, the results obtained have important meaning; they will give us an approach to explore the dynamical behavior of biochemical systems. The surface roughness and irregularity of proteins and enzymes can be described by fat fractal and multifractal analysis. The surface fractal dimensions calculated by the variation method are different from those obtained by other methods, since the former is applicable to self-affine systems. Thus, the results are reliable for the surface. The fat fractal and multifractal features of proteins and enzymes are studied by simulation. The surface mass exponents are regarded as another kind of scaling exponent, and the f-α/spectrum provides further detailed information about the surfaces of proteins and enzymes. The f-α/spectrum and the computer simulation have also been used to study the enzymatic reaction. The results indicate that the active site probability distribution of enzyme surface and the "poison" time are all related to the fractal dimension. The catalytic reaction kinetics of proteins and enzymes have been investigated by the statistical methods of random walks on fractal structures. The relations between Hill coefficents and the fractal, as well as fracton dimensions, have been obtained. The results are reasonable and suggest a new mechanism for the allosteric effects of proteins and enzymes. Finally, we have introduced five fractal theorems to investigate the models of imitative enzymes. The enzyme model design is related not only to the fractal geometry, but also to the topological indices and information parameters. Using them, we have given a satisfactory explanation for the catalytic activity sequence. In addition, the linear relation between fractal dimensions and structural information parameters has been discovered and prediction of the reverse phenomena of activity order has been verified by experiments.

Fractal and chaotic theories are powerful tools to study protein conformation and enzymatic kinetics; however, there are also some weaknesses, e.g., their physical meaning is not clear. We hope that more studies of the interpretation of the observed power laws will become available.

Acknowledgments

The authors wish to thank Drs Z. Tang, X. Zheng, and M. Xiang for the computer simulations. This project is supported by the National Natural Science Foundation of China, the Sichuan Youth Science and Technology Foundation, and the Chinese Educational Committee Foundation.

References

1. Mandelbrot, B. B., *The Fractal Geometry of Nature*, W. H. Freeman, San Francisco, 1982.
2. Elber, R., Fractal analysis of protein, in *The Fractal Approach to Heterogeneous Chemistry*, (Ed., Avnir, D.), John Wiley & Sons, New York, 1989, 407.
3. Li, H. Q., Chen, S. H., and Zhao, H. M., Fat fractal and multifractal for protein and enzyme surfaces, *Intl. J. Biol.*, 13, 210, 1991.
4. Eykholt, R. and Umberger, D. K., *Physica D.*, 30, 43, 1988.
5. Feder, J., *Fractals*, Plenum Press, New York, 1988.
6. Li, H. Q., Chen, S. H., and Zhao, H. M., Fractal mechanisms for the allosteric effects of proteins and enzymes, *Biophys. J.*, 58, 1313, 1990.
7. Alexander, S. and Orbach, R., Density of states on fracton, *J. Physique (Paris) Lett.*, 43, L625, 1982.
8. Rammal, R. and Toulouse, G., Random walks on fractal structures and percolation clusters, *J. Physique (Paris) Lett.*, 44, L13, 1983.
9. Cates, M. E., Brownian dynamics of self-similar macromolecules, *J. Physique*, 46, 1059, 1985.
10. Muthukumar, M., Dynamics of polymeric fractals, *J. Chem. Physiol.*, 83(6), 15, 3161, 1985.
11. Dewey, T. G., Fractal aspects of protein structure and dynamics, *Fractals*, 1, 179, 1993.
12. Adam, M. and Lairez, D., Fractal conformation of polymers, *Fractals*, 1, 149, 1993.
13. Miyaki, Y., Einaga, Y., and Fujita, H., Excluded volume effect in dilute polymer solutions, *Macromolecule*, 11, 1180, 1978.
14. Davidson, N. S., Fetters, L. J., Fund, W. G., Hadjichristids, N., and Graessley, W. W., Measurement of chain dimensions in dilute polymer solutions, *Macromolecules*, 20, 2, 2614, 1987.
15. Bauer, B. J., Fetters, L. J., Graessley, W. W., Hadjichristids, N., and Quack, G. F., Chain dimensions in dilute solutions, *Macromolecules*, 22, 2337, 1989.
16. Richter, D., Farago, B., Fetters, L. J., Huang, J. S., and Even, B., On the relation between structure and dynamics of star polymers in dilute solution, *Macromolecules*, 23, 1845, 1990.
17. Stanley, H. E., *Introduction to Phase Transitions and Critical Phenomena*, Clarendon Press, Oxford, 188, 1971.
18. Dewey, T. G., Protein structure and polymer collapse, *J. Chem. Physiol.*, 98, 2250, 1993.
19. Pfeifer, P. and Obert, M., Fractals: basic concepts and terminology, in *The Fractal Approach to Heterogeneous Chemistry*, (Ed., Avnir, D.), John Wiley & Sons, New York, 1989, 22.
20. Helman, J. S., Coniglio, A., and Tsallis, C., Fractals and the fractal structure of proteins, *Phys. Rev. Lett.*, 53(12), 1195, 1984.
21. Wagner, G. C., Colvin, J. T., Allen, J. P., and Stapleton, H. J., Fractal models of protein structure, dynamics and magnetic relaxation, *J. Am. Chem. Soc.*, 107, 5589, 1985.
22. Colvin, J. T. and Stapleton, H. J., Fractal and spectral dimensions of bio-polymer chains, *J. Chem. Physiol.*, 62(10), 4699, 1985.
23. Stapleton, H. J., Protein as fractals, *Phys. Rev. Lett.*, 54(15), 1734, 1985.
24. Li, H. Q. and Wang, F. Q., The fracton dimension of proteins, *Chin. Sci. Bull.*, 38(8), 706, 1993.
25. Wako, H. and Gō, N., *J. Comp. Chem.*, 8(5), 625, 1987.
26. Go, N. et al., *Proc. Natl. Acad. Sci. U.S.A.*, 80, 3696, 1983.
27. Li, H. Q., Chen, S. H., and Zhao, H. M., Fractal structure and conformational entropy of protein chain, *Intl. J. Biol. Macromol.*, 12(6), 374, 1990.

28. Aharony, A. et al., Scaling approach to phonon fracton crossover, *Phys. Rev.*, B31(4), 2565, 1985.
29. Alexander, S. et al., *Phys. Rev.*, B34, 2726, 1986.
30. Isogai, Y. and Itoh, T., Fractal analysis of tertiary structure of protein molecule, *J. Phys. Soc. Jpn.*, 53.(6), 2162, 1984.
31. Wang, C. X., Shi, Y. Y., and Huang F. H., Fractal study of tertiary structure of proteins, *Phys. Rev.*, A41(12), 7043, 1990.
32. Havlin, S. and Ben-Avraham, D., *J. Phys.*, A15, L311, 1982.
33. Li, H. Q. and Zhao, H. M., Fractal studies on protein molecular chains, *J. Bioactive Comp. Polymer*, 9, 318, 1994.
34. Allen, J. P., Colvin, J. T., Stinson, D. G, Flynn, C. P., and Stapleton, H. J., *Biophys. J.*, 38, 299, 1982.
35. Lewis M. and Rees, D. C., *Science*, 230, 1163, 1985.
36. Liebovitch, L. S., Fischbarg, J., Koniarek, J. P., Todorova, I., and Wang, M., Fractal model of ion — channel kinetics, *Biochem. Biophys. Acta*, 896A, 173, 1987.
37. Yang, Y.S., *Fractals in Physics*, (Eds., Pietronero, L. and Tosatti, E), North-Holland, Amsterdam, 1986, 119.
38. Krumhans, J. A., Vibrational anomalies are not generally VUA to fractal geometry, *Phys. Rev. Lett.*, 56(25), 2696, 1986.
39. LI, H. Q., Li. Y., and Zhao, H. M., Fractal analysis of protein chain conformation, *Intl. J. Biol. Macromol.*, 12, 6, 1990.
40. Fichthorn, K. A., Ziff, R. M., and Gulari, E., *Catalysis*, (Ed., Ward, J. W.), Elsevier, Amsterdam, 1988, 883.
41. Binder, K., *Monte Carlo Methods in Statistical Physics*, Springer-Verlag, Berlin, 1979.
42. MacDonald, M. and Jan, N., Fractons and the fractal dimension of proteins, *Can. J. Physiol.*, 64(10), 1353, 1986.
43. Pfeifer, P., Wetz, U., and Wippermann, H., Fractal surface dimension of proteins, *Chem. Phys. Lett.*, 113 (6), 535, 1985.
44. Blundell, J. L. and Johnson, L. N., *Protein Crystallography*, Academic Press, New York, 1976.
45. Li, H. Q. and Wang, F. Q., *The Fractal Theory and Its Applications in Molecular Seience*, Science Press, Beijing, 1993, 33.
46. Wang, F. Q. and Li, H. Q., *The Fractal Geometry and Dynamical Systems*, Heilongjiang Education Press, Harbin, 1993, 189.
47. Rackovsky, S. and Seheraga, H. A., Differential geometry and polymer conformation. 1. Comparison of protein conformation, *Macromolecules*, 11(6), 1168, 1978.
48. LeGuillon, J. C. and Zinn-Justin, *J. Phys. Rev.*, B21, 3976, 1980.
49. Zhao, D. L. and Huang, Y., *Acta Polym. Sin.*, 3, 310, 1 989.
50. Ottinger, H. C., *Macromolecules*, 18, 93, 1985.
51. Grassberger, P. and Proeaeeia, I., Characterization of strange attractors, *Phys. Rev. Lett.*, 50(5), 346, 1983.
52. Grassberger, P. and Procaceia, I., Estimation of the Kolmogrov entropy from a chaotic signal, *Phys. Rev.*, A28(4), 2591, 1983.
53. Wang, F. Q., Luo, C. S., and Chen, G. X., An improvement of G-P algorithms and its applications, *Chin. J. Comp. Physiol.*, 10(3), 345, 1993.
54. Wang, F. Q., Luo, C. S., and Chen, G. X., Quantitative characterization of chaotic systems, in *ICSSSE '93*, (Ed., Zheng, W. M.), International Academic Publishers, Beijing, 1993, 603.
55. Kubota, Y., Takahashi, H., Yoshino, T., and Tsuchiya, T., Chaos-theoretical analysis of possible structural quantities for globular proteins, *Biophysica Acta*, 1079, 73, 1991.
56. Nishikawa, K and Ooi, T., *J. Biochem.*, 100, 1043, 1986.
57. Bernstein, F. C., Koetzle, T. F., Williams, G. J. B., Meyer, E. F., Jr., Briee, M. D., Rodgers, J. R., Kennard, O., Shimanouehi, T., and Tasumi. M., *J. Mol. Biol.*, 112, 535, 1977.
58. Grantham, R., *Science*, 185, 862, 1974.
59. Dunki, R. M., The estimation of the Kolmogorov entropy from a time series and its limitations when performed on EEG, *Bull. Math. Biology*, 53(5), 665, 1991.

60. Wolf, A., Swift, B., Swinney, H. L., and Valtano, J. A., Determining Lyapunov exponents from a time series, *Physica*, 16D, 285, 1985.
61. Hess, B. and Markus, M., Order and chaos in biochemistry, *Trends Biochem. Sci.*, 12, 45, 1987.
62. Markus, M. and Hess, B., Transitions between oscillatory modes in a glycolytic model system, *Proc. Natl. Acad. Sci. U.S.A.*, 81, 4394, 1984.
63. Goetze, T. and Brickmann, J., Self-similarity of protein surfaces, *Biophys. J. Biophys. Soc.*, 61, 109, 1992.
64. Connolly, M. L., *Science*, 221, 709, 1983; *J. Appl. Crystallogr.*, 16, 548, 1983; *Biopolymers*, 25, 1229, 1986.
65. Aqvist, J. and Tapia, O., *J. Mol. Graphics*, 5, 30, 1987.
66. Farin, D., Peleg, S., Yavin, D., and Avnir, D., Applications and limitations of boundary: line fractal analysis of irregular surfaces, *Langmuri*, 1(4), 399, 1985.
67. Farin, D. and Avnir, D., *Characterization of Porous Solids*, (Eds., Unger, K. K. et al.), Elsevier, Amsterdam, 421, 1988.
68. Dubuc, B., Zucker, S. W., Tricot. C., Quiniou, J. F., and Wehbi, D., *Proc. R. Soc. Lond.*, A425, 113, 1989.
69. Dubuc, B. and Tricot, C., *Acad. Sci. Paris*, 306, 531, 1988.
70. The protein data bank: a computer-based archival file for macromolecular structures, *J. Mol. Biol.*, 112, 535, 1977 (Brookhaven National Laboratory, Protein Data Bank Newsletter, January 1990.)
71. Coniglio, A. and Stanley, H. E., *Phys. Rev. Lett.*, 1X. 52, 1068, 1984.
72. Meakin, P., Coniglio, A., and Stanley, H. E., Scaling properties for the surfaces of fractal and nonfractal objects, *Phys. Rev.*, A34(4), 3325, 1986.
73. Matsushita, M., Hayakawa, Y., Sato, S., and Honda, K., *Phys. Rev. Lett.*, 59, 86, 1987.
74. Ohta, S. and Honjo, H., *Phys. Rev. Lett.*, 60(7), 611, 1988.
75. Leyvraz, F., The "active perimeter", *J. Physiol.*, A18(15), 1491, 1985.
76. Turkevich, L. and Scher, H., Occupancy probability scaling in diffusion limited aggregation, *Phys. Rev. Lett.*, 55(9), 1026, 1985.
77. Li, H. Q., Wang, F. Q., and Zhao, H. M., Fractal reactions on enzyme surfaces, *Chin. J. Chem. Phys.*, 8(2), 162, 1995.
78. Meakin, P., Simulation of effects of fractal geometry on the selectivity of hetergeneous catalysts, *Chem. Phys. Lett.*, 123(5), 428, 1986.
79. Witten, T. A., Jr., and Sander, L. M., *Phys. Rev. Lett.*, 47, 1400, 1981.
80. Lee, J., Havlin, S., and Stanley, H. E., *Phys. Rev.*, A45(2), 1035, 1992.
81. Anacker, L. W. and Kopelman, R., Fractal chemical kinetics: simulation and experiments, *J. Chem. Physiol.*, 81(12), 6402, 1984.
82. Kopelman, R., Rate processes on fractals theory, simulations and experiments, *J. Stat. Phys.*, 42(1/2), 185, 1986.
83. Monod, J., Wyman, J, and Changeux, J. P., On the nature of allosteric transitions: a plausible model, *J. Mol. Biol.*, 12, 88, 1965.
84. Koshland, D. E., Nemethy, G., Jr., and Filmer, D., Comparison of experimental binding data and theoretical models in proteins containing subunits, *Biochemistry*, 5, 365, 1966.
85. Schweitzer-Stenner, R. and Dreybrodt, W., An extended-monod-Wyman-Changeux-model expressed in terms of the Herzfeld-Stanley formalism applied to oxygen and carbon monoxide binding curves of hemoglobin trout IV, *Biophys. J.*, 55, 691, 1989.
86. Dewey, T. G. and Datta, M. M., Determination of the fractal dimension of membrane protein aggregates using fluorescence energy transfer, *Biophys. J.*, 56, 415, 1989.
87. Li, H. Q., Chen, S. H., and Zhao, H. M., Fractal of hybrid orbitals and their applications in the enzyme models, *Chin. Chem. Lett.*, 1(3), 257, 1990.
88. Kopelman, R., Fractal reaction kinetics, *Science*, 241, 1620, 1988.
89. Li, H. Q., Chen, S. H., and Zhao, H. M., Fractal approach for the investigation of the enzyme model, *Chin. Sci. Bull.*, 36(15), 1319, 1991.
90. Farin, D. and Avinr, D., *J. Am. Chem. Soc.*, 110, 2039, 1988.
91. Kier, L. B. and Hall, L. H., *Molecular Connectivity in Structure-Activity Analysis*, Research Studies Press, Letchworth, U.K., 1986.

92. Lall, R. S. and Srivastava, V. K., *Math. Chem.*, 13, 325, 1982.
93. Barysz, M. et al., *Stud. Phys. Theor. Chem.*, 28, 222, 1983.
94. Zhang, H. G. and Xin, H. W., *Chin. J. Chem. Phys.*, 2(6), 413, 1989.
95. Ashby, W., *An Introduction to Sybernetics*, John Wiley & Sons, New York, 1956.
96. Mowshovitz, A., *Bull. Math. Biophys.*, 30, 1750, 1968.
97. Wu, D. C., *Introduction to Conformational Statistical Theory of Macromolecules*, Sichuan Education Press, Chengdu, 1985, 370.

Section III

Cells

chapter five

Morphogenesis by Bacterial Cells

Tohey Matsuyama and Mitsugu Matsushita

> *Cloud boundaries, mountain skylines, coastlines, forked lightning, … these, and many other natural objects, have a form much better described in fractal terms than by straight lines and smooth curves of classical geometry. Fractal mathematics ought, therefore, to be well suited to modelling and making predictions about such phenomena.*
>
> K. Falconer

5.1 Introduction

As Robert Koch and his colleagues clearly showed with many successful identifications of bacterial pathogens, bacteriological studies must be done with pure cultures. For preparation of such a pure culture, two technical steps are indispensable: (1) isolation of a single colony on a solid medium and (2) confirmation of homogeneity in the descendant colonies growing after spreading inoculation of cells in the single colony. In these processes, colony properties are examined precisely from various points (color, size, morphology, etc.) and have been regarded as important characteristics expressed by each bacterial isolate. Since failure in this initial step of bacteriological practice causes formidable mistakes in subsequent studies, most bacteriologists are extremely nervous about colony properties. Their interests in bacterial colonies, however, did not extend beyond the technical requirements to obtain pure cultures and differential properties useful in determinative bacteriology.

In fact, the characteristic morphology of a bacterial colony indicates the presence of specific biological and physico-chemical processes. It is obvious that the colony is not a simple mass collection of multiplying cells.[1,2] It means that bacterium inoculated on a solid medium exerts controlled multiplication responding specifically to internal and external factors. Species-specific colony morphologies may be related to genetically coded cell activities.

A bacterium is a single-celled organism and is able to live independently. Such a single-celled living style is, however, unusual in nature. Most bacteria prefer to habituate on some surface environment and have close contact with their brothers and sisters. In such a situation, a few cells may take off from the surface and translocate freely to new environments, and most cells staying on the surface will develop intimate interactions with their neighbors. We think that a lot of advantages and disadvantages of living together may be also present in the microbial world. Evolution from single-celled organisms to multicellular organisms might have occurred through the colonial life of single-celled organisms.

We have noted that some colony patterns made by bacteria are quite similar to patterns developed in nonliving systems. Is this a coincidental event or are there common mechanisms in the morphogenic processes? In the present research we will analyze a bacterial pattern formation by using various experimental systems and will try to show some basic principles underlying morphogenesis in nature.

5.2 Fundamentals in bacterial colony formation

After fractal growth of a bacterial colony was demonstrated,[3,4] studies on microbial colony patterns increased among scientists in the field of physics and the related sciences. This chapter is written for such scientists with limited microbiological backgrounds. Some special factors for experiments with colony morphogenesis are also presented here.

5.2.1 Bacterial strains

Every bacterial isolate has its individual character. Even though they have the same species name (e.g., *Bacillus subtilis*), their characteristics are never identical one to the other. So, we usually give a strain name to each isolate to distinguish it from others and keep a record of the isolation source (e.g., sputum of a patient suffering from bronchitis). A bacterium without such a strain name is outside the realm of experimental science. The real bacterium with the strain name always has its own specific properties. If a particular property (such as normal or defective in flagellation) is critical in the experiments, it must be examined and described clearly in the published manuscript.

The strain properties must be described precisely and unambiguously. If there are too many strains or too many characteristics to describe by workable sentences, abbreviations for these properties will be necessary. To avoid confusion arising from the use of such abbreviated terms, there are standardized rules for abbreviations, designations, and their usage.[5] For example, the abbreviation consists of three-letter symbols set in Roman type with the first letter capitalized: Mot$^-$ designates the immotile phenotype. The phenotype must be clearly discriminated from the genotype which corresponds to a firmly identified gene. The gene names are generally described by three lowercase italic letters, and one capital italic letter is used to indicate the locus: *motA*.

Bacterial strains tend to change their properties insidiously during repeated cultivation, so we must be careful to maintain a strain having the original properties. In addition, bacterial strains used in the published studies must be available for examination by other scientists who want to confirm the reproducibility of the published results. Stocks of strains are made by suspending bacterial cells in 10% (wt/wt) skim milk and kept at –80°C. Sterilization of skim milk is carried out by autoclaving at 113°C for 15 min. Sterility of the prepared skim milk is tested by the plating method.

5.2.2 Culture conditions

Bacteria are small and have seemingly simple anatomical structures. As a result, they often are thought of as simple organisms that are easy to manipulate and have a straightforward life style that is easy to elucidate. We do not think so. Their life strategy must be sophisticated, as they are creatures that have thrived throughout an extremely long (3,500,000,000 years) evolutionary history. Their behavior in various environments is made quite sensitive and variable due to their response to biological or physico-chemical factors (mostly not identified). So, the precise setting of experimental conditions and a description of these details are quite important to obtain reproducible findings. Without such attention to detail, experiments can be rendered meaningless.

Although we can obtain the same lot of commercial products for agar media, properties of a prepared agar plate will be a little different from those prepared in other laboratories. The most likely difference will be in the content of water in an agar plate. It depends on various factors (plate drying temperature, time, etc.), and these routine procedures usually are not mentioned in manuscripts; however, morphogenic behavior of bacteria is profoundly influenced by the dryness of the surface, so it is important to find appropriate preparation methods for the experiments. Atmospheric humidity in an incubator is also important, especially when cultivation time is longer than 2 to 3 days.

Agar is a crude commercial product made from marine algae and contains many kinds of unidentified materials. Thus, the properties of agar depend profoundly on the manufacturer. Swarming (surface translocation behavior of bacterial population) of *Escherichia coli* (ATCC 25922) on soft agar media is prominent on media containing Japanese Eiken agar, but difficult to observe on the media containing Difco agar.[6]

Glucose is added to some media as a nutrient for bacterial growth. Glucose, however, is not only a source of carbon and energy, but it sometimes works as a promoter for morphogenesis of a colony. Bacterial behavioral changes that may be related to unknown bacterial responses to glucose have been reported.[7] Glucose (0.1 to 0.4%, the concentration range which produces no increase in the growth rate of *Salmonella typhimurium* in a liquid culture) is shown to promote fractal colony growth of the bacteria on Vogel-Bonner agar media.[7] Swarming of *E. coli* and *S. typhimurium* on semi-solid media is also remarkably enhanced in the presence of glucose.[6] On the other hand, swarming of *Proteus mirabilis* (NPC 3007) on a peptone agar is inhibited by glucose.[8]

Amino acids evoke an unexpected colony morphogenesis in addition to their role as nutrients for cell multiplication. *P. mirabilis* is a gram-negative rod, and the wild type strain is able to grow without an external supply of amino acids. However, drastic morphological changes in swarming colonies of the bacteria are inducible by the addition of some amino acid.[8,9] Thus, the colony pattern depends on added amino acid (Figures 1A, B, and C). In the case of amino acids, optical configuration of the amino acid is also critical (Figures 2A and B). Morphogenesis promoted by D-enantiomer progresses differently from that promoted by L-enantiomer (biologically natural isomer).

Effects of salts on the morphogenic behavior of bacteria is quite mysterious. The promoting effects of NaCl on swarming of *P. mirabilis* is well known,[10] but we cannot explain this outcome. *P. mirabilis* exhibits characteristic swarming responses to various salts. In the presence of 0.1 M of NaH_2PO_4 or KH_2PO_4, the bacteria make a spiral swarming[8] (Figure 3).

Morphogenic behavior of some microbial species depends on cultivation temperature. *Serratia marcescens* cultivated at 30°C forms a red-pigmented giant fractal colony, but one cultivated at 37°C forms a nonpigmented round colony.[3,11] Although a bacterium has many genes (more than 1000), these are not always expressed. Temperature is one of many environmental signals which can switch bacterial gene expression.

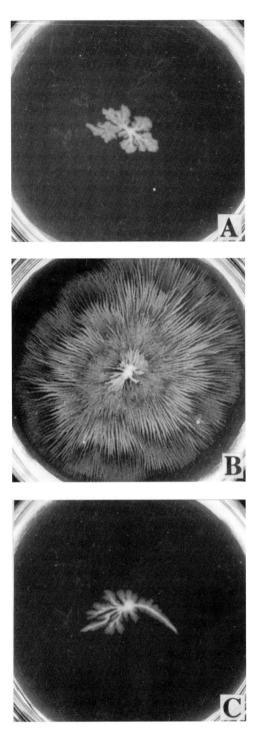

Figure 1 *Proteus mirabilis* NPC 3007 colony on a minimum growth medium[9] (1.5% Eiken agar) supplemented with nothing (A), 10 μM L-asparagine (B), and 10 μM L-aspartic acid (C). Cultivation at 37°C for 48 h.

Incubation time and observation timing are also quite important. Most bacteriologists discard a culture plate after recognizing a well isolated colony and after sampling bacterial cells from the colony. At that culture stage, bacterial colonies are

chapter five: Morphogenesis by Bacterial Cells 131

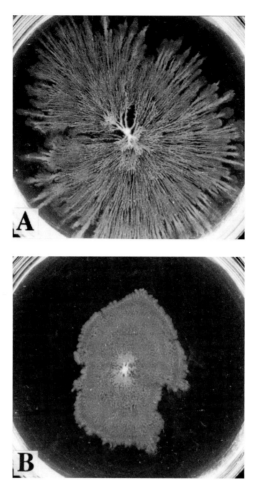

Figure 2 *Proteus mirabilis* NPC 3007 colony on a minimum growth medium (1.5% Eiken agar) supplemented with 1 μM L-alanine (A) and 1 μM D-alanine (B). Cultivation at 37°C for 48 h.

Figure 3 *Proteus mirabilis* NPC 3007 colony on a peptone medium (1.5% Bactoagar) containing 0.1 μM NaH$_2$PO$_4$. Cultivation at 37°C for 32 hr.

not large enough to develop a fully characteristic morphology. Our discovery of fractal growth of a bacterial colony was made by incubating the plate for an unusually long time. From the model of diffusion-limited growth, this long incubation time was shown to be reasonable. A careful approach to bacterial behavior is fairly recent; therefore, little is known that can predict how bacteria will behave from moment to moment. Continuous recording of colony growth with a video machine will inform us of critical steps of morphogenesis.

During long incubation times permitting extensive cell multiplication, it is natural that colony-forming populations become heterogeneous in their cellular physiology and hereditary traits. Mutant cells will emerge through extensive cell multiplication. Development of a new pattern from the margin of an original colony is a usual phenomenon (Figure 4). It is necessary to isolate a bacterial cell from such a new pattern and to reinoculate it on the same kind of plate. Bacteriological characterization of the isolate also will be necessary to compare it to original strains. New properties exhibited in the variant may be worthy as a clue for elucidation of morphogenic mechanisms in a modified colony.

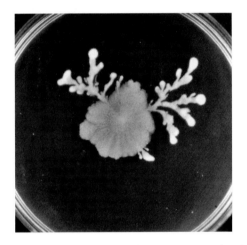

Figure 4 *Serratia marcescens* NS 38 (producer of serrawettin W1) colony on a Vogel-Bonner medium (1.4% Bactoagar). Cultivation at 30°C for 7 days.

5.3 *Characterization of colony morphology*

5.3.1 *Diversity in colony morphology*

Bacterial colonies are remarkably rich in morphological diversity; however, it is not due to morphological diversity of a unit constructing a colony. The morphogenic unit, i.e., a bacterium, is known as an organism having relatively simple shapes (coccal or rod-like). The origin of the diversity rather may be related to the smallness of bacteria (approximately 0.5 to 2.0 µm). Thus, they are living under the direct influence of intermolecular forces. This means that the specificity in activities (interactions with other cells, substrates, etc.) of a bacterium is determined on the molecular level. Diversity in the molecular world may be nearly infinite.

As living creatures, bacteria are sensing and releasing various kinds of information. As a consequence, some specific decisions will be made through complex processes in each bacterium. Such biological responses are mostly controlled by genetic systems. Genetic information is part of a huge stock of molecular diversity collected through the extremely long evolutionary history of bacteria. Thus, such diversity derived from the expression of specific nucleotide sequences in DNA is another source of colony heterogeneity. A combination of diversity in the present environment and diversity from accumulated genetic traits may produce various colony forms.

While this is generally held to be true by bacteriologists, it is noteworthy that some fractal colony morphology has been explained without regard to diversity of either the material world or the genetic world. Observing characteristic patterns of a colony, we usually tend to look for something responsible for it. This does not necessarily mean that responsible factor(s) must be a specific material (living or nonliving). By using computer simulations we know that a simple setting of repetitive processes generates characteristic random patterns. For analysis of colony patterns, simulation approaches considering key specific processes offer an important approach to solving the mystery of morphology. We must realize that colony morphogenesis is an expanding process composed of a number of repetitive operations.

Here we provide a brief review of various patterns of bacterial colonies. With regard to three-dimensional forms, we have noted various types of vertical sections as shown in Figure 5. These all have emerged after a point inoculation of the bacteria on the surface of an agar medium. Since most present studies on colony morphology are focused on horizontally spreading patterns on the agar surface, we show details from actual examples. All colonies shown here were made in plastic dishes with 85-mm diameter.

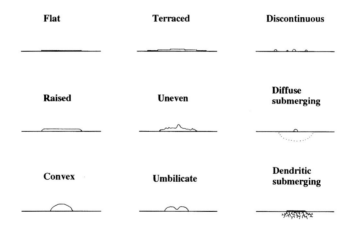

Figure 5 Illustrated vertical sections of various bacterial colonies.

A compact round form is the most familiar morphology of a bacterial colony. It is small but large enough for recognition and isolation for pure culture work. Such colonies are so familiar that no one has tried to explain why the form is round. It is strange, though, that the colony keeps the roundness even after prolonged cultivation (Figure 6). During horizontally extended surface growth, bacterial colonies usually develop characteristic morphology. In this situation, simple round forms are rather rare. Some species develop scattered colonies (Figure 7) or successively produce wavy expanding clusters (Figure 8). Continuous spreading growth also produces various patterns. It is noteworthy that spreading growth is not always radial from the point-inoculated center. *Salmonella* in a specific culture condition spread in quite unpredictable directions (Figure 9). The random pattern shown is complex and difficult to describe, but has a self-similar fractal property as reported previously.[11]

As described already, compact expansions producing monotonous architecture (such as Figure 6) are rare. Expansion usually progresses by uneven generation and extension of spurs followed by their branching.[11,12] The mode of extension and

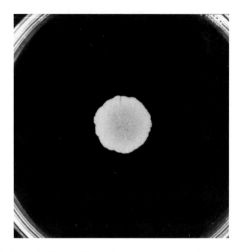

Figure 6 *Micrococcus luteus* ATCC 9343 colony on a tryptic soy agar (Difco, 1.5% agar). Cultivation at 37°C for 28 days.

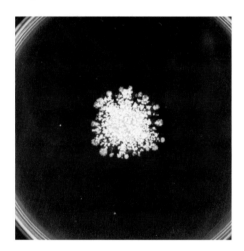

Figure 7 *Staphylococcus epidermidis* ATCC 90120 colony on a tryptic soy agar (Difco, 1.5% agar). Cultivation at 37°C for 42 days.

branching is dependent on bacterial species and culture conditions.[7,11] This process seems to be one of the actual points responsible for colony morphologies. Straight extension of a spur will result in a colony morphology as shown in Figure 10. Results of reeling extension are shown in Figure 11. It is interesting that spur extension has some tendency for turning to the right. Figure 12 is a representative pattern. This tendency, however, is not exclusive; counterclockwise spiraling occurs sometimes (Figure 13). Width of a branch is also controlled biologically, and remarkable differences can be observed (Figures 14 and 15). Circular outline of the overall expansion is also noteworthy in several figures. Mutually restrained growth among neighboring branches seems to persist. Appearance of rings in the internal part of a colony (Figure 16) also suggests the presence of synchronized growth of the branches. Figure 17 is an example contrasted with the colony with circular outline.

Lastly, it should be mentioned that a given colony is not always composed of the same patterns. Representative examples of such heterogeneous colonies are given in Figures 4 and 18. Characteristic giant colonies of *Clostridium tetani* (pathogenic

chapter five: Morphogenesis by Bacterial Cells   135

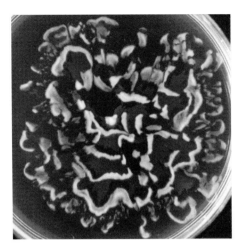

Figure 8   *Salmonella typhimurium* LT2-3γ (flagella-less mutant of LT2 strain) colony on a Vogel-Bonner medium (1.0% Eiken agar, 0.8% glucose). Cultivation at 37°C for 48 hr.

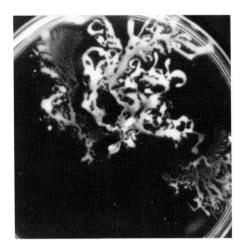

Figure 9   *Salmonella typhimurium* ATCC 14028 colony on a Vogel-Bonner medium (0.5% Eiken agar, 0.8% glucose). Cultivation at 37°C for 72 hr.

Figure 10   *Serratia marcescens* NS 45 (producer of serrawettin W3) colony on a Luria-Bertani medium (0.75% Bactoagar). Cultivation at 30°C for 24 hr.

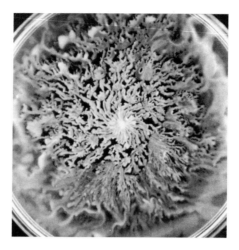

Figure 11 *Serratia marcescens* NS 38 colony on a nutrient agar (0.5% Eiken agar). Cultivation at 30°C for 12 hr.

Figure 12 *Serratia marcescens* NS 38-09 (serrawettin-less mutant of NS 38 strain) colony on a nutrient agar (0.5% Eiken agar). Cultivation at 30°C for 72 hr.

agent of tetanus) always generate secondary faint spreadings (indicated by an arrow in Figure 18).

5.3.2 Examination of fractal property

Let us imagine cauliflower in a salad served at dinner, displaying a short and thick trunk-like part with a compact and fleshy head. Now split a branch out of it and take a close look at it. Then you will find that the shape is very similar to that of the original one, a smaller but short trunk-like part with a compact and fleshy head again. If you take an even closer look at the branch, it will be found to consist of still smaller parts with similar shape again. Namely, that characteristic shape of cauliflower comprises a nested structure of branches with repeatedly reduced scales.

We see lots of random patterns in our surroundings that do not seem to have any regularity at first glance. For instance, a chain of mountains, clouds, lightning, coasts, river networks, cracks of the earth surface and rocks, soot or dust aggregates, and so on. They are diverse and too numerous to mention. However, an outline

chapter five: Morphogenesis by Bacterial Cells 137

Figure 13 *Serratia marcescens* NS 45 colony on a nutrient agar (0.5% Eiken agar). Cultivation at 30°C for 15 hr.

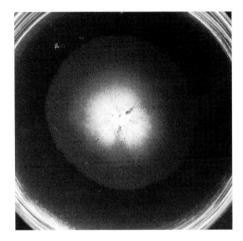

Figure 14 *Bacillus subtilis* ATCC 21331 colony on a nutrient agar (0.5% Eiken agar). Cultivation at 37°c for 14 hr.

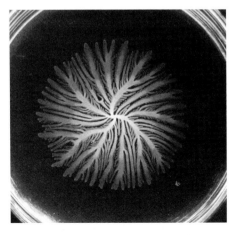

Figure 15 *Bacillus subtilis* 21331-b3 (surfactin-less mutant of ATCC 21331 strain) colony on a nutrient agar (0.5% Eiken agar). Cultivation at 37°C for 19 hr.

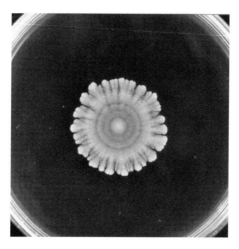

Figure 16 *Serratia marcescens* NS 38 colony on a tryptic soy agar. Cultivation at 37°C for 21 days.

Figure 17 *Salmonella anatum* KS 200 colony on a Vogel-Bonner medium (1.4% Eiken agar, 0.4% glucose). Cultivation at 37°C for 28 days.

Figure 18 *Clostridium tetani* TV 82 colony on a GAM medium (Nissui, 1.5% agar). Cultivation at 37°C for 40 days. An arrow indicates secondary spreading growth with different density.

obtained from a chain of mountains can be clearly distinguished from a white (uncorrelated) noise curve obtained from, say, an electronic circuit. Also, river networks and railroad networks are a different kind. Many patterns seen in nature seem to be, in spite of apparent randomness, formed under some unique regularity, such as the nested structure of cauliflower described above. Mandelbrot recognized this fact very early, and has pointed out, with many concrete examples, that very simple and beautiful regularity or symmetry called *scale invariance* is hidden in apparently featureless random patterns. He has tried to describe them by introducing a new concept, *fractal*.[13,14,15] In this section we will present a brief review of fractals.

5.3.2.1 Self-similar fractals and fractal dimension

5.3.2.1.1 Self-similar fractals. Let us examine a very complicated coastline, such as the Norwegian fjords, by maps on variously reduced scales. Along with the decrease of scale reduction of the maps, i.e., zooming in on the coastline, we sometimes have the situation in which smaller irregularities due to peninsulas and bays that could not be seen before continue to appear and the pattern complexity does not change so much. Just like this example, among various complicated patterns there are characteristic ones whose part, when taken out from the original pattern and enlarged, cannot be distinguished from the original. This kind of symmetry or invariance on the isotropic enlargement or reduction of scales is called *self-similarity*, and patterns with this property are said to be *self-similar fractals*.

Many Rias coasts are random patterns that are known to be self-similar in the statistical sense. Here, however, we first will take regular geometric examples that satisfy rigorous self-similarity. Figure 19 shows a (triadic) Koch curve. As indicated in Figure 20, the construction is very simple. Let us initially have a segment of a line of unit length, as shown in Figure 20 (0). In general, this kind of an initial or zeroth generation pattern is called an *initiator*. At the first stage we divide the segment into three equal segments and replace the central one by two sides of an equilateral triangle, the bottom side of which is exactly that central segment. The first generation pattern obtained from the zeroth generation is, in general, called a *generator*. We now have a curve with four segments of length 1/3 as a generator, as seen in Figure 20 (1). At the second stage, we carry out the same procedure for the four segments as we did for the first stage. Namely, by replacing each segment by the generator of the same size, we obtain the second generation curve, as seen in Figure 20 (2). It is the Koch curve shown in Figure 19 that is the *infinitie*th generation curve obtained by continuing this procedure *ad infinitum*.

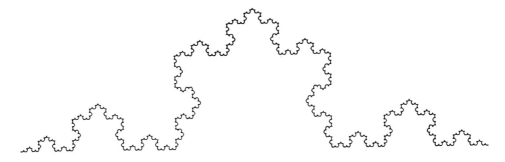

Figure 19 A Koch curve. $D = \ln 4 / \ln 3 \approx 1.26$.

As seen from Figure 21, it is easy to imagine from the construction that, whichever part of a Koch curve one enlarges, one can always obtain exactly the same curve

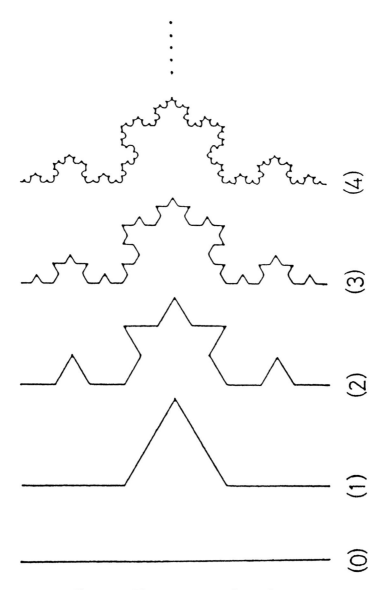

Figure 20 The construction of a Koch curve.

as the original. Namely, a Koch curve is self similar. One can say intuitively that a Koch curve consists of irregularities of various sizes and has no characteristic length scales except its whole size. Particularly interesting is the fact that in spite of the finiteness of the end-to-end distance of a Koch curve (the whole size), the length along the curve is infinite. In other words, in addition to the nondifferentiability due to its irregularity, a Koch curve has the pathological property that has no length at all. This can be easily understood from the construction shown in Figure 20, because the length of the zeroth generation curve, i.e., the initiator, is 1, that of the first generation curve is 4/3, that of the second generation curve is $(4/3)^2$, and so on, increasing unlimitedly.

5.3.2.1.2 Fractal dimension. Let us examine another self-similar fractal shown in Figure 22a. A square of unit side length is taken as the *zero*th generation ($k = 0$) pattern or an initiator. It is divided into nine equal squares of side length 1/3, out

chapter five: Morphogenesis by Bacterial Cells 141

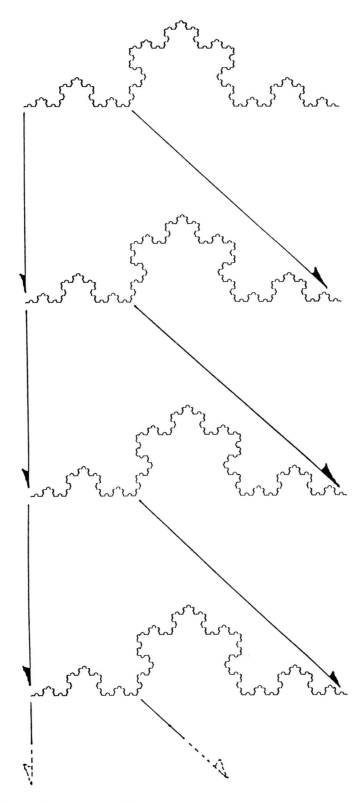

Figure 21 Schematic illustration of the self-similarity of a Koch curve. A fourth part of a Koch curve enlarged three times isotropically looks exactly the same as the original.

of which five squares constituting the diagonals are left and the other four are discarded. The resultant pattern is the first generation ($k = 1$) or a generator. The second generation ($k = 2$) pattern is obtained by carrying out exactly the same procedures for each of five smaller squares of the first generation ($k = 1$) pattern. It is understandable that the infinitieth generation ($k = \infty$) pattern obtained by continuing the procedures *ad infinitum* is self similar.

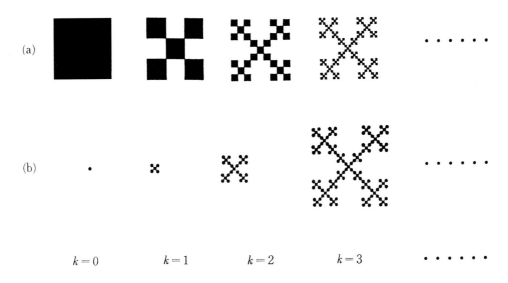

Figure 22 Two ways of constructing a self-similar fractal called Vicsek's snowflake. $D = \ln 5/\ln 3 \approx 1.46$.

The side length ε of smaller squares constituting the kth generation pattern is $\varepsilon = 3^{-k}$, because the side length of the initiator ($k = 0$) square is 1. Total number N of the squares for this kth generation pattern is given by $N = 5^k$, as is easily seen from Figure 22a. Making logarithms of the two equations and eliminating k, one can find that N is scaled with ε as

$$N = \varepsilon^{-D}, \quad D = \frac{\ln 5}{\ln 3} \approx 1.46. \tag{1}$$

It is this exponent D that is called the *fractal dimension* of the self-similar pattern obtained from Figure 22a. We will later discuss why this quantity implies the pattern dimensionality. It should be noted here that the fractal dimension corresponds to the degree of roughness and fineness or complexity of self-similar patterns and is the most important quantity characterizing self-similar fractals.

As for regular self-similar fractals such as those shown in Figures 19 and 22, the fractal dimension D can be represented by generalizing Equation (1) as

$$D = \frac{\ln N}{\ln(1/\varepsilon)}. \tag{2}$$

Here N is the number of constitution units at the present generation which have emerged from one constitution unit at the previous generation when proceeding with the pattern construction one stage ahead, or the number of reduced initiators constituting the generator, and ε is the reduction rate. $N = 5$ and $\varepsilon = 1/3$ for the pattern obtained from Figure 22a, which reproduces the results of Equation (1). It is easily understood from Figure 20 that a Koch curve in Figure 19 has $N = 4$ and $\varepsilon = 1/3$, and the fractal dimension $D = \ln 4/\ln 3 \approx 1.26$.

At each stage of the construction in Figure 22a, if only three small squares on a diagonal (e.g., one from upper left to lower right) are left instead of five on two diagonals, a single diagonal segment (one-dimensional pattern) is obtained at the *infinitie*th generation ($k = \infty$).

This corresponds to the case in which the generator consists of three squares on a diagonal. If instead all the nine small squares are left, the resultant pattern remains the same square as the original (two-dimensional pattern), no matter how many times generations are repeated. This is the case in which the generator coincides exactly with the initiator. Applying Equation (2) for each case, one has $N = 3$, $\varepsilon = 1/3$, and $D = 1$ for the former, and $N = 9$, $\varepsilon = 1/3$, and $D = 2$ for the latter. These are consistent with the intuitive results. As seen from these results, the fractal dimension is a natural extension of the ordinary dimension that takes only an integer value. It should, however, be noted that there can be ways to leave, for example, three small squares other than those three on a diagonal ($N = 3$). This means that the pattern of $D = 1$ is not necessarily an ordinary one-dimensional curve, but a collection of very scattered points. This also implies that there can be innumerable, qualitatively different self-similar patterns, even if the fractal dimension alone is specified. It is very important to understand that the fractal dimension alone is insufficient to characterize self-similar fractals and that other characteristic quantities or topological properties such as the degree of ramification are necessary.

5.3.2.1.3 Determination of fractal dimensions. **Box-counting method:** One can generalize Equation (2) in order to examine the self-similarity of various random patterns and to determine the fractal dimension D. The space that contains a given pattern is divided as a lattice into cells (or pixels, as in a computer) of side length ε, and the number $N(\varepsilon)$ of pixels that carry any part of the given pattern is counted. This corresponds to investigating a given pattern through coarsening with the resolution ε. The number $N(\varepsilon)$ is then measured by varying the value of ε and plotting double-logarithmically against ε. If the points fit a straight line, one can conclude that the given pattern is self similar, and fractal dimension D is obtained from the absolute value of the line slope. This is called the box-counting method, which is a handy way of checking the self-similarity of one given pattern obtained, e.g., in a well resolved photograph. Figure 23 shows an example of fractal analysis by making use of the box-counting method for the colony pattern shown in Figure 17. As to the example shown in Figure 22a, the kth generation pattern corresponds exactly to the one obtained through coarsening the self-similar pattern of the *infinitie*th generation ($k = \infty$) by the pixel size $\varepsilon = 3^{-k}$. In this case, therefore, N takes discrete values of $N(\varepsilon) = 1, 5^1, 5^2, \ldots$ for $\varepsilon = 1, 3^{-1}, 3^{-2}, \ldots$, respectively. When plotting them double-logarithmically, one can easily find that they are exactly on a line with the slope of $-D$ ($D = \ln 5/\ln 3$).

Radius-of-gyration method: Patterns defined rigorously in mathematics such as the Koch curve shown in Figure 19 and the one shown in Figure 22 keep their self-similarity over infinitely small scales. For real patterns seen in nature, however, there emerge characteristic lengths such as branch thickness of a ramified cluster (exemplified by a DLA-like colony shown later in Figure 37) due to some physical cause such as surface tension, before going to the scale of molecules and atoms. This means that the self-similarity breaks down around such a length scale. In other words, there are upper and lower limits in the range of length scales in which the self-similarity

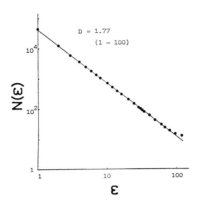

Figure 23 An example of fractal analysis by the box-counting method for a colony pattern shown in Figure 17. The least-mean-squares fitting yields the fractal dimension (absolute value of the line slope) $D = 1.77$.

of real patterns holds. Bearing this in mind, let us construct in Figure 22b the same pattern as the one shown in Figure 22a by using constituent units, or "particles" of some small length scale. That is, we regard the size of the particles as the lower limit described above. The pattern of the first generation ($k = 1$) in Figure 22b is obtained by arranging five particles of the zeroth generation ($k = 0$) with just the same arrangement of five small squares of the first generation in Figure 22a. In the second generation ($k = 2$), a pack of five particles obtained in the first generation is regarded as a cluster, and the five clusters are arranged in the same way as in the first generation. This procedure is repeated *ad infinitum*. In Figure 22b, therefore, the pattern grows large every generation, unlike in Figure 22a.

Let the diameter of constituent "particles" be a. Then the total number N of particles contained in the pattern of the kth generation is $N = 5^k$, and the side length R of the square which touches the pattern externally is $R = 3^k a$. R can be regarded as the size of the kth generation pattern. Let us take the logarithm of both equation and eliminate k under the condition $R \gg a$ ($k \gg 1$). Then N is found to be scaled with R as

$$N \sim R^D. \tag{3}$$

(The symbol $\sim$ represents the proportionality relation.) D here coincides with that which appeared in Equation (1). This is a natural result, because the contraction of the kth generation pattern in Figure 22b to the size of the kth generation pattern in Figure 22a results in exactly the same pattern as in Figure 22a.

The relation of Equation (3) is very important and useful when examining the self-similarity and determining the fractal dimension D of growing patterns such as dendritic crystals and bacterial colonies. For instance, let us take as an example a trajectory of a random walk with a constant step length, which is the idealization of Brownian motion. After taking N steps, the random-walker (or "drunken man") weaves away to some distance R from the starting point. (Here one step of the walk can be regarded as one "particle" and an N-step trajectory as an N-particle linear polymer of size R.) According to Einstein's theory of Brownian motion, the leaving distance R is given on average by $R \sim N^{1/2}$, irrespective of the dimension d of space where the random walk takes place. Since R can be regarded as the average size of the trajectory, this relation implies from Equation (3) that the random-walk trajectory is self similar and its fractal dimension is $D = 2$, independent of d.

chapter five: Morphogenesis by Bacterial Cells

A sequence of snapshots of a growing pattern is divided into pixels of the same size. The pixel size *a*, once decided, is never changed; this point is essentially different from the box-counting method. The number *N* of pixels that contain any part of the pattern is then counted. Some small characteristic length such as branch thickness of dendritic crystals should be taken as the unchangeable pixel size *a*. The distance across can be taken as the measure of the size of a growing pattern at each growth stage. The radius of gyration *R* of a pattern, however, is better to use because this is a statistical quantity that has less statistical scattering. It is calculated from

$$R^2 = \frac{1}{N}\sum_{i=1}^{N}(r_i - r_c)^2, \quad r_c = \frac{1}{N}\sum_{i=1}^{N} r_i, \qquad (4)$$

where *N* is the number of pixels which contain any part of the pattern, r_i is the position vector of the *i*th pixel among them, and r_c represents the center-of-mass coordinate of the pattern. Many sets of values of *R* and *N* calculated thus are plotted double-logarithmically. If these points are on a line with some statistical error, the growing pattern under examination can be concluded to be self similar, and the slope of the line yields the fractal dimension *D* according to Equation (3). This is called the *radius-of-gyration* method. When a pattern grows like a line, a disk, or a sphere, it is easy to understand that the value of *D* gives, respectively, *D* = 1, 2, or 3. This method is also useful for the case in which clusters of various sizes are obtained at one time, such as colloidal aggregates obtained from colloidal suspension through salting-out. In such cases *R* and *N* are evaluated for each cluster, and sets of values of (*R*,*N*) for many clusters of various sizes are plotted double-logarithmically.

There are other ways to determine fractal dimensions, such as divider method and density-density correlation function method.[14,15] The box-counting and radius-of-gyration methods described so far, however, are representative ways to do so.

5.3.2.2   Self-affine fractals and scaling indices

5.3.2.2.1   Self-affine fractals.   As described before, the length of a Koch curve is infinite, although the size or the length of the zeroth generation segment (initiator) is finite. This cannot occur unless infinite number of superpositions take place when a Koch curve is projected onto the initiator segment. Let us take an outline of a chain of mountains as an example. It looks like a self-similar curve, since it is composed of rugged parts of various sizes. It cannot be imagined, however, that there is an infinite number of overhangs of various sizes for a chain of mountains, unlike the case of a Koch curve. The outline of mountains is not self-similar. Then an interesting question arises immediately: What kind of regularity is hidden in such rugged curves as an outline of mountains?

Figure 24 shows a plot of the displacement x_H of the usual one-dimensional Brownian motion (*H* = 1/2) as a function of time *t*. The subscript *H* is called the Hurst exponent (0 < *H* < 1) which indicates the persistence in variation of a curve. A curve with *H* > 1/2 is persistent, i.e., when the curve happens to increase, it shows the tendency to continue increasing and vice versa. Hence, the curve looks relatively smooth locally, but rougher globally. On the other hand, a curve with *H* < 1/2 is antipersistent, which means that when the curve happens to increase, it shows the tendency to decrease afterwards, and vice versa. Hence, the curve looks more rugged locally, but flatter globally. A curve *H* = 1/2 is marginal. In fact, usual Brownian motion is marginal for increase and decrease. Time-series curves $x_H(t)$ with general *H* (0 < *H* < 1) are called *fractional Brownian motion* (fBm) curves. One-dimensional

Figure 24 A typical example of one-dimensional Brownian motion traces $x_H(t)$ ($H = 1/2$).

Brownian curves have $H = 1/2$ and are a special case of fBm (see Reference 14 for more detail).

Figure 24 looks like an outline of rugged mountains such as the Alps or the Rocky Mountains. Let us regard this curve as a given pattern and the length of the curve from time 0 and time T to be N. Since each step in the time–space (t–x_H space) has a constant and unit length, N is proportional to T. That is, the horizontal width of the pattern T is scaled with the length N as $T \sim N^{\nu_t}$, $\nu_t = 1$. On the other hand, let us take a standard deviation $X = \sqrt{\langle x_H^2(t) \rangle}$ as a measure of the size of the displacement of Brownian motion within time interval T ($\langle \cdots \rangle$: ensemble average). Then Einstein's theory of Brownian motion described in the previous section gives $X \sim T^{1/2}$. The vertical width of the pattern X is, therefore, scaled with the length N as $X \sim N^{\nu_x}$, $\nu_x = 1/2$. An important point here is that for any partial curve taken from the one-dimensional Brownian curve $x_H(t)$ ($H = \nu_x = 1/2$), the way for the vertical width X to be scaled with the length N is different from that for the horizontal width T, i.e., $\nu_x \neq \nu_t$.

The size R of a self-similar pattern such as a Koch curve (Figure 19), a DLA (diffusion-limited aggregation) cluster (Figure 38), and a Brownian motion trajectory described in the previous section is isotropically scaled with N (the number of units or particles constituting the pattern) as $R \sim N^{\nu}$, $\nu = 1/D$ (D is the fractal dimension of the pattern), independent of the directions. On the other hand, the scaling property of the pattern shown in Figure 24 is dependent on the directions, i.e., the horizontal and vertical directions. This kind of more general, anisotropic scaling symmetry is called *self-affinity*, and patterns with self-affinity are called *self-affine fractals*. Such patterns cannot be characterized by a single exponent ν (or D), but by more than one ν_i. For instance, the curve shown in Figure 24 is characterized by two exponents, ν_x and ν_t, although the latter is trivial in this case. Self-similarity is, therefore, a special case of self-affinity in which values of all the scaling exponents ν_i are equal to each other.

Let us investigate properties of self-affine fractals by analyzing one-dimensional Brownian curve $x_H(t)$ with $H = 1/2$. Figure 25a is an example of $x_H(t)$ for a long time interval. It looks very flat, i.e., the vertical change looks very small. The reason is that from the results described above the relative change can be represented as $X/T \sim T^{-1/2}$, which becomes smaller as time interval T becomes larger. On the other hand, Figure 25b shows a pattern obtained by enlarging 32 times the initial 1/32 of Figure 25a isotropically. If the curve $x_H(t)$ were self similar, the curve of Figure 25b would look similar to the curve of (a), just as a Koch curve exhibits its self-similarity in Figure 21. The fact is that the curves (a) and (b) of Figure 25 look very different. The latter looks much more rugged than the former. This means that the curve $x_H(t)$ is not self similar. Let us now enlarge the initial 1/32 of the curve of Figure 25a 32 times horizontally and $\sqrt{32}$ times vertically. The result of this anisotropic enlargement is shown in Figure 25c, which looks very similar to the curve

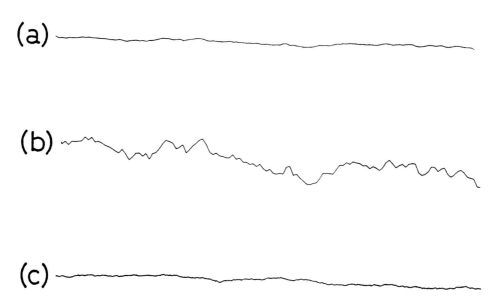

Figure 25 (a) An example of one-dimensional Brownian motion curves $x_H(t)$ ($H = 1/2$) for a long time interval. (b) Initial 1/32 of curve (a) is isotropically enlarged, i.e., 32 times horizontally and 32 times vertically. Note that the resultant curve is very different from the original curve. (c) Initial 1/32 of curve (a) is now anisotropically enlarged, i.e., $\sqrt{32}$ times horizontally and 32 times vertically. Note that now the resultant curve looks very similar to the original curve.

Figure 25a. This similarity is the concrete meaning of the anisotropic, self-affine scaling.

The change of patterns seen in Figures 25a and 25b implies a serious problem for the following reason. Since convenient ways to determine fractal dimensions, such as the box-counting method, presuppose isotropic scaling of patterns under examination, they cannot be applied to self-affine fractals. In fact, their forcible application to self-affine fractals leads to the unreasonable result that the fractal dimension obtained varies, depending on scales used. Of course, one can define some special dimensions such as local and global dimensions for a known, well defined curve such as an fBm curve.[14,15] These quantities, however, are determined from the limiting properties of a given curve. When a curve of finite length, whose characteristics we want to know, is given, these dimensions cannot be determined in general.

5.3.2.2.2 Determination of self-affine scaling exponents. Suppose, for instance, that the curve shown in Figure 24 is given as an unknown one. This may have some scaling invariant property, if not self similar. How can one extract it from this curve of finite length? This can be done simply by determining the anisotropic scaling exponents v_i described above, instead of fractal dimensions D.[16]

Suppose we have a curve in two-dimensional space, $y = f(x)$, as shown in Figure 26. The extension to a curve or surface in higher dimensions is straightforward. Let us first define the smallest length scale or unit length scale a (= 1) and measure by this scale the curve length Na (= N) between two arbitrary points A and B on the curve. (This is equivalent to regarding the curve as consisting of particles of diameter a and counting the number of particles between A and B; in this sense, the present method is a natural extension of the radius-of-gyration method described before.)

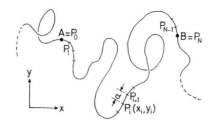

Figure 26 Measuring the curve length Na between a pair of points A and B on a given curve by the smallest fixed length scale a. (From Matsushita, M. and Ouchi, S., *Physica D*, 38, 246, 1989. With permission.)

We then calculate x- and y-variances X^2 and Y^2 of all measured points on the curve between the two points A and B:

$$X^2 = \frac{1}{N}\sum_{i=1}^{N}(x_i - x_c)^2, \quad Y^2 = \frac{1}{N}\sum_{i=1}^{N}(y_i - y_c)^2 \tag{5}$$

with the center-of-mass coordinates of the curve

$$x_c = \frac{1}{N}\sum_{i=1}^{N}x_i, \quad y_c = \frac{1}{N}\sum_{i=1}^{N}y_i, \tag{6}$$

where (x_i, y_i) is the coordinate of the ith measured point P_i on the curve. The standard deviation X and Y indicate, respectively, the approximate size of the part between A and B on the curve. Repeat the measurement procedures described above for many pairs of points on the curve and examine by double-logarithmic plots of X and Y vs. N whether they are scaled as

$$X \sim N^{v_x}, \quad Y \sim N^{v_y}. \tag{7}$$

The exponents v_x and v_y, even when Equation (7) holds, are in general different. In that case, the curve is self affine. If and only if they are equal to each other, the curve is self similar.

Figure 27 shows the plot of results obtained by applying the method described above, i.e., the horizontal and vertical standard deviation T and X_H as a function of curve length N for the one-dimensional Brownian motion curve $x_H(t)$ ($H = 1/2$) shown in Figure 24.[16] It gives $v_t = 1.00$ and $v_x \approx 0.50$, which are expected values for $x_H(t)$ with $H = 1/2$.

5.4 Pursuit of random pattern growth

In the analysis of mechanisms of generation of a specific pattern, it is important to follow steps which actually occur precisely. Since each bacterium is visible under a microscope, it is possible to trace the generation process at the level of the smallest morphogenic unit. On the other hand, examination of the morphogenic processes from a global point of view is also possible by just observing an inoculated culture plate macroscopically. Time scales of processes under examination are at the level of seconds–minutes in the microscopic world, and at the level of hours–days in the

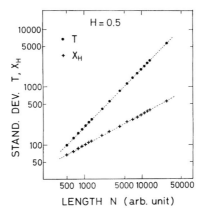

Figure 27 Dependence of the standard deviations of horizontal and vertical coordinates T and X_H, respectively, on the curve length N between various points on the one-dimensional Brownian motion curve $x_H(t)$ ($H = 1/2$) shown in Figure 24. The slopes yield the self-affine scaling exponents $v_t = 1.00$ and $v_x \approx 0.50$. (From Matsushita, M. and Ouchi, S., *Physica D*, 38, 246, 1989. With permission.)

macroscopic world. So, combined study is quite informative by covering a broad range of morphogenic processes.

5.4.1 *Macroscopic generating process*

It is easy to see the developing colonies at intervals. Many characteristic morphologies are observable, and unique processes generating such morphologies have been noted. By comparing sets of photographs taken at different incubation times (Figures 28A and B and Figures 29A and B), differences in developmental steps become evident in addition to the morphology itself. In Figures 28A and B, random branch extension is continuously progressing and leaves fixed patterns behind. In Figures 29A and B, random groups of bacterial mass are splitting intermittently from a preformed patterns. So, the preformed pattern is not stable. We know that the former (dendritic branching) is the process carried out by flagellated motile cells. The latter (wave-like row advancing) is carried out by nonflagellated immotile cells. Microscopic profiles of these processes will be described later.

For quantitative analysis of the morphogenic time course, a time-lapse video with a charge-coupled device (CCD) camera targeting a growing colony in a see-through incubator is convenient. A diagram of our recording system is illustrated in Figure 30.

Video recording sometimes informs us of unexpected events in the processes of colony morphogenesis. For example, recently we discovered a peculiar behavior of *P. mirabilis*. After the bacteria completed swarming on the whole surface of a semisolid medium, they abruptly began to swirl multifocally and developed a mobile and transient moiré-like pattern;[17] they then settled into a patchy pattern as shown in Figure 31.

5.4.2 *Microscopic generating process*

Pursuit of the bacteria engaged in characteristic colony formation is quite intriguing. Real time behavior of flagellated bacteria such as *P. mirabilis* or *B. subtilis* in the microscopic field under an objective lens (100×, Nikon nonimmersion lens with a long working distance) reveals the bacteria to have the same magnitude of relative

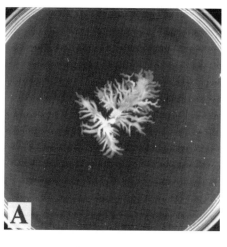

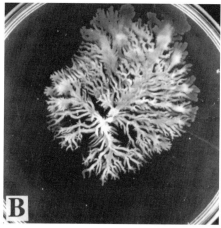

Figure 28 *Proteus mirabilis* NPC 3007 colony on a nutrient agar (0.5% Eiken agar, 0.4% glucose). Cultivation at 37°C for (A) 12 hr and (B) 18 hr. (From Matsuyama, T. and Matsushita, M., *Appl. Environ. Microbiol.*, 58, 1227, 1992. With permission.)

speed as the motion of human beings. Since there are many moving cells and they all look similar, it is quite difficult to trace individual cells without a video monitor/recorder.

Let us examine an extending tip of the branch shown macroscopically in Figures 28A and B; two video-recorded figures under a microscope are given in Figures 32A and B. There are four important findings in the two photographs:

1. Bacterial cells are unusually elongated.
2. At the extruding site (toward the right), they are making a cluster by neighboring side by side.
3. Numbers and morphologies of cells in clusters are changing between the two photomicrographs.
4. There are free spaces among bacterial clusters.

The bacterium, *Proteus mirabilis*, is famous for the ability of dimorphic transition. In a liquid culture or at the center of a colony, this bacterium has a short vegetative form with several flagella (Figure 33A). However, at the swarming margin of a colony on a solid medium, the bacteria exerts remarkable differentiation as shown in Figure

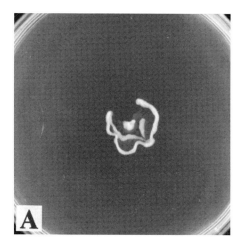

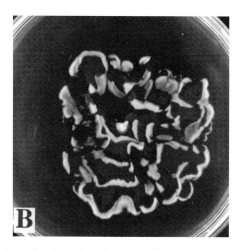

Figure 29 *Salmonella typhimurium* LT2-3γ colony on a Vogel-Bonner medium (1.0% Eiken agar, 0.8% glucose). Cultivation at 37°C for (A) 24 hr and (B) 36 hr.

33B. These hyperflagellated and elongated cells are quite active in forward and backward translocation and move together or apart by mutual sliding on their elongated bodies (member cells of a cluster are exchanging frequently, as seen in video analysis). Free space among bacteria indicates a microscopically discontinuous extension profile. Thus, the bacterial population gradually occupies a new outer space (compare Figures 32A and B). A single bacterium seems to be unable to translocate freely into new space. Bacteria "explore" a new space only as a bundled mass. On the other hand, in the backward area (on the surface once occupied by bacteria), a single bacterium will translocate forward and backward at random. This seems to suggest some modification of the surface by bacterial products which varies according to the position of the bacteria.

An elongated cell may be suitable for a long-lasting sliding contact with other cells. Such bacterial differentiation for group translocation (swarming) is shown to be common among many species of bacteria.[18,19] Recently we have reported that *S. typhimurium* and *E. coli* also exert swarming-associated differentiation.[6] A schematic illustration of the dimorphism of *P. mirabilis* is given in Figure 34.

152   *Fractal Geometry in Biological Systems: An Analytical Approach*

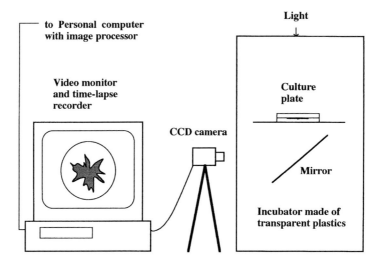

Figure 30   A macroscopic system for colony growth tracing.

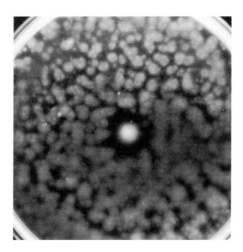

Figure 31 *Proteus mirabilis* NPC 3007 colony on a nutrient agar (0.5% Eiken agar). Many spots are final results of a point inoculation at the center of the plate. Cultivation at 30°C for 4 days.

Next, let us examine microscopic processes occurring in the wavy extension of *S. typhimurium* on a semi-solid medium (macroscopic Figures 29A and B). In contrast to *P. mirabilis* described above, this *Salmonella* mutant is lacking flagella. So, each bacterium is not moving independently in the microscopic field. Then, how can the colony split and extend outward? There was a slow stream of cell population suggesting that bacteria were passively translocated. Although the patterns formed are different, flagella-less *S. marcescens* also demonstrates such spreading.[20,21] We have not yet obtained data which completely explains the mechanisms of this stream. We do know, however, that tilting the plate has no effect on the direction of the spreading growth which suggests involvement of some translocating mechanisms strong enough to resist the gravitational force.

Microscopic examination of dense branches of a giant colony (e.g., Figures 14 and 15) revealed the presence of specific structure in the branch extension.[22] As seen in a photomicrograph of a branch tip (Figure 35), bacterial cells located on the

chapter five: Morphogenesis by Bacterial Cells 153

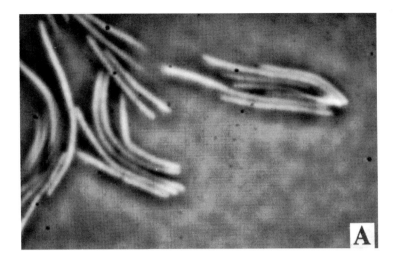

Figure 32 Photomicrographs of *Proteus mirabilis* NPC 3007 translocating on a nutrient medium (0.5% Eiken agar, 0.4% glucose). Figure B was taken 2 seconds after Figure A.

outermost margin seem to be reluctant to move (clear images in the figure). In contrast, cells a little behind the tip are quite active in random swirling (blurred images). Consequently, the outermost cells forming a shell are gradually pushed outward by the pressure made by swirling inside cells. This is a real extension process of this distinct type of branch. Just behind the extending tip, there are always hot spots composed of busily moving cells. After analysis of many video records, we summarized the findings as illustrated in Figure 36. Thus, a structured cell population is necessary for an effective occupation of the surface environment. It is important to consider this mass behavior in theoretical simulations of colonial pattern formation by bacteria.

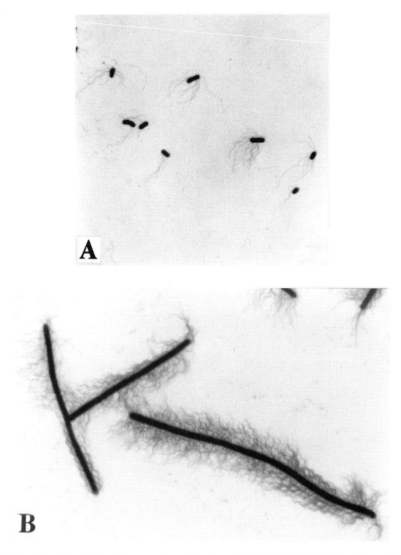

Figure 33 Flagella stain photomicrographs of *Proteus mirabilis* NPC 3007 taken from the center (A) and the margin (B) of a swarming colony on a nutrient agar (1.5% Eiken agar).

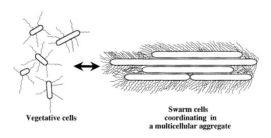

Differentiation of *Proteus mirabilis*

Figure 34 Schematic illustration of two different phases of *Proteus mirabilis*.

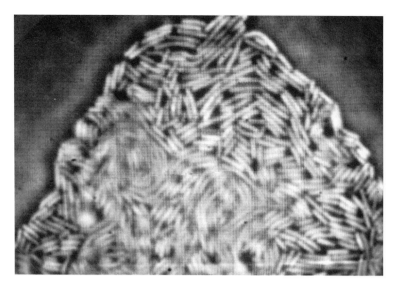

Figure 35 Photomicrograph of a growing branch tip of *Bacillus subtilis* OG-01 colony on a Vogel-Bonner medium (0.5% Eiken agar, 0.8% glucose). Random swirlings of inside bacterial cells are remarkable in this photograph taken by 1/4-second shatter speed.

Figure 36 Schematic illustration of a growing branch. Open box indicates an actively moving bacterial cell. Closed box indicates an inactive cell. Inside cells locating near to the tip are crowded and vigorously swirling. In contrast, outermost cells are mostly inactive and making a shell-like arrangement. Extending direction is indicated by an arrow.

5.5 Experimental approach for elucidation of morphogenic mechanisms

5.5.1 Designs of experiments and revealed mechanisms

As we have shown, the same bacterial strain forms different types of colony pattern when the culture conditions are different. So it is usual to design comparative experiments to identify the factor(s) responsible for such pattern differences. Modification of one experimental condition, however, will produce multiple effects on bacteria and environment. For instance, glycerol added as a nutrient, will change various physico-chemical factors (e.g., surface hydrophobicity) of the medium, so it is not easy to identify the factors actually responsible for colony morphogenesis. Practically speaking, responsible factors have been identified by summing up many experiments differing in various aspects. While the real factors remain to be unambiguously identified, it is important to know the range of availability of those factors empirically suggested.

5.5.1.1 Nutrients

Nutrients are indispensable for multiplication of bacteria and subsequent colony formation. A colony will be large on a rich medium and small on a poor medium.

While this is commonly known among bacteriologists, the concept must be modified now.[11,23] By continuing culture for a long time, we have indicated that the effect of nutrient concentration has different features. Consumption of nutrients by a bacterial colony growing at a fixed site will produce a gradient in a nutrient diffusion field. When the concentration of nutrients is high, effects of the gradient may be too small to affect the multiplication of bacteria. On the other hand, effects of this nutrient gradient will become profound for the bacteria if the medium contains a limited amount of nutrients. Thus, bacterial multiplication will depend on the accessibility of nutrients in the diffusion field. This diffusion field may be a real example of a so-called Laplacian field in which diffusion length l is extremely long. Here, diffusion length l is defined as $l = 2D/v$ (D, diffusion constant of nutrient; v, the growth rate of a colony frontline, in our case). By preparing a thin solid medium in a plastic dish and inoculating bacteria at the center of the plate, we are able to induce the development of a two-dimensional nutrient diffusion field. A colony of *B. subtilis* point-inoculated on a poor nutrient (peptone, 0.1%) medium continues to grow slowly on the diffusion field generated by the growing colony itself. Occurrence of these dynamic processes is realized by the appearance of a characteristic colony pattern (Figure 37). This colony pattern is quite similar to the pattern (Figure 38) formed by computer simulation adopting the diffusion-limited aggregation (DLA) model.[24] This DLA model describes a prototype of random pattern growth as an irreversible cluster aggregation in a Laplacian field. In a computer simulation, Brownian particles are released from random positions far from the central aggregation site one at a time to keep the concentration of Brownian particles low. The pattern produced by this model is an outwardly branching self similar fractal, with the fractal dimension of about 1.71. A fractal dimension of a *B. subtilis* colony formed in the experimental condition was each determined by the box-counting method, an example of which is shown in Figure 23. The mean value was shown to be 1.72 ± 0.02 (number of samples, n = 24).[23] When we consider that bacteria have a real surface area (i.e., not a simple point of dimension 0), the obtained fractal dimension of the colony is quite reasonable.

Figure 37 *Bacillus subtilis* OG-01 colony on a low nutrient agar (0.1% peptone, 0.9% Bactoagar). Cultivation at 35°C for 30 days. (From Matsuyama, T. and Matsushita, M., *Crit. Rev. Microbiol.*, 19, 117, 1993. With permission.)

As shown in a phase diagram (Figure 39), we could induce a pattern shift from the DLA type to a reaction-limited growth model, or Eden-like pattern (B in the

chapter five: Morphogenesis by Bacterial Cells

Figure 38 Example of two-dimensional DLA cluster obtained by computer simulations. (From Matsuyama, T. and Matsushita, M., *Crit. Rev. Microbiol.*, 19, 117, 1993. With permission.)

diagram), by increasing the concentration of nutrients.[25] Thus nutrient concentration in the medium can be shown to be important not only for a colony size but also for fractal morphogenesis of a *B. subtilis* colony.

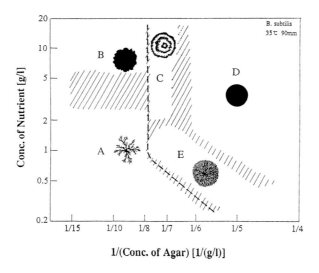

Figure 39 Phase diagram of pattern change in colonies of *Bacillus subtilis* OG-01 (wild type). Thick broken line indicates the boundary of observing the active movement of bacterial cells inside colonies. (From Wakita, J. et al., *J. Phys. Soc. Jpn.*, 63, 1205, 1994. With permission.)

5.5.1.2 Spatial arrangement

Presence of a Laplacian diffusion field around the growing DLA type colony was also demonstrated by experiments observing spatial factors affecting the colony growth. First, the screening effect was confirmed by a video recording of the colony growth. This effect appears as the growth inhibition of inner branches by surrounding more protruding branches.[7,23] That is, inner branches have less chance to trap Brownian moving nutrients in this diffusion field. The screening effect described above is the phenomenon predicted by Meakin through extensive computer simulation of DLA growth.[26,27] Second, Meakin also predicted the repulsion effect between neighboring DLA patterns as shown in Figure 40. So, experiments were simply done by inoculating bacteria on two neighboring sites. The results obtained (Figure 41) are quite consistent with Meakin's prediction and indicate the involvement of a DLA-like process in this fractal colony morphogenesis.

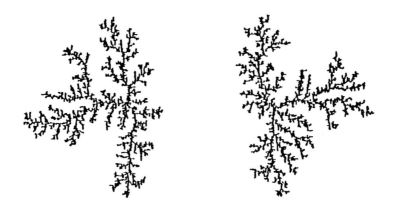

Figure 40 Repulsion effect between two-dimensional DLA clusters grown from two seed particles. (From Matsuyama, T. and Matsushita, M., *Crit. Rev. Microbiol.*, 19, 117, 1993. With permission.)

Figure 41 Two neighboring *Salmonella typhimurium* ATCC 14028 colonies on a Vogel-Bonner medium (1.4% Eiken agar, 0.2% glucose). Cultivation at 37°C for 21 days. (From Matsuyama, T. and Matsushita, M., *Appl. Environ. Microbiol.*, 58, 1227, 1992. With permission.)

5.5.1.3 Substrate construction

As shown in Figure 39, modification of substrate gel by changing the concentration of agar in the medium causes remarkable pattern changes in a *B. subtilis* colony.[25] This effect is presumably due to enabling the bacteria to translocate by themselves. On a hard agar, most bacteria are not able to translocate by themselves and the colony growth is just the result of volume increase due to cell multiplication. On a semi-solid agar, bacteria can multiply and spread (if they are motile). So most colonies develop in a short time and form patterns dependent on the mode of bacterial spreading behavior. There are plenty of spreading patterns to be analyzed with reference to diverse biological activities of bacteria.[11]

5.5.1.4 Genetic approach

For analysis of biological activities responsible for colony morphogenesis, genetic approaches are quite useful. Generally speaking, there are two types of genetic approaches. One approach starts by obtaining many independent mutants defective in the formation of the specific colony pattern to be analyzed. Each mutant is examined to determine steps which are altered. Thus, one analyzes morphogenesis by isolation of a series of mutants each of which corresponds to a unit process in the pattern forming pathway. After characterization of the activity of each unit process, the total morphogenic processes can be reconstructed step by step. Such studies are now in progress with the Liesegang-ring-like morphogenesis by *P. mirabilis*.[28,29]

The second approach is carried out by isolating mutants defective in an already known activity. Then the mutation is examined for its effect on the colony morphogenesis. By this method, the role of that specified activity in the colony morphogenesis will be put forward. Introduction of flagella-less mutation in *B. subtilis* resulted in failure of the bacteria to form three patterns (C, D, and E in Figure 39), so these patterns are shown to be formed by flagella-dependent morphogenic processes.[25] *B. subtilis* is famous for production of a surface-active exolipid called *surfactin*. The role of surfactin in colony morphogenesis was also examined by obtaining a surfactin-less mutant.[11,22] Inability of the mutant to produce surfactin is confirmed by thin-layer chromatography (Figure 42). The mutant thus isolated was unable to form a DLA-type colony on a hard agar.[11,22] This suggests some essential role of the surfactin in the colony morphogenesis of *B. subtilis*. In addition, on semi-solid media, the surfactin-less mutant formed a different type of spreading the colony as shown in Figures 14 and 15.[11]

The role of surface active bacterial products in colony morphogenesis was studied more extensively in *S. marcescens*.[3,30] The bacteria have been shown to produce wetting agents called *serrawettins*. Extracellular secretion of these lipids is confirmed by various experiments. Extracellular vesicles advancing before bacterial cells (Figure 43) has been shown to contain serrawettins as the major component.[31] By using a newly devised mutant screening method,[32] we could isolate many mutants defective in the production of these serrawettins. Three kinds of serrawettins have been reported so far. Each strain of *S. marcescens* produces only one kind of serrawettin.[33] In contrast to wild type strains which develop giant self similar fractal colonies (Figure 44) on a hard agar, all of the mutants defective in the production of the serrawettins were equally unable to form a self-similar fractal colony.[3] Since back mutation from serrawettin-negative to serrawettin-positive restores the fractal morphogenic activity, involvement of other unknown mutations is improbable.[3] So, surface-active exolipids such as surfactin and serrawettins seem to have critical roles in the colony morphogenesis of these bacteria.

Spreading behavior of these mutants of *S. marcescens* on a semi-solid medium is also quite similar to that of *B. subtilis* surfactin-less mutant. Pattern changes resulting from the introduction of the mutation (compare Figures 45 and 46) include a shift of branch width from thin to broad type and strengthening of the tendency to turn to the right. By the inoculation of a serrawettin-producing but flagella-less strain nearby the serrawettin-less strains, a pattern-changing complementation effect is clearly visualized (Figure 47). There seems to be some common pathway in the working processes of these surface active exolipids.[33]

5.5.1.5 Terrestrial factor

On a semi-solid medium, radial branching patterns of most bacterial colonies so far examined demonstrated the tendency of turning to the right. Why is it so? We are going to examine possible explanations one by one through experiments. Since our laboratory is located in the northern hemisphere, the effect of the Coriolis force, known to be responsible for the whirlpool direction of a typhoon, is worthy of examination. Through collaboration with a bacteriologist working in the southern

Figure 42 Thin-layer chromatograms of surfactin (lane 1), lipids from *Bacillus subtilis* ATCC 21331 (lane 2), and 21331-b3 (lane 3). In a lane 3, there is no spot corresponding to surfactin.

hemisphere (Dr. P. R. Fisher, LA Trobe University, Australia), this effect was examined by setting up the same experimental conditions, the only exception being the location of the laboratory. The resulting *S. marcescens* spreading is shown in Figure 46. The bacteria (*S. marcescens* and *B. subtilis*) showed a clear tendency to turn to the right in the laboratory in Australia. So, the effect of the Coriolis force is negligible.

5.5.2 Identification of chemical factors

From the mutational analysis described above, the roles of surface-active exolipids in colony morphogenesis are suggested. Therefore, it is necessary to isolate active agents and to examine the effect of purified agents on morphogenesis. We have succeeded in the purification of serrawettin W1, W2, and W3, and in the determination of the complete chemical structures of W1 and W2 and the partial structure of W3.[30,34,35] They were shown to have similar structures to each other and all were classified as cyclodepsipeptides. It is noteworthy that surfactin, the surface-active exolipid of *B. subtilis*, also belongs to the cyclodepsipeptide family. In Figure 48, the chemical structure of serrawettin W2 which was revealed by us is given as a representative molecule of a cyclodepsipeptide.

chapter five: Morphogenesis by Bacterial Cells 161

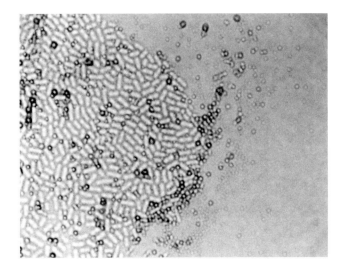

Figure 43 Phase-contrast photomicrograph of growing tip of a *Serratia marcescens* NS 38 colony on a Davis medium just after inoculation of a bacterial mass. In addition to rod-shaped bacterial cells, round extracellular vesicles are scattering on the agar surface.

Figure 44 *Serratia marcescens* NS 38 colony on a Vogel-Bonner medium (1.4% Eiken agar, 0.1% glucose). Cultivation at 30°C for 28 days.

Effects of the purified bacterial product were examined by placing a paper disk containing the serrawettin W2 near the colony of the bacteria defective in the production of surface active agent. The effects were evident both on hard and semi-solid agar plates as seen in Figures 49 and 50. It is noteworthy that serrawettin W2 produced by *S. marcescens* is active on *B. subtilis* in spite of the difference in bacterial species. Other surface active exolipids described herein have all been ascertained to be active in promotion of spreading growth for fractal morphogenesis and modification of branching patterns.[11] Molecular and physico-chemical mechanisms of their function in morphogenesis are intriguing subjects to be investigated in the future.

5.6   *Modeling and population dynamics of colony formation*

It should be noted that in Figure 39 the colony formation in the region A can be explained by the DLA model. Unfortunately, it is still difficult to understand even

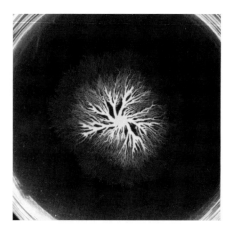

Figure 45 *Serratia marcescens* NS 38 colony on a nutrient agar (0.5% Eiken agar). Cultivation at 30°C for 12 hr. (From Matsuyama, T. and Matsushita, M., *Crit. Rev. Microbiol.*, 19, 117, 1993. With permission.)

Figure 46 *Serratia marcescens* NS 38-09 colony on a nutrient agar (0.5% Eiken agar). Cultivation at 30°C for 18 hr. (Courtesy of Dr. P. R. Fisher.)

phenomenologically the formation of concentric-ring-like colonies in the region C and dense-branching morphology (DBM)-like colonies in the region E. If, however, colonies look macroscopically very simple, then we may be able to understand the formation of the mechanism. Let us here discuss the growth mechanism of homogeneously spreading, disk-like patterns seen in the region D, an example of which is shown in Figure 51, from another viewpoint.[36]

5.6.1 Population dynamics approach

As seen in Figure 51, a colony in the region D looks macroscopically like a perfect disk with a clearcut interface. We noticed by microscope observations, however, that the growing front is obscure microscopically, as seen in Figure 52A, compared with that in regions C and E. In other words, the interface cannot be defined clearly in this D region. On the other hand, as seen in Figure 52B, the distribution of bacterial cells inside a colony looks almost homogenous, and they are moving around actively even deep inside the colony. We also noticed that the population density of bacterial cells inside a colony in the region D is lower even in a monolayer of cells than in the region E. This is the reason why colonies in D looked transparent and were hard

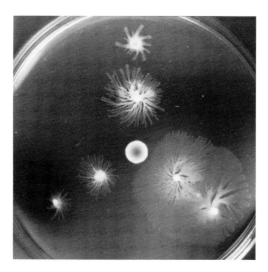

Figure 47 Extracellular complementation by serrawettin producing live cells. Serrawettin W1 producing *Serratia marcescens* NS 38-45 (flagella-less mutant) was point inoculated at the center of the plate. Three different *S. marcescens* mutants defective only in production of serrawettins were point inoculated nearby and apart from the center. Compare inside colonies (near the central colony) to outside colonies. The inside colonies are modified and extending thin branches.

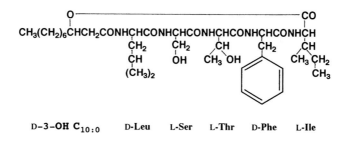

Figure 48 Chemical structure of serrawettin W2. (From Matsuyama, T. and Matsushita, M., *Crit. Rev. Microbiol.*, 19, 117, 1993. With permission.)

to observe. Inside the colony, active but erratic movement of individual cells is seen almost everywhere. Hence we assume as the *zeroth* approximation that the bacterial cell movement in the region D can be described in terms of diffusion in two dimensions. According to close observations, however, their motion does not look completely Brownian but rather somewhat collective, especially when their population is dense. They tend to make a group of several cells parallel with each other, as seen in Figure 32, which move around together. Of course, one cell or two join a group and another or two leave it from time to time. Our intention is to regard the colony formation in the region D as a combination of the multiplication of cells inside the colony and their outward diffusion.

The spatio-temporal variation of population density of bacterial cells $b(r,t)$ and concentration of nutrient $n(r,t)$ is then represented by reaction-diffusion-type equations:

$$\frac{\partial b}{\partial t} = \nabla \cdot (D_b \nabla b) + f(b,n), \qquad (8)$$

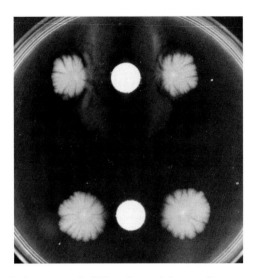

Figure 49 Effect of purified serrawettin W2 on bacterial spreading growth on a solid medium. An upper paper disk contains 100 µg of serrawettin W2 and a lower disk is a control. Serrawettin-less mutants *Serratia marcescens* NS 25-03 (left) and NS 25-04 (right) were point inoculated nearby the disk. Promotion of spreading growth is recognizable near the serrawettin W2 disk. Cultivation at 30°C for 21 days.

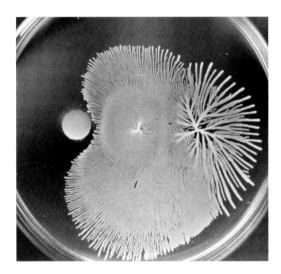

Figure 50 Effect of purified serrawettin W2 on bacterial spreading growth on a semi-solid medium. *Bacillus subtilis* JH 642 (nonproducer of surfactin) were point inoculated at sites apart from and nearby the paper disk (left) containing 100 µg of serrawettin W2. The exolipid from *Serratia marcescens* is active to *B. subtilis* beyond the species difference. Cultivation at 37°C for 24 hr.

$$\frac{\partial n}{\partial t} = D_n \nabla^2 n - \nu b n, \tag{9}$$

where D_b and D_n are the diffusion coefficients of bacterial cells and nutrient, respectively; ν is the consumption rate of nutrient by bacteria; and $f(b,n)$ denotes the reaction term due to local bacterial growth. D_b is in general dependent on both b and n. In the following, however, we regard D_b as a constant, since we are investigating here

chapter five: Morphogenesis by Bacterial Cells 165

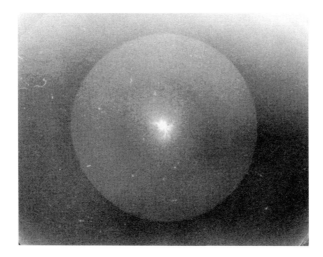

Figure 51 An example of simple spreading disk-like colonies of *Bacillus subtilis* OG-01 observed in region D. The initial concentration of nutrient peptone is 3 g/l and the initial concentration of agar is 5 g/l. The photograph was taken after a 13-hr incubation. (From Wakita, J. et al., *J. Phys. Soc. Jpn.*, 63, 1205, 1994. With permission.)

only the region D. Moreover, we assume that $f(b,n)$ can be described by the following logistic-like form:

$$f(b,n) = [\varepsilon(n) - \mu b]b, \qquad (10)$$

where the term $\varepsilon(n)$ represents the rate of Malthusian growth of individual cells, while μ is the coefficient of competition among cells that describes the suppression of population increase. This form is plausible because too many bacterial cells restrain themselves from increasing their population.

Let us now consider the limiting case in which nutrient is so rich that n can be put spatially constant, i.e., $\varepsilon(n) = \varepsilon$. It should be noted that this does not mean that ε is independent of n. When Equations (8) and (9) are decoupled, Equation (8) becomes

$$\frac{\partial b}{\partial t} = D_b \nabla^2 b + (\varepsilon - \mu b)b. \qquad (11)$$

This is known as the Fisher equation.[37] This equation and its travelling wave solutions have been widely studied. In particular, this equation asymptotically yields isotropically spreading, homogeneous solutions with stable travelling wavefronts of constant speed in two dimensions. The population density inside and wavefront speed, respectively, are given by $b = \varepsilon/\mu$ and $v = 2(\varepsilon D_b)^{1/2}$.

This Fisher equation may describe the simple, homogeneously spreading colony pattern observed in the region D in Figure 39. Let us here discuss experimental confirmation for this conjecture. It turned out to be very difficult to estimate the value of ε and especially μ experimentally. Here we assume that μ is independent of the nutrient concentration n. This assumption is plausible because basically μ describes how strongly two bacterial cells are repelled when they encounter each other. This is in clear contrast to the case of ε, the rate of the Malthusian population increase of bacteria, which may be strongly dependent on n. We will, therefore, argue

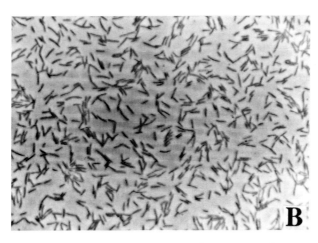

Figure 52 Photomicrographs of a disk-like colony in the region D shown in Figure 39. (A) A snapshot of fuzzy interface of the colony. (B) A snapshot inside the colony. The scales of the figures are both 0.2 mm in width. (From Wakita, J. et al., *J. Phys. Soc. Jpn.*, 63, 1205, 1994. With permission.)

the consistency in describing experimental data by the Fisher equation (11) under the assumption of constant μ.

5.6.2 *Experiments for verification of a proposed model*

We first let a pair of colonies inoculated simultaneously collide with each other. They were then observed to fuse together, as shown in Figure 53. This is a striking contrast to the case of DLA-like colonies in region A (Figure 41) or even DBM-like colonies in region E, which were found to repel each other. Numerical calculations of the Fisher equation, Equation (11), started from two points in two dimensions, were found to yield exactly the same behavior. This is the first (but qualitative) evidence of the conjecture.

Secondly, we estimated the population density b of bacterial cells. We measured the occupation rate of area by cells in homogeneously populated region just inside the colony interface for various nutrient concentrations n in region D. We obtained

chapter five: Morphogenesis by Bacterial Cells   167

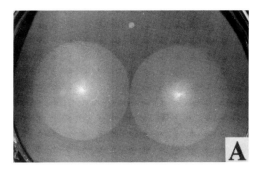

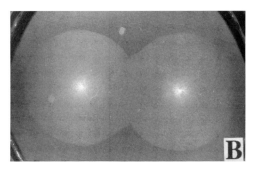

Figure 53 Two disk-like colonies in region D. Bacteria were inoculated simultaneously at two points and then incubated for about a half day. (A) They came close together independently and (B) fused together. (From Wakita, J. et al., *J. Phys. Soc. Jpn.*, 63, 1205, 1994. With permission.)

the result that b is approximately proportional to n. This implies that under the assumption of the Fisher equation with constant μ, ε should be proportional to n, i.e., $\varepsilon = \varepsilon_1 n$ (ε_1 is some constant). This proportionality may not be true in general. It should be noted, however, that here we are concerned with the growth rate of bacterial cells only in the specific region D.

Next we measured the diffusion coefficient of bacterial cells D_b for various nutrient concentrations n. As described before, the movement of individual cells does not look like Brownian motion in shorter periods and for higher nutrient concentrations because of their elongated shape and high population density, respectively. Nevertheless we tried to estimate D_b from the ensemble average of squared displacements of individual cells $<R^2>$ during time interval t by using the relation $<R^2> = 2D_b t$. We found that D_b does not depend so much on the nutrient concentration n, i.e., $D_b \sim n^0$.

Finally we measured the interface growth speed v for various nutrient concentrations n. The diameters of growing colonies are not quite proportional to the incubation times. There always seems to be a dormant period of 7 or 8 hours after incubation during which colonies do not grow. It looks as if bacteria need this period to adjust themselves to a new environment and to initiate their new life. After this period, however, colonies grow rather quickly; hence, we estimated the growth speed υ when the colony diameter reached 5 cm. The results clearly showed that v is proportional to $n^{1/2}$, as shown in Figure 54. This is compatible with the speed v = $2(\varepsilon D_b)^{1/2} \sim n^{1/2}$ that the Fisher equation gives.

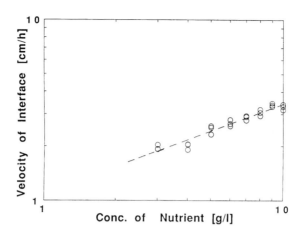

Figure 54 The growth speed v of the interface of colonies in region D, measured as a function of the nutrient concentration n. Note the double-logarithmic plotting. The slope of a broken line determined by the least-mean-squares fitting is 0.49, clearly implying $v \sim n^{1/2}$.

All these experimental results are consistent with the behaviors derived from the Fisher equation. We, therefore, can conclude that the colony formation in region D in Figure 39 essentially can be described by the Fisher equation.[36] An interesting theoretical problem in the near future may be to determine if this population dynamics approach can be generalized to other regions such as C and E in order to explain the colony formation observed there.

5.6.3 Fractal growth of colonies — revisited

When agar plates are hard, bacterial cells cannot move around on the surface of the plates. They only grow and perform cell division locally by feeding on the nutrient. Under this condition one can observe the effect of nutrient diffusion on the colony growth (see Sections 5.1.1 and 5.1.2). Let us fix the initial nutrient (peptone) concentration at some low value, say 0.1%. We are now under the cultivation condition of poor nutrient and hard agar medium. Colony patterns with characteristically branched structure are normally observed after 3 or 4 weeks of incubation. They look like dendritic crystals without any rotational symmetry such as the sixfold one seen in snowflakes.

Under this condition the colony growth is governed by diffusion-limited processes: a protruding part on the interface of a growing colony tends to grow more because it faces medium richer in nutrient compared with surrounding interfacial parts. This tendency eventually results in the dendritic or fractal growth of the colony interface. The prototype model that describes this process is the diffusion-limited aggregation (DLA) first proposed by Witten and Sander.[24] The DLA model is known to produce randomly branched self-similar fractal patterns with the fractal dimension of about 1.71 in two-dimensional space. It is known that the colony growth of common rod-like bacteria such as *B. subtilis*, *E. coli*, and *S. typhimurium* under the diffusion-limited condition can be described in terms of DLA.[7,11,22] In order to check this, one should confirm not only the values of the fractal dimensions but also the existence of screening and repulsion effects characteristic to DLA processes as already described in the previous section.

Even when colony patterns are compact in its bulk structure instead of being ramified, i.e., nonfractal in the ordinary sense, there is still a possibility that their interfaces may be self-affine fractals.[15,38] An interface is a boundary between two different phases such as a colony and the medium. It must then have some spatial correlation along it according to how it was made. However simple or rough it looks, a growing interface of a colony may in many cases conceal the anisotropic scaling symmetry known as self-affinity. This can be examined by the method described in Section 3.2.2.

As for the population dynamics approach to the colony formation, it is at this stage too early to describe the fractal growth of colonies. Let us hope that this interesting theoretical approach will soon become useful to understand many aspects of colony formation.

5.7 Prospects

The variety of morphology is a visible aspect of diversity in nature. Variation is an enjoyable aspect of daily life; however, since these various forms in nature are so familiar, we tend to simply regard them as they are. As a result, the causes behind these usual morphologies have not been queried or resolved properly. Until recently the morphology of bacterial colonies has not been of interest. After we noticed the strange colony pattern of *S. marcescens*, we then found that there is a variety of colony patterns in the microbial world and that it is possible to reproduce these patterns by controlling biological and culture conditions. This means that an experimental approach is possible in the study of patterns. Furthermore, patterns similar to those made by bacteria are everywhere in nature. Thus, we hope this experimental study will provide novel insight for the elucidation of the general question about the patterns in nature.

Acknowledgments

We are grateful to M. Sogawa, M. Ohgiwari, J. Wakita, Y. Shimada, H. Itoh, and A. Nakahara for assistance with the experiments.

References

1. Shapiro, J. A., Bacteria as multicellular organisms, *Sci. Amer.*, 256, 82, 1988.
2. Shapiro, J. A. and Hsu, C., *Escherichia coli* K-12 cell- cell interactions seen by time-lapse video, *J. Bacteriol.*, 171, 5963, 1989.
3. Matsuyama, T., Sogawa, M., and Nakagawa, Y., Fractal spreading growth of *Serratia marcescens* which produces surface active exolipids, *FEMS Microbiol. Lett.*, 61, 243, 1989.
4. Fujikawa, H. and Matsushita, M., Fractal growth of *Bacillus subtilis* on agar plates, *J. Phys. Soc. Jpn.*, 58, 3857, 1989.
5. *ASM Style Manual for Journals and Books*, American Society for Microbiology, Washington, D.C. 1991, 25.
6. Harshey, R. M. and Matsuyama, T., Dimorphic transition in *Escherichia coli* and *Salmonella typhimurium*: surface-induced differentiation into hyperflagellated swarmer cells, *Proc. Natl. Acad. Sci. U.S.A.*, 91, 8631, 1994.
7. Matsuyama, T. and Matsushita, M., Self-similar colony morphogenesis by gram-negative rods as the experimental model of fractal growth by a cell population, *Appl. Environ. Microbiol.*, 58, 1227, 1992.
8. Matsuyama, T. and Matsushita, M., Abstract, *9th Annu. Meet. Jpn. Soc. Microb. Ecol.*, 1993, 49 (in Japanese).
9. Allison, C., Lai, H.-C., Gygi, D., and Hughes C., Cell differentiation of *Proteus mirabilis* is initiated by glutamine, a specific chemoattractant for swarming cells, *Mol. Microbiol.*, 8, 53, 1993.

10. Williams, F. D., Nature of swarming phenomenon in *Proteus, Annu. Rev. Microbiol.,* 32, 101, 1978.
11. Matsuyama, T. and Matsushita, M., Fractal morphogenesis by a bacterial cell population, *Crit. Rev. Microbiol.,* 19, 117, 1993.
12. Cooper, A. L., Dean, A. C. R., and Hinshelwood, C., Factors affecting the growth of bacterial colonies on agar plates, *Proc. Roy. Soc.,* B171, 175, 1968.
13. Mandelbrot, B. B., *The Fractal Geometry of Nature,* W. H. Freeman, San Francisco, 1982.
14. Feder, J., *Fractals,* Plenum Press, New York, 1988.
15. Vicsek, T., *Fractal Growth Phenomena,* 2nd ed., World Scientific, Singapore, 1992.
16. Matsushita, M. and Ouchi, S., On the self affinity of various curves, *Physica D,* 38, 246, 1989.
17. Shimada, Y., Nakahara, A., Matsushita, M., and Matsuyama, T., Spatio-temporal patterns produced by bacteria, *J. Phys. Soc. Jpn.,* 64, 1896, 1995.
18. Harshey, R. M., Bees aren't the only ones: swarming in gram-negative bacteria, *Mol. Microbiol.,* 13, 389, 1994.
19. Allison, C. and Hughes, C., Bacterial swarming: an example of prokaryotic differentiation and multicellular behavior, *Sci. Prog.,* 75, 403, 1991.
20. O'Rear, J., Alberti, L., and Harshey, R. M., Mutations that impair swarming motility in *Serratia marcescens* 274 include but are not limited to those affecting chemotaxis or flagellar function, *J. Bacteriol.,* 174, 6125, 1992.
21. Matsuyama, T., Bhasin, A., and Harshey, R. M., Mutational analysis of flagellum-independent surface spreading of *Serratia marcescens* 274 on a low agar medium, *J. Bacteriol.,* 177, 987, 1995.
22. Matsuyama, T., Harshey, R. M., and Matsushita, M., Self- similar colony morphogenesis by bacteria as the experimental model of fractal growth by a cell population, *Fractals,* 1, 302, 1993.
23. Matsushita, M. and Fujikawa, H., Diffusion-limited growth in bacterial colony formation, *Physica A,* 168, 498, 1990.
24. Witten, T. A. and Sander, L. M., Diffusion-limited aggregation, a kinetic critical phenomenon, *Phys. Rev. Lett.,* 47, 1400, 1981.
25. Ohgiwari, M., Matsushita, M., and Matsuyama, T., Morphological changes in growth phenomena of bacterial colony patterns, *J. Phys. Soc. Jpn.,* 61, 816, 1992.
26. Meakin, P., A new model for biological pattern formation, *J. Theor. Biol.,* 118, 101, 1986.
27. Meakin, P., The growth of fractal aggregates and their fractal measures, in *Phase Transitions and Critical Phenomena,* vol. 12. Domb, C. and Lebowitz, J. L., Eds., Academic Press, New York, 1988, 335.
28. Belas, B., Erskine, D., and Flaherty, D., *Proteus mirabilis* mutants defective in swarmer cell differentiation and multicellular behavior, *J. Bacteriol.,* 173, 6279, 1991.
29. Allison, C. and Hughes, C., Closely linked genetic loci required for swarm cell differentiation and multi-cellular migration by *Proteus mirabilis, Mol. Microbiol.,* 5, 1975, 1991.
30. Matsuyama, T., Kaneda, K., Nakagawa, Y., Isa, K., Hara- Hotta, H., and Yano, I., A novel extracellular cyclic lipopeptide which promotes flagellum-dependent and -independent spreading growth of *Serratia marcescens, J. Bacteriol.,* 174, 1769, 1992.
31. Matsuyama, T., Murakami, T., Fujita, M., Fujita, S., and Yano, I., Extracellular vesicle formation and biosurfactant production by *Serratia marcescens, J. Gen. Microbiol.,* 132, 865, 1986.
32. Matsuyama, T., Sogawa, M., and Yano, I., Direct colony thin-layer chromatography and rapid characterization of *Serratia marcescens* mutants defective in production of wetting agents, *Appl. Environ. Microbiol.,* 53, 1186, 1987.
33. Matsuyama, T., Bacterial wetting activity and fractal colony growth, *Surface,* 31, 114, 1993 (in Japanese).
34. Matsuyama, T., Fujita, M., and Yano, I., Wetting agent produced by *Serratia marcescens, FEMS Microbiol. Lett.,* 28, 125, 1985.
35. Nakagawa, Y. and Matsuyama, T., Chromatographic determination of optical configuration of 3-hydroxy fatty acids composing microbial surfactants, *FEMS Microbiol. Lett.,* 108, 99, 1993.

36. Wakita, J., Komatsu, K., Nakahara, A., Matsuyama, T., and Matsushita, M., Experimental investigation on the validity of population dynamics approach to bacterial colony formation, *J. Phys. Soc. Jpn.*, 63, 1205, 1994.
37. Murray, J. D., *Mathematical Biology*, Springer-Verlag, Berlin, 1989.
38. Vicsek, T., Cserzo, M., and Horvath, V. K., Self-affine growth of bacterial colonies, *Physica A*, 167, 315, 1990.

chapter six

Fractal Studies of Neuronal and Glial Cellular Morphology

Thomas G. Smith, Jr. and G. D. Lange

> *Self-similarity, or invariance against changes in scale or size, is an attribute of many laws of nature and innumerable phenomena in the world around us. Self-similarity is, in fact, one of the decisive symmetries that shape our universe and our efforts to comprehend it.*
>
> M. Schroeder

6.1 Introduction

6.1.1 Fractal geometry

There are two aspects of fractal geometry — the synthetic and the analytic. Synthesis is concerned with generating images or forms by following certain mathematical rules.[1-3] These images can be either idealized or natural looking (i.e., fractal forgeries).[4] Analysis on the other hand is concerned with extracting quantitative information from data sets, often images. In this chapter, the focus will be on analysis.

6.1.1.1 The fractal dimension

Fractal geometry is a generalization of Euclidean geometry which allows for the concept of nonintegral dimension, i.e., (units)^D, where D may not be an integer but is a fraction, and hence the name "fractal". We are interested in describing the shapes of objects quantitatively so that we can associate numbers with complexity of form. Analytic Euclidean geometry does not easily lend itself to this goal, but fractal geometry does. This is largely because complexity and scaling are intimately related. For example, to the microscopist an object that continues to reveal more and more detail as it is magnified is morphologically complex. It is precisely this property of fractal objects (self-similarity or scale invariance) that suggests that fractal geometry might be a good model to provide measures of complexity. Indeed, the fractal dimension measures the rate of addition of structural detail with increasing magnification. Fractal dimension (D), therefore, serves as a quantifier of complexity. There are several related Ds. When these quantities take on integer values they correspond to the common, Euclidean definition of dimension.

6.1.1.2 Range of D

In fractal geometry, fractal objects that range from a point to a line (e.g., broken-line, Cantor sets) have a D between 0 and 1; those between a line and a plane (irregular lines, borders, etc.) have values between 1 and 2; those between a plane and a solid (surfaces of hills, valleys, etc.) have values between 2 and 3. Because we will be dealing with borders of objects (images) in a plane, we will be concerned with Ds between 1 and 2. The numbers of interest are those that indicate the degree of complexity and are found in the fractional part of D.

6.1.1.3 Empirical nature of D

There is a considerable literature on the theoretical aspects of fractal dimension. We have chosen here to emphasize the empirical nature of D. That is, we assume that the D is strictly descriptive. Thus, D is a statistic, like a mean. In general, we will not attempt to connect it to possible underlying mechanisms, which is in line with our own work in this field. In any case, connections between the empirical values of D and any specific growth or developmental mechanisms require the answering of specific experimental questions and not statistical or mathematical ones. We view D in much the same way that an optician might view refractive index — that is, as a descriptor of the properties of some object or material even though, in the case of the latter measure, a good deal is known about the underlying mechanisms leading to its value.

6.1.1.4 Methods of determining D

For visual clarity, the illustrations in this paper are binary silhouettes of nerve cells, glial cells, and idealized (Koch) figures. But the measurements and calculations were based on the borders of these and similar objects, and the borders were unbroken sets of black on white pixels. An image was, therefore, an array of pixels which were either white or black. The border pixels (Boolean ones) were restricted to a closed continuous, one-pixel-wide set of contiguous square pixels surrounding a compact set of Boolean zero pixels. There were two kinds of measures applied to this set of pixels. The first was distances between points on the border. The second was count of border pixels in grids of various spacing or in randomly positioned discs of various diameter. A transition between these is exemplified by the box-counting and dilation methods. When a distance measure was applied, the side of a pixel became a unit of length. When a counting measure was applied, the border pixel became a unit of mass.

6.1.1.5 Grey scale to binary images

Of course, the cells are first recorded as grey-scale images either from scanning of photographs or from direct video microscopy. There are several ways this grey scale-to-binary conversion can be achieved. The most obvious (but tedious and time-consuming) is to trace the image and then digitize the trace. An easier method employs edge-detecting algorithms. We have found that a method using a convolution of the image with a so-called Mexican Hat kernel, followed by thresholding, and some occasional human intervention produces an acceptable one-pixel-thick boundary.[5,6] This method produces some inevitable smoothing of the boundary. When we are comparing several cells, we compensate for this distortion by carefully applying exactly the same algorithm to each sample in the hope that differences, results, and conclusions will not be seriously altered.

6.1.1.6 Length methods

There are several ways of measuring the D related to length. The classical method of Richardson[1] involves measuring the perimeter of an object with various spans of calipers or various lengths of rulers (*trace* method). When the log of the perimeter is plotted against the log of the ruler lengths, a fractal object gives a straight line with a negative slope S. From this measurement, D is calculated as D = 1 – S.

A second method is based on the concept of covering the border and is called a *box-counting* or *grid* method. Here, sets of square boxes (i.e., grids) are used to cover the border. Each set is characterized by a box size. The number of boxes necessary to cover the border is noted as a function of the box size. The log of the number of covering boxes of each size times the length of a box edge is plotted against the log of the length of a box edge. Again, a straight line results, with slope S, and D is calculated as before.

A third method, developed by Flook,[7,8] is called the *dilation* method. Dilation, in this case, means a widening and smoothing of the border. It is accomplished by convolution with a binary disk, i.e., all the components of all the kernels have a unitary value (Boolean one). The result is a thickened, but grey-scale border. To return this border to Boolean one values, all non-zero pixels are thresholded to a Boolean one. The rate at which the total mass (or surface area) of the border grows as a function of the diameter of the convolution kernel depends on D. The log of the resulting area divided by the kernel diameter is plotted against the log of the kernel diameter. Yet again, a straight line results with a negative slope S, and D is calculated as above. This method, while a length-related measure, provides a bridge between the length- and the mass-based methods in that every pixel in the border is involved in the measurement and the area is merely a count of the pixels or mass points.

It should be noted that, with these length-related methods, the magnitude of the resultant measure (perimeters or counts/diameter) increases as the measuring element decreases in size. In a deterministic or true fractal, such as Koch figures[1] (see Figure 1), this continues without limit. This is an illustration of the important and defining property of fractals: self-similarity or scaling symmetry. That is, fractal objects look qualitatively the same at all scales or magnifications. Such deterministic self-similarity exists only as mathematical formulas or in computers. Real world or natural fractals are only self-similar in a statistical sense and have straight lines in log-log plots over a limited extent.[9] It should also be noted that the above measuring operations have the characteristic of a low-pass filter, with increased filtering with increases in the size of the measuring elements. That is, they progressively remove the high spatial frequencies of an image, while leaving the lower spatial frequencies. The final value calculated is an average property of the whole object and has no spacial locality. All three of the methods employed to calculate D are basically plots of log length vs. log length.

Given sufficient detail and magnification of a border, the results of all three operations give similar Ds; see Figure 2.[8] They can be tested by using the borders of fractals of known D — for example, the Koch figures, as in Figure 1.[8] All methods consistently underestimate the values of these deterministic fractals by a small percentage. This is a consequence of the fact that a finite, digitized image with a limited number of pixels cannot realize the detail implicit in a deterministic fractal. This error probably is found in the measurements of natural fractals also, but since it is a consistent, and not random, error and since most results are used comparatively, it probably does not significantly affect the conclusions drawn.

The first two methods can readily be implemented on a computer or can even be done (tediously) by hand. The third method is rather more difficult to program on a computer, but is somewhat superior to the others in that it is less sensitive to the location of the image in a frame or to pixelization effects. For example, if the border is not centered in the image frame and does not significantly fill the frame, the grid method will give too few points with large grids and erroneously increase

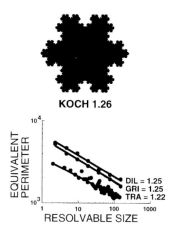

Figure 1 (Upper panel) Measuring the fractal dimension (D) of a Koch triadic island of D = 1.26 by the dilation (DIL), grid (GRI), and perimeter trace (TRA) methods. (Lower panel) Log-log plots of the equivalent perimeters with measuring elements (resolvable size) from size 2 to 128, in powers of 2. See text.

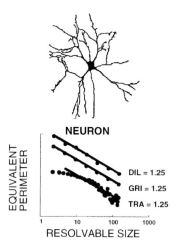

Figure 2 (Upper panel) Measuring the fractal dimension (D) of a cell-cultured spinal cord neuron by the dilation (DIL), grid (GRI), and perimeter trace (TRA) methods. (Lower panel) Log–log plots of the equivalent perimeters with measuring elements (resolvable size) from size 2 to 128, in powers of 2. See text.

the slopes in the log-log plots. In addition, since most borders become straight (Euclidian) lines with small rulers in perimeter measurements, the slope tends towards zero (D = 1) with small rulers and leads to nonlinearities in the log-log plots (see Figure 2, trace method). The dilation method is superior because it measures at every point on the border and, hence, generates more data.[8,10]

6.1.1.7 Mass method

The mass-oriented measure of D is different, but is similar to the box-counting method above (see Figure 3). Here one randomly centers boxes or circles (the results will be the same regardless of the shape used) of different sizes at many points along the border and counts the number of border pixels contained within the box or circle. Then, the log of the number of pixels within each box or circle is plotted against the log of some measure of the measuring element (edge size, diameter). A fractal object *chapter six: Fractal Studies of Neuronal and Glial Cellular Morphology*   177

gives a line with a positive slope, which is the D for that object. The power relationship plotted is

$$\mu(d) = A\, d^D \qquad (1)$$

where μ(d) is the number of pixels in a box of size d, d is the diameter or box-edge length, A is a variable, and D is the fractal dimension (see Figure 7).

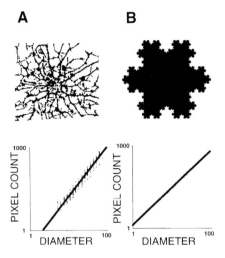

Figure 3 Measuring the fractal dimension, D, of (A) a cell-cultured oligodendrocyte and (B) a Koch triadic island of D = 1.26 by the mass-related measuring method. Below are log-log plots of pixel counts of the number of pixels in each box of diameter d vs. the diameters. See text.

6.1.1.8 Factors affecting D

The characteristics of cellular morphology that most influence the magnitude of D are the profuseness of branching and the ruggedness or roughness of the border, with increases in either leading to a larger D.[8] This means, of course, that two cells that look very different (one with a smooth border and many branches; the other with few branches and a rugged border) may have the same D (see Figure 4). This result emphasizes that, with such global measures, D provides no unique morphological specification.

The exact method one uses to determine the D does not depend on whether an image is self similar. That is determined by the log-log plots. The range of the linear slope of those plots identifies the range of self-similarity. In the work considered here, it is only the border that is fractal. The structureless interior is Euclidian, not fractal. One can show this by doing an analysis of a filled vs. a border of a Koch figure with known D. Only the border image gives the correct result (except for the trace method).

6.2   Uses of fractal geometry and D

Now that we have defined D and explained the methods of obtaining it, the question naturally arises as to whether knowledge of the magnitude of the dimension has any utility. The answers come from several experiments where it has been employed to quantify some aspects of cellular morphology.

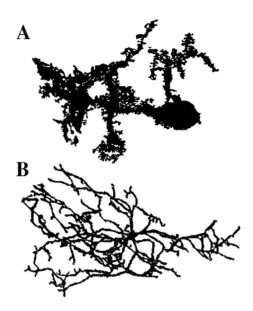

Figure 4 Two cell cultured neurons with the same D but with very different morphologies. (A) Cerebellar Purkinje cell. (B) Spinal cord cell. See text.

6.2.1 Quantification and classification

Some of the earliest uses of fractal geometry in the study of the cellular morphology of individual cells were simply to quantify their complexity and to see if the numbers agreed with intuitive estimates. The general results were deemed positive.[8,11] As can be seen in Figure 5, where the Ds of various glial cells are indicated by the verticle line from the silhouette to the D axis, as the cells become more complex, their D value increases.

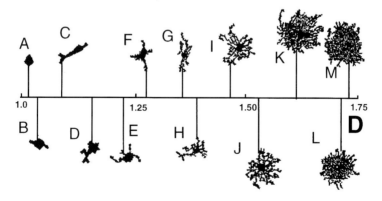

Figure 5 Silhouettes of cell-cultured glial cells. Lines connect each cell with its D value displayed on the D-axis. See text.

Caserta et. al.[12] compared the Ds of mature vs. immature retinal ganglion cells and suggested diffusion limited aggregation (DLA) as a possible mechanism for neuronal growth; DLA is traditionally associated with fractal growth (see below). Other studies of retinal ganglion cells suggested other mechanisms (L-systems growth[13]) for neuronal growth.[14-16] Both hypotheses have appealing features, but have yet to be verified by experiments. In two papers,[17,18] the property of self-similarity of certain cell types was demonstrated by optical magnification changes of up to tenfold.

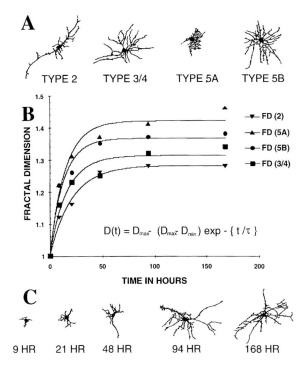

Figure 6 Data from cell-cultured spinal cord neurons at different stages and times of development. (A) Cells are classified on basis of number of primary dendrites emanating from cell soma (2, 3 or 4, 5) and on dendritic characteristics (5A and 5B). (C) Cells increase in size and complexity over time in culture. (B) Plot of the changes in complexity as a function of time in culture for the four cell types. Curves are derived from equation shown above the time axis in (B). See text.

6.2.2 Growth and differentiation

A significant use of the concepts and measures of fractal geometry has been in the area of the study of the morphological growth and differentiation of individual nerve and glial cells in cell culture. Several studies[10,19,20] have measured the Ds of individual cells from the time their progenitors were planted in culture dishes until they matured to some plateau value of complexity. This is illustrated in Figure 6. First, cells were identified on the basis of the number of primary dendrites emanating from their somas, namely two primary dendrites, three or four dendrites, and five or more denrites. Cells with five or more dendrites were further subdivided on the basis of their dendritic branching patterns (e.g., density, length, total branches; Figure 6A).

All cell types are virtual spheres when plated onto culture dishes, and, thus, when observed on a plane have a D near 1. As these cells grow and extend branches or dendrites, their D increases monotonically to some final or plateau value between 1 and 2 (see Figures 6B and C and Table 1). In all cell types, this monotonic growth often can be well fit by a simple mathematical model which employs the concept of a time-constant:

$$D(t) = D_{max} - (D_{max} - D_{min})\exp(-t/\tau), \qquad (2)$$

where $D(t)$ is D as a function of time, D_{max} is the asymptotic value at infinite time, D_{min} is the value (usually near 1) at zero time, t is time in culture, and τ is the time constant, as shown in Figure 6B. In a few cases, there has appeared to be two stages of growth, and a two-time-constant model has given a somewhat better fit (see Figure 8).[19] It is truly remarkable that so simple a model fits as much as a fourfold increase in complexity (D = 1.1 to 1.4). We should point out that each cell phenotype tends to have a characteristic range in both plateau value of D and time constant (see Table 1). These several papers also suggest that D is a useful quantitative measure of cellular morphological differentiation, which is the *Webster's Dictionary* definition of differentiation: "... a change from the simple to the complex".[21] In these studies, in which the plating densities were quite low and with little chance for intercellular communication, it was concluded that the growth and complexity determinants were largely intrinsic to the individual cells.[19]

Table 1 Maximum Fractal Dimensions (D_{max}) and Time Constants (τ) for Four Cell Types

Cell type	D_{max}	Time constant τ (hr)
Type 2	1.28	20.4
Type 3/4	1.32	16.4
Type 5A	1.41	14.5
Type 5B	1.37	12.6

The recognition that some cells differentiate faster or to a greater perceived complexity than others is not new. The significance of the results reviewed here is that they specify quantitatively how much faster or slower and how much more or less cells differ from one another in their complexity. For example, the ratios of time constants quantify the fractional rate differences, and the ratios of the fractional parts of the Ds quantify the complexity differences (e.g., a change from 0.1 to 0.4 represents an fourfold change in complexity). Moreover, these experiments provide a baseline for future attempts to quantify the effects of factors which are known to modify morphological growth and differentiation. Measurement of D can, in principle, be done on living cells. Therefore, in growth and differentiation experiments it may be particularly useful to measure D rather than depend on immunological markers to define stages of development.[10,18]

Some authors[22] have held that the magnitude of D in a growth process is related to the number of degrees of freedom involved in that process. If this is the case, then the increase in D during development implies that the number of degrees of freedom increases as morphologic differentiation proceeds and eventually reaches a plateau.

The question arises as to how these findings of changing Ds relate to the concept of fractal growth. Fractal growth usually refers to growth of some property (length, area, etc.) at constant D. Clearly, the growth discussed above is not fractal growth but is more likely related to something resembling the L-systems growth mentioned previously.[13] Indeed, if one generates deterministic fractals, such as Koch islands or snowflakes, with an L-systems regime and plots the increases in D as functions of growth iteration or stage, a curve is obtained that is analogous to that described by Equation (2) (see Figure 7).[20] Objects that have a D less than their final D are called prefractals.[1]

chapter six: Fractal Studies of Neuronal and Glial Cellular Morphology 181

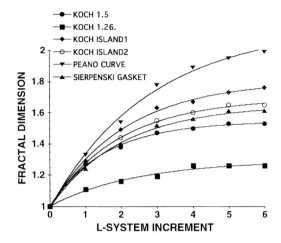

Figure 7 Plots of changes in D vs. L-systems iteration of several deterministic fractals. See text.

In some of these papers,[19] other more conventional cytological measures (membrane area, total dendritic length, etc.) were undertaken in parallel with the D measurements on the same cells (see Figure 8). Most often, these measures grew linearly with time and some continued to increase after D had reached a plateau. Presumably, these latter stages represent fractal growth.

In addition, Sholl analyses[18,19,23] were performed on skeletonized, individual cells. Sholl analysis measures the rate at which branches form minus the rate at which they terminate as a function of distance from the soma. Certain plots (semi-log or log-log) of these changes produce a Sholl coefficient, which correlates with D (see Figure 8). This is understandable since the Sholl coefficient relates to branching, as does D, but it is not sensitive to the border roughness. D, therefore, is thought to be a superior measure of complexity.[10,19,20] On the other hand, the two coefficients together might be of value in resolving the inability of D to distinguish between branching and border roughness, when reliable and consistent Sholl coefficients can be measured.[18,19]

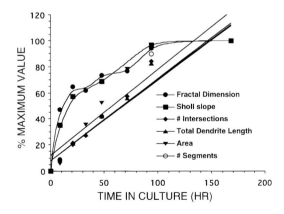

Figure 8 Plots of several measures (fractal dimension, Sholl slope, total dendritic length, cell area, etc.) as a function of time in culture. See text.

6.3 Other studies

6.3.1 Vertebrate central nervous system cortical pyramidal neurons

In one study,[17] it was found that the motor cortical pyramidal cells of the brains of cats were significantly more complex than those of the monkey. This somewhat surprising result was interpreted as follows. The size of the brain and the number of neurons in the cat are both smaller than those of the monkey. That paper suggested "... that the fractal dimension and the degree of morphological complexity may be related to the requirement for the number of separable functions to be accommodated within one neuron. For example, as the size of the cortex and the number of neurons in a region increase, the opportunity exists, within a given cortical zone, for individual functions to be segregated and for functional specialization to be accommodated with less morphological complexity of the neurons performing each of those functions."[17] The implication here is, that cat neurons, being more complex, integrate many functions in a kind of serial mode of information processing; whereas monkey neurons may be more specialized, with fewer functions, and operate in a parallel mode of information processing and perhaps at a faster rate.

6.3.2 Bergmann glial cells

Similar results were found when comparing the Ds of Bergmann glial cells (also known as Golgi epithelial cells) across three species in two mammalian orders: rat and monkey and man. There, the rank order was: rat > monkey > man. The smaller and phylogenetically "lower" animals had larger Ds[24] and may represent more complex functions for the smaller animals' glia; for example, more extensive spatial buffering of extracellular potassium ion concentration by individual cells.

6.3.3 Surface-to-volume ratio

A technically efficient use of D derives from its correlation with the surface-to-volume (S/V) ratios of individual glial cells across several species. The S/V ratio is important in quantitative studies, such as transmembrane flux experiments but is difficult to measure. Reichenbach et al.[25] found that, for any given cell type and structure, there was linear relation between the S/V ratio of a given cell type and its D. Thus, when the S/V ratio is required, the technically simpler and faster calculation of D can be measured and the S/V ratio can be read from a previously calibrated curve.

6.3.4 Miscellaneous research

There has been a number of papers in which there were "obvious" morphological differences in the results from different experimental causes and where D has been used to demonstrate quantitatively the magnitude of these differences and, sometimes, to illustrate their statistical differences. McKinnon et al.[26] showed that two growth factors produced statistically different morphologies when used on cultured glial cells . Siegel et al.[24] showed that D correlated with structural and functional differences in Bergmann glial cells in cerebella of various species. They also showed that the complexity of these cells increased progressively throughout life. Senitz et al.[27] used D in a clinical pathology study to compare human neocortical astroglial cells during development, aging, and dementia.

Finally, an important usage of D was a recent study of the human retina. For some time there had been a dispute over the number of types of horizontal cells in the human retina. A number of conventional measures (membrane surface, cell volume, soma diameter, etc.) did not resolve the issue. By the use of fractal analysis with other novel and conventional measures, Kolb et al.[28] showed that, in fact, there are three statistically different horizontal cell types.

6.4 Lacunarity and moments of mass distributions

As mentioned, a value of D does not uniquely specify a cellular morphology and very different looking cells can have the same or very similar Ds (see Figure 4). This is a general problem in fractal geometry. What is needed is at least one other measure that distinguishes such cells. Lacunarity is potentially such measure. Lacunarity is a measure of the lack of rotational or translational invariance or symmetry in an image. Lacunarity is a neologism from the Latin *lacuna* for lack, gap, or hole. Hence, a fractal is said to be lacunar if the gaps are wide, i.e., if they have large intervals. High lacunarity is suggested if the scatter of mass-related points around each disc size is large in the log-log plots (see Figure 3A). Mandelbrot[1] notes that the prefactor, A in Equation (1), is roughly inversely related to lacunarity. He notes, however, that this definition has limited validity and application.

For these studies of lacunarity the mass-oriented measure of Ds was used, and all the results discussed below were based on measures of this type. Some 60 cells from five cell types in nine different species were employed. In addition, one type had measurements taken at four time periods during growth. Originally, the variance for each cell at each circle size was calculated and was normalized by dividing its associated mean. Then, these normalized variances were averaged to obtain the number called the lacunarity. When the entire 60 cells lacunarity vs. their D were plotted, an unwanted correlation of –0.75 was found between the two measures (see Figure 9). This suggested that the two measures were not sufficiently independent to be distinctive.

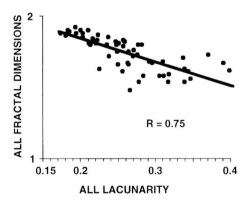

Figure 9 Plot of D vs. lacunarity of each cell for a number of cells from several different cell types. See text.

Taking a different approach, the original pixel counts (mass) were rescaled, using the power function relation.

$$\mu_{scaled} = (A-1)d^D \mu(d) - 1 \qquad (3)$$

where μ, A, d, and D have the meaning shown in Equation (1). This forces the mass distributions for the various measuring diameters (d) into a single mass distribution with a mean of zero. The variance and third moment of the distribution were calculated by standard methods.

The fractal dimension and the two moments were then taken as coordinates in a three-dimensional description of the shape of each cell. The pairwise (partial) correlations among the three parameters were D and second moment, 0.41; D and third moment, 0.22; second and third moment, 0.64. Pairwise plots of the parameters did not separate the points into recognizable cell groups. Cluster analysis, factor analysis, and other multivariate statistical techniques were then applied to this three-dimensional space. Principle component analysis led to two predominant eigenvalues with all three components contributing to the eigenvector associated with the largest eigenvalue. This would seem to indicate that the three measures do improve the resolution of the shapes of cells as opposed to using D alone. Inspection of the data does not, however, at this time, lead to revelation of new categories of cells. On the other hand, despite the lack of orthogonality, a close inspection of the data shows that lacunarity can often distinguish between two cells with the same length-related D, on a case by case basis. Thus, lacunarity can have some utility.

6.5 Discussion and conclusions

From the results and conclusions of previous research reported in this review, it should be abundantly clear that the concepts of fractal geometry and the use of the notion of a fractal dimension are helpful analytical tools for quantitative studies of the morphology of individual nerve and glial cells.

There is some concern, however, about the stability of D and τ values for any given cell type across experiments over a period of time, even when the cells or tissues are prepared in the same manner. In two reports[10,18] from the same laboratory and apparently following the same methods on the same preparation, the D and τ values were not exactly the same. This may reflect some (unkown) variability in the methods and/or measurements. Comparative studies of the effects of given factors may require simultaneously run controls to measure D in every experiment, unless it is found with identical techniques that D and τ are stable for each phenotype over a number of experiments and with time.

The notion that biological tree-like, fractal structures represent an optimal design for a particular function is generally accepted. For example, such ideas have been proposed for the lung for the flow of air[29] and the vascular bed for the flow of blood.[30] It may well be that the fractal, dendritic trees of neurons are also optimally designed, but in this case for the flow of their most important commodity: information.

Indeed, there is a considerable body of theory and data which indicates that some neurons are optimally structured for the integration of their graded input signals (synaptic potentials) which results in a pulsatile output signals (action potentials).[31-34] In cases where the organization is known, excitatory synaptic inputs are located mainly on distal dendritic branches, whereas inhibitory synaptic inputs are located on the cell body and proximal dendrites. Since the action potential-generating membrane is located at the axon hillock or the axon itself, the excitatory synaptic potentials must depolarize that membrane to a threshold value in order to generate an action potential. If, however, prevention of such action potentials is required, then a simultaneous increase in inhibitory conductances can shunt the excitatory currents flowing from the distal dendrites toward the axon hillock and block the generation of an action potential. It is not clear, at this time, how the physiology and D are related, but an analysis of substructure (e.g., dendrites vs. somata) at a higher magnification might lead to excitatory areas having a different D from inhibitory areas.

It already has been shown, for example, that dendritic arborizations have different Ds than axon terminal arborizations.[17]

In addition, several papers[24,25] involving fractal measurements along with other factors suggest that glial cells are optimally structured to perform one of their known functions, namely, the regulation of extracellular potassium ion concentration in the nervous system.

We believe that it is clear from the examples in this review that the use of fractal geometry in microscopic anatomy is now well established. It would seem that calculation of fractal dimensions and other related parameters are useful as quantitative morphological and developmental descriptors. Perhaps more promising from our point of view, is the possibility that they will be increasingly useful in establishing links between structure and function.

References

1. Mandelbrot, B. B., *The Fractal Geometry of Nature*, W. H. Freeman, New York, 1982.
2. Peitgen, H.-O. and Richter, P. H., *The Beauty of Fractals*, Springer-Verlag, New York, 1986.
3. Peitgen, H.-O. and Richter, P. H., *The Science of Fractal Images*, Springer-Verlag, New York, 1988.
4. Voss, R. F., Random fractal forgeries, in *Fundamental Algorithms for Computer Graphics*, Earnshaw, R. A., Ed., Springer-Verlag, Berlin, 1985, 805.
5. Marr, D. and Hildreth, E., Theory of edge detection, *Proc. Roy. Soc. Lond. B*, 207, 187, 1980.
6. Smith, T. G., Jr., Marks, W. B., Lange, G. D., Sheriff, W. H., Jr., and Neale, E. A., Edge detection in images using Marr-Hildreth filtering techniques, *J. Neurosci. Methods*, 26, 75, 1988.
7. Flook, A. G., The use of dilation logic on the quantimet to achieve fractal dimension characterisation of textured and structured profiles, *Powder Technol.*, 21, 295, 1978.
8. Smith, T. G., Jr., Marks, W. B., Lange, G. D., Sheriff, W. H., Jr., and Neale, E. A., A fractal analysis of cell images, *J. Neurosci. Methods*, 27, 173, 1989.
9. Baumann, G., Barth, A., and Nonnenmacher, T. F., Measuring fractal dimensions of cell contours: practical approaches and their limitations, in *Proc. First Intl. Symp. on Fractals in Biology and Medicine*, Nonnemacher, T. F., Losa, G. A., and Weible, E. R., Eds., Birkhäuser-Verlag, Basel, 1993, 182.
10. Smith, T. G., Jr., Behar, T. N., Lange, G. D., Marks, W. B., and Sheriff, W. H., Jr., A fractal analysis of cultured rat optic nerve glial growth and differentiation, *Neuroscience*, 41, 159, 1991.
11. Cutting, J. E. and Garvin, J. J., Fractal Curves and Complexity, *Percept. Psychophys.*, 42, 365, 1987.
12. Caserta, F., Stanley, H. E., Eldred, W. D., Dacord, G., Hausman, R. E., and Nittman, J., Physical mechanisms underlying neurite outgrowth: a quantitative analysis of neuronal shape, *Phys. Rev. Lett.*, 64, 95, 1990.
13. Prusinkiewicz, P. and Lindenmayer, A., *The Algorithmic Beauty of Plants*, Springer-Verlag, New York, 1990.
14. Pellionisz, A. J., Neural geometry: towards a fractal model of neurons, in *Models of Brain Function*, Coterill, R. M. J., Ed., Cambridge University Press, Cambridge, 1989, 453.
15. Montague, P. R. and Friedlander, M. J., Expression of an intrinsic growth strategy by mammalian retinal neurons, *Proc. Natl. Acad. Sci., U.S.A.*, 86, 7223, 1989.
16. Montague, P. R. and Friedlander, M. J., Morphogenesis and territorial coverage by isolated mammalian retinal ganglion cells, *J. Neurosci.*, 11, 1440, 1991.
17. Porter, R., Ghosh, S., Lange, G. D., and Smith, T. G., Jr., A fractal analysis of pyramidal neurons in mammalian motor cortex, *Neurosci. Lett.*, 130, 112, 1991.
18. Smith, T. G., Jr., and Behar, T. N., Comparative fractal analysis of cultured glia derived from optic nerve and brain demonstrate different rates of morphological differentiation, *Brain Res.*, 634, 181, 1994.

19. Neale, E. A., Bowers, L. M., and Smith, T. G., Jr., Early dendrite development in spinal cord cell cultures: a quantitative study, *J. Neurosci. Res.*, 34, 54, 1993.
20. Smith, T. G., Jr., and Neale, E. A., A fractal analysis of morphological differentiation of spinal cord neurons in cell culture, in *Proc. First Intl. Symp. on Fractals in Biology and Medicine*, Nonnemacher, T. F., Losa, G. A., and Weible, E. R., Eds., Birkhäuser-Verlag, Basel, 1993, 210.
21. *Webster's New Collegiate Dictionary*, G & C Marriam Company, Springfield, MA, 1976.
22. Mayer-Kress, G., Introductory remarks, in *Dimensions and Entropies in Chaotic Systems*, Mayer-Kress, G., Ed., Springer-Verlag, New York, 1985.
23. Sholl, D. A., Dendritic organization in the neurons of the visual and motor cortices of the cat, *J. Anat.*, 87, 387, 1953.
24. Siegel, A., Reichenbach, A., Hanke, S., Senitz, D., Brauer, K., and Smith, T. G., Jr., Comparative morphometry of Bergmann glial (Golgi epithelial) cells. A Golgi study, *Anat. Embryol. (Berlin)*, 183, 605, 1991.
25. Reichenbach, A., Siegel, A., Senitz, D., and Smith, T. G., Jr., A comparative fractal analysis of various mammalian astroglial cell types, *NeuroImage*, 1, 69, 1992.
26. McKinnon, R. D., Smith, C., Behar, T., Smith, T. G., Jr., and Dubois-Dalcq, M., Distinct effects of bFGF and PDGF on oligodendrocyte progenitor cells, *Glia*, 7, 245, 1993.
27. Senitz, D., Reichenbach, A., and Smith, T. G., Jr., Surface complexity of human neocortical astrocytic cells: changes with development, aging, and dimentia, *J. Brain Res., J. Hirnforschung*, 36, 531–537, 1995.
28. Kolb, H., Fernandez, E., Schouten, J., Ahnelt, P., Linberg, K. A., and Fisher, S. K., Are there three types of horizontal cell in the human retina?, *J. Comp. Neurol.*, 343, 370, 1994.
29. Bassingthwaighte, J. B., Physiological heterogeneity: fractals link determinism and randomness in structures and functions, *News Physiol. Sci.*, 3, 5, 1988.
30. West, B. J. and Goldberger, A. L., Physiology in Fractal Dimensions, *Am. Sci.*, 75, 354, 1987.
31. Rall, W., Burke, R. E., Smith, T. G., Jr., Nelson, P. G., and Frank, K., Dendritic location of synapses and possible mechanisms for the monosynaptic EPSP in motoneurons, *J. Neurophysiol.*, 30, 1169, 1967.
32. Jack, J. J. B., Noble, D., and Tsien, R. W., *Electric Current Flow in Excitable Cells*, Clarendon Press, Oxford, 1975.
33. Rall, W., Core conductor theory and cable properties of neurons, in *Handbook of Physiology*, Brookhart, J. M. et al., Eds., Bethesda, MD, 1977, 39.
34. Rall, W., Time constants and electrotonic length of membrane cylinders and neurons, *Biophys. J.*, 9, 1483, 1969.

Section IV

Tissues

chapter seven

Mosaic Pattern in Tissues from Chimeras

Mustafa Khokha and Philip M. Iannaccone

> *Surely it is evident that in living pattern and form, nature has provocatively concealed some essential underlying simplicity in an excess of ornament.*
>
> **L. G. Harrison**

When a mammalian embryo develops *in utero*, a few progenitor cells must grow and organize themselves into functional tissues if the organism is to survive. This process of tissue organization and growth is a critical and prerequisite step in the generation of an organ. A number of methods to gain an understanding of this process have been undertaken, and in this chapter we will briefly discuss a few and then discuss in detail how fractal geometry has increased the understanding of tissue organization in the mosaic pattern of chimeric rats. But first it will be necessary to discuss some aspects of developmental biology and the nature of mosaic pattern.

7.1 Mammalian development

Following the release of a mature egg, fertilization triggers embryonic development. Sperm must deliver its chromosomes to the egg in order to complete the genetic material necessary for development to begin. Fertilization in most mammalian species occurs in the ampulla (distal 1/3) of the oviduct. Sperm capacitate as they migrate to this location and through a series of steps acquire the ability to fertilize the mature egg. The sperm becomes hypermotile and releases proteolytic enzymes which allow it to gain access to the zona pellucida of the egg. The zona pellucida contains a number of complex glycoproteins, including, in the case of the mouse, ZP3 which binds sperm to the egg. Nearly instantaneously, the sperm head gains access to the oolema (egg cell membrane) and the ooplasm. Cortical degranulation of the ooplasm occurs and the egg is rendered impervious to fertilization by additional sperm, although more sperm may bind the egg. The midpiece of the sperm head binds the oolema in a fusion event and the nuclear material (highly condensed DNA) swells to form a male pronucleus. This structure migrates toward the center of the egg to meet the female pronucleus which formed following the second meiosis.[1] The end result is a cell called the zygote (Figure 1) that has acquired a haploid

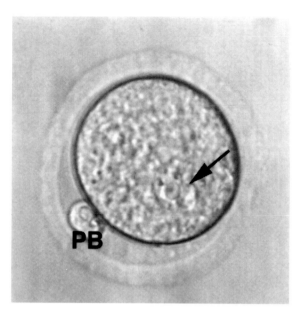

Figure 1 Photomicrograph of a living one-celled fertilized rat embryo. PB indicates a polar body which is enclosed by the zona pellucida; the arrow indicates the pronucleus of the cell. The embryo at this stage is approximately 100 µm in diameter.

set of paternal genes and a haploid set of maternal genes and is ready to begin the next step in development: cleavage.

The zygote heads toward multicellularity by first dividing its already massive cytoplasm (the fertilized mouse, rat, and human one-cell zygotes are about 100 µm in diameter) amongst replicating nuclei. The zygote does not grow in size in this period but rather rapidly replicates its genome and cleaves itself into smaller and smaller cells called *blastomeres*. The first division is in half, quarters, then eighths, until eight blastomeres are formed which form a solid aggregation of cells called a *morula*. The blastomeres compact and undergo further divisions until there are approximately 32 cells. A cavity forms in the mass of cells by a combination of programmed (apoptotic) cell death and active accumulation of fluid. This is the blastocyst stage of development and the zygotes at this stage comprise two populations of cells with different developmental potentials. The outside cells are known as *trophectoderm* cells and make extraembryonic membranes and elements of the chorio-allantoic placenta in the mouse.[2,3] The inside cells are called *inner cell mass* cells and they will develop into the fetus. Perhaps as few as three to four of these cells are required to give rise to all of the cells necessary to construct the entire mouse fetus.[4] The formation of these inner cell mass cells in the blastocyst from the compact morula is a critical step requiring cellular reorganization and is prerequisite to the next step in development: gastrulation.[5]

Once the inner cell mass is formed, the cells organize themselves into two sheets: the hypoblast and the epiblast. The hypoblast (or primitive endoderm) forms a sheet-like lining around the blastocyst cavity. The epiblast (or primitive ectoderm) is a group of cells that lie in contact with the hypoblast and will eventually form the fetus. After the blastocyst implants in the uterine wall, the primitive ectoderm proliferates and develops into two distinct populations: embryonic ectoderm and extra-embryonic ectoderm. The embryonic ectoderm of the mouse is totipotent,[6] capable of forming all tissues of the adult mouse including functional gametes. The embryonic ectoderm is surrounded by visceral endoderm at this stage, which in the mouse is about 6 days after fertilization. Embryonic ectoderm cells enter a region on the posterior aspect called the *primitive streak* and, as a result of specific signaling events, differentiate into mesodermal cells which grow and divide and then migrate to lateral aspects of the embryo. The process is a complex coordination of induction, which instructs cells to form into a particular germ cell layer, and cellular migration, which places cells into correct positions within the developing embryo.

With the establishment of the three germ layers — ectoderm, endoderm, and mesoderm — the embryo under genetic control undergoes a coordinated process of three cellular forces to form pattern. These forces are cellular division, cell movement, and cell death. Organogenesis, a critical developmental phase during which organs form, begins with the allocation of a few cells to form a primordium, and then growth, organization, and differentiation of those cells into functional tissues ensues. Pattern formation requires that genetic information be translated into cellular forces that produce particular forms in a regulated manner that is conserved from individual to individual.[7]

7.2 *Proposed mechanisms of pattern formation*

Positional information models of pattern formation suggest that in a field of cells pattern is determined by allocating a cell with a positional value from some reference and that this positional value then determines the fate of the cell in creating a pattern.[8] For example, in order to create a pattern of stripes in a section of tissue, first cells along one length of the tissue could be given incrementally increasing positional values indicating their distance from the edge of the tissue. In this way each cell would know how far it was from the reference edge of the tissue. Let us say that the tissue is assigned values from one to ten from one edge of the tissue to the other. Then starting from the reference edge where the positional values are low, a range of positional values (for example zero to one) could be given an instruction to change into one cell type (type A). Adjacent to these, another small range of positional value cells (from one to two) could be assigned the alternate cell type (type B). Then, in the next range from two to three, the first cell type (in this case type A) could be repeated, generating a stripe of alternate cell type (type B) between the two stripes of type A cells. Repeating this process would eventually produce a striped pattern across the field of cells.

A few aspects of this model of pattern formation are critical. First, there must be a way that cells can "know" what their position is. A commonly proposed mechanism is that a diffusible substance, a morphogen, is produced in such a manner that its concentration varies across the field in a manner that relates position. For example, in the striped pattern a source of morphogen being produced at the reference edge diffuses toward the other end and consequently decreases in concentration across the field of cells. The cells detect the concentration of morphogen and from that they can determine their positional value and consequently whether they should become a type A cell or type B. In this way, cells that may appear similar by histology or morphology are in fact nonequivalent. By manipulating the morphogen gradient, different patterns can be produced.[9-13] However, another important aspect of this model is that prior to the generation of positional information, the cells are capable of following either differentiation pathway. It is the process of generating positional information which then restricts the developmental potential and therefore creates pattern. For example, given a normal morphogen gradient, two outcomes are possible if a cell in a region that would normally have become type A were transplanted into a region that would normally produce type B. First, if the cell were transplanted before gaining positional information, then it would form a type B cell and acquire the positional information of the location to which it was transplanted.[14] On the other hand, if the cell already had obtained its positional cue and therefore became nonequivalent, it would become a type A cell despite its new location and positional value. In this way, by changing the positional values prior to the point of nonequivalence, different patterns may be generated. In our example this would entail altering the morphogen gradient along the length of the tissue.

Another theory of pattern formation, called the *kinetic theory*, is representative of a broad field which emphasizes that, through the processes of activation, inhibition, and communication, pattern which is correct both spatially and temporally may be generated. Kinetic theory is true to its name in that it attempts to explain how pattern may be generated from an unpatterned state in a dynamical way.[15] Kinetic theories of development imply the generation of form with a process that occurs over time rather than suddenly without due cause.

Although kinetic theory potentially encompasses a large number of specific ideas, reaction-diffusion theory has been the most extensively elaborated and referenced as an example of a kinetic theory of pattern formation. In fact, Turing[16] first described a reaction-diffusion model in 1952 to describe certain biological patterns including gastrulation. The form of this model is that two morphogens are found in a system, the concentrations of which are affected by the fact that local areas of increased concentration of either will tend to diminish due to diffusion and that the two morphogens react with each other and can alter each other's concentration at particular rates of reaction. By proper selection of certain reaction rate parameters, a nonhomogeneous pattern of morphogens results with distinctive similarities to biological patterns. How could the developing cells choose the right reaction rate parameters? Presumably, this would fall on the genome by the proper production of protein catalysts or other regulators of the reaction.

How does the reaction-diffusion model of kinetic theory then differ with a positional information model? After all, a positional information model which generated stripes as discussed above also used a morphogen to create pattern. In a positional information model, the determination of position is paramount. That is, a coordinate system must be developed which will include a location from which all other locations will be determined relatively (the edge of the tissue in the stripe example) and polarity, which will determine the direction in which position is measured. The cells then recognize where they are in this coordinate system and assume the correct form consequently. A diffusible substance is not even necessary. For example, we could have generated the same pattern of stripes if our tissue section grew from the reference edge outward so that once cells left the leading edge of the tissue they stopped dividing. In this way only cells at the outer edge of the tissue divide and grow. Position could have been determined then by how long the cell had been out of the growth region. Different ranges of time out of the growth region would determine type A or type B cells, and stripes would result.

On the other hand, in the kinetic model or more specifically in the reaction-diffusion model, the production of pattern by the interplay of morphogens is the critical aspect. It is inherent in Turing's definition of a morphogen as a "form producer". A certain number of morphogens react with each other, altering their respective concentrations, and those concentrations are also dependent on the processes of diffusion. The culmination of all of these interactions produces a pattern that is not homogeneously distributed, but in fact resembles biological pattern.[17] The cells interpret their surrounding local concentration milieu of morphogens and alter their state to form the necessary pattern. In this way, if we imagine a source of signaling substance and a local sink which together produce a gradient of this substance, this substance would be a perfectly fine positional information morphogen since it could indicate position from a reference along a certain direction between source and sink. However, it would fail as a reaction-diffusion morphogen since the generation of the pattern is dependent on the proper localization of source and sink. The substance by itself is incomplete to produce pattern, and it requires specification of the manner of localization of source and sink before we can consider it as a reaction-diffusion model.

Another theory of pattern formation which can be classified as a kinetic theory is the mechanochemical model. Unlike in the reaction-diffusion model where a chemical pattern is first generated and then cells interpret their signaling environment followed by appropriate morphogenesis with differentiation and cell movement, in the mechanochemical model these steps of pattern formation and morphogenesis are inseparable processes interacting continuously. In the mechanochemical model a chemical signal may cause cells to move toward the source of this signal and therefore alter that signal either by increasing the signal (thus being autocatalytic) or by decreasing the effectiveness of the signal by depleting the surrounding area of cells. A number of mechanical processes are possible. For example, there may be chemotaxis whereby cells migrate toward a chemical attractant arrayed in the substratum.[18-22] Alternatively, in haptotaxis cells move along an adhesive gradient, or the cells may move by convection where the cell is moved passively either by transportation due to other cells or by forces generated within the extracellular matrix when deformed by other cells or osmotic changes. These cellular mechanics can be measured by actually measuring cellular forces, watching cellular movements and rearrangements, and determining cellular densities. In each of these cases, chemical signals coordinate with cellular mechanics to produce pattern represented by cellular differentiation, cell shape changes, and cell movement. The process is an interactive one where cellular mechanics and cellular morphology affect the chemical signal, and the altered chemical signal in turn affects cell mechanics and morphology. The patterns obtained by either reaction-diffusion or mechanochemical means can be very similar, although the method of producing those patterns may be very different. Both models require some prepattern in order to begin the patterning process and ultimately morphogensis, but the reaction-diffusion models require a much more elaborate prepattern than mechanochemical models.

7.3 Mosaic pattern

There is a twofold purpose to the discussion of models of pattern formation before any mention of fractal geometry. The first purpose is to create a sense of familiarity with current general theories of pattern formation in development and the background in which a fractal model will be introduced. The second is to show how the fractal model in certain cases proves to be superior in describing the resultant pattern. Edelman[7] defines pattern formation as "processes by which ordered arrangements of cells or their products are attained." In order to illustrate the fractal model, we need first to describe one way of examining some aspects of pattern formation, namely mosaic pattern analysis.[23-25] Then, by way of this example, we shall show how fractal geometry may prove helpful in understanding the biology of development and pattern formation.

Mosaic pattern can be visualized in the organs of chimeras. Chimeras are produced by first generating at least two strains of animals that differ in some cellular marker. Then, embryos from the two strains are harvested and cells from one strain are amalgamated with the cells of an embryo from the other strain (Figure 2). The resulting embryo is then composed of cells from both strains. This embryo is allowed to fully grow into an adult which also has cells from both strains distributed within

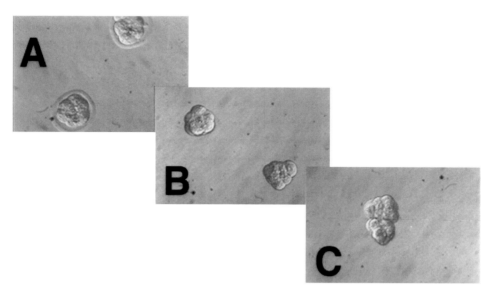

Figure 2 Photomicrograph of living rat cleavage stage embryos. (A) Eight-cell embryos with zonae pellucidae. (B) Eight-cell embryos after zonae a pellucidae are removed. (C) The eight-celled embryos in B are apposed to form an embryo chimera. This amalgamated embryo is transferred to the uterus of a surrogate mother, and the resulting offspring is a chimera whose tissues comprise progeny cells from both embryos. All embryos are approximately 100 µm in diameter.

its tissues, and thus its tissues are mosaic composites comprising progeny cells from the genetically distinguishable strains.[26,27] By various histochemical staining techniques, the cells of each strain can be localized within the tissue. The distribution of the distinguishable cells results in a histologic image which is called *mosaic pattern*. It consists of patches which are contiguous areas of like cell type bordered by cells of the other cell type.

We must ask at this point whether studying this mosaic pattern is useful in attempting to understand development. Mosaic pattern is both conserved and regulated in the sense that the patterns in a single organ (such as the liver) from many different chimeras appear similar (conserved), while the pattern in a given individual will vary from organ to organ (regulated). Therefore, livers from different chimeric rats would all have mosaic patterns that are similar to each other (Figure 3), and different adrenal glands from different chimeras would also have mosaic patterns similar to each other. The mosaic pattern in the adrenal gland, on the other hand, is different from that seen in the liver in every individual chimera examined (Figure 4). We argue that, since mosaic pattern possesses the properties of regulation and conservation, the processes that form mosaic pattern, i.e., cell division, cell movement, and cell death, must also be regulated and of consequence to normal development.[25]

It has also been shown[28] that the patches which comprise the mosaic pattern of the livers of rat chimeras are fractal (Figure 5). This was done by measuring the perimeter and the area of the patches observed in livers from rat chimeras at several different magnifications. The log of the measurement (uncorrected for scale) was plotted against the log of $1/m$ (mismagnification), and the scale of observation and fractal dimensions (Hausdorff-Besicovitch dimension) were determined. All of the dozens of patches examined were shown to be fractal objects. Because of the properties of self-similarity and scaling which fractals possess (see Chapters 1, 8, and 11), the possible mechanisms that could produce a fractal mosaic pattern are constrained.

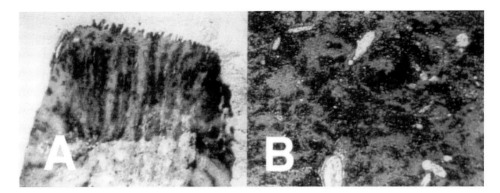

Figure 3 Photomicrograph of frozen sections of adrenal gland (A) and liver (B) from a rat chimera. Progeny cells of one parental lineage (from one of the two embryos in Figure 2) are stained black, while progeny cells of the other parental lineage (from the other of the two embryos in Figure 2) are unstained. The patches of stained cells in the liver look like islands in a sea of the unstained cells, while patches in the adrenal cortex are contiguous stripes.

The mosaic pattern can be considered a mere footprint of the processes that are responsible for its formation: cell proliferation, cell movement, and cell death. So models of the generation of fractal mosaic pattern are models of those processes. If we consider the models of pattern formation addressed earlier, limitations in those models with respect to constraints on patch formation become obvious. Fractal geometry displays tremendous complexity, with details nested within details. In order to determine the exact location for cell placement such that fractal patches could be obtained with a positional information model would require a great deal of genomic information. Moreover, as cells divide, any error in the placement of a cell from its correct position would be multiplied through succeeding generations, requiring an error-correction mechanism in order to maintain robustness. In the case of the reaction-diffusion model, generating a chemical prepattern with the necessary complexity that can withstand perturbations in the environment by any sort of convective current would be difficult. The mechanochemical model would not require such an elaborate prepattern but would require complex cellular movements and the necessary expenditure of energy to perform such movements in a coordinated way. A fractal model of mosaic pattern formation eliminates many of these difficulties with its relative simplicity and consequent robustness. Before describing this model we review how well described mathematical fractals develop in order to make the parallels with the biological fractals more obvious.

Fractals are made by an iterative set of rules applied to a set of initial conditions. Despite the resultant complexity of a fractal manifested in its self-similarity and scaling, generating a fractal need not require either complex rules nor a complex set of initial conditions. For example, in order to create the triadic Koch curve, the initial condition is simply a line segment (see Figure 6). The generator, the rule applied repetitively, is to divide every line segment into thirds and form an equilateral triangle in the middle third of each line segment. This generator is applied time and time again, and each new line segment formed from creating a triangle in the middle third becomes a new line segment for the next application of the generator. Therefore, at iteration $t = 0$ we have our initial condition of a single line segment. At $t = 1$, the generator rule is applied once and the middle third of the line segment is bent into an equilateral triangle. The curve is now composed of four equally long parts. Two parts of the curve at the ends are line segments unchanged from the original line segment, and two parts of the curve in the middle are formed by the process of the

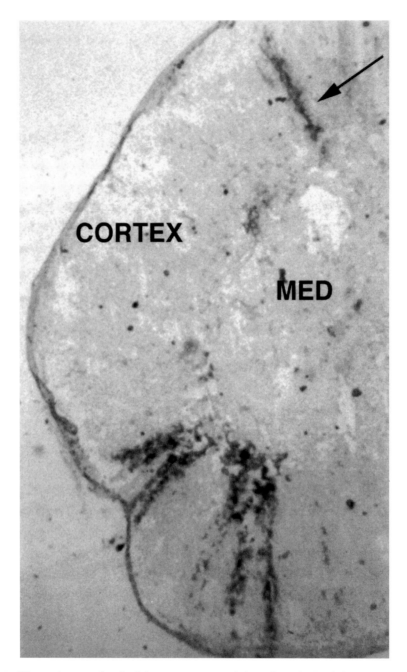

Figure 4 Photomicrograph of of frozen sections of the adrenal gland from a rat chimera, prepared as described in Figure 3. Notice that the patches are stripes even when the chimera is highly unbalanced (i.e., most of the tissue is derived from the parental lineage which does not stain). Arrow indicates one such stripe. MED indicates the medulla of the gland which does not stain in either lineage; CORTEX indicates the cortex of the gland.

generator. Then at t = 2, the generator can be applied again this time to four line segments each of which is divided into thirds, and the middle third is bent into an equilateral triangle. The result is 16 line segments. At t = 3, each of these line segments is again acted upon by the generator and the process is repeated in this manner an infinite number of times, generating a fractal (Figure 6). Therefore, when t = ∞, the

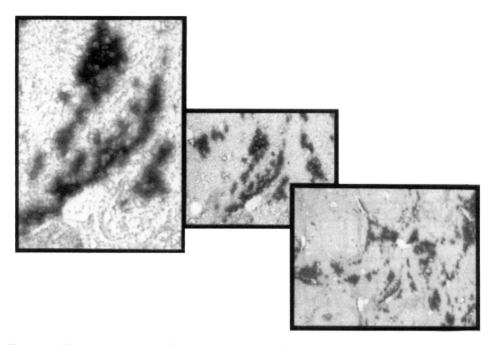

Figure 5 Photomicrographs of frozen sections of the liver from a rat chimera at three different magnifications (lowest magnification at lower right, highest magnification at upper left). Notice that as the magnification, and hence scale of observation, increases more and more detail is revealed. In these patches we demonstrated that more detail (i.e., more perimeter and more area) is revealed than predicted by lower magnification observation. These patches are fractal.

curve has an infinite number of line segments and therefore an infinite length and exhibits properties of scaling and self-similarity. To the mathematician, it is this final curve that is so fascinating because of these properties. Further, the triadic Koch curve is a fractal with infinite length that we can draw a circle around and contain in a finite area. However, to the developmental biologist it is not just the beauty of the curve, but also how it was formed that is so interesting. It required only a simple set of initial conditions and a single generator rule to produce the rich characteristics of detail.

7.4 Models of organogenesis

With the knowledge that the patches in the mosaic pattern of the chimeric rat liver are fractal,[28] it seems reasonable to postulate that an iterative set of rules applied to a set of initial conditions could also produce the mosaic pattern seen in the chimeric rat liver. Theoretical evidence suggesting that this postulate may be true came from computer simulation studies of mosaic fields. In these studies, cells in an anlagen were randomly chosen to divide, and then the daughter cells were placed in a random orientation. Appropriate cells were shuffled in order to make room for the newly generated cells. The simulations showed that complex geometric patterns could be generated using only this iterated program of cellular division. One study also found that the mosaic fields generated exhibited fractal properties similar to those seen in biological tissue.[29,33] Thus, the generation of mosaic pattern in the liver need not require cell movement or cell death.

It seems reasonable, therefore, to postulate that an iterated program of cellular division could generate mosaic pattern. The initial condition would be the anlagen

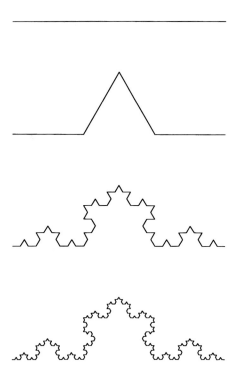

Figure 6 Koch curve generated as described in the text. The top figure is the intiator, next is the effect of applying the generator, next is after several applications of the genrator, and bottom is after many applications of the generator.

of cells that are the organ primordia, and the simplest of rules would be an iterative program of cellular division with random placement of daughter cells after cell division. A cell in the anlagen would be randomly chosen to divide and its daughter cells would be placed in an orientation chosen at random. Cells around the dividing cell would be shuffled in a manner that would provide room for the newly generated daughter cells. In this way, each cell in the anlagen would be chosen to divide, daughter cells would be placed in a random orientation, and then surrounding cells would be shuffled in order to generate the necessary room (Figure 7).

Initially, small patches would form as cells of the same parental lineage are positioned contiguously. These small patches would tend to grow and become larger because of random placement of daughter cells after cellular division. Eventually, however, cells from the other parental lineage would occur in the patch, also because of the random placement of daughter cells, causing the patch to fragment. This force would oppose the outward expansion of the patch. However, separate patches could aggregate forming a single patch. Eventually these forces would balance and the patches would oscillate, maintaining a relatively fixed average size.[30]

This oscillatory behavior also would constrain certain measures of the geometric complexity of the patches as determined by fractal dimension. By slicing into the organ, the mosaic pattern can be visualized in two dimensions. Also the proportion of tissue that is of one marker type can be quantified and then related to the fractal dimensions to be measured. Three different fractal dimensions have been measured. First, the surface fractal dimension of a patch can be measured by various methods including the yardstick method, the box-counting method, and the dilation method of Minkowsky-Bouligand (see Chapter 8). All of these techniques apply to the perimeters of the patch that is cross-sectioned. Second, the mass fractal dimension can be

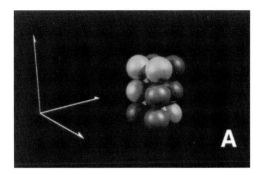

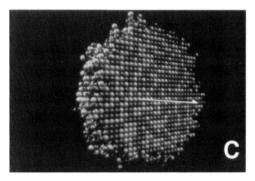

Figure 7 Computer simulation of the generation of mosaic pattern. (A) Red and green cells are mixed in an intial condition simulating the allocation of organ primordial cells (anlagen). (B) Following several applications of a generator rule set: cells to divide are choosen randomly and the daughter cell is placed in a random adjacent position forcing cells in those positions (if present) to move. (C) After ten "generations" or applications of the rule set. The organ parenchyma mass has been cut in half revealing in section the mosaic pattern as it would be seen in sectioned tissue from the liver of a rat chimera.

calculated by box-counting over a section of mosaic tissue where each box that contained any marked tissue would be counted. Finally, the fractal fragmentation can be determined by measuring the sizes of each of the patches in the section and determining the number of patches that are greater than a certain size.[31] A brief description of the methodology follows. Mass fractal dimension was estimated using the box-counting method. A grid with boxes of different sizes ranging from 1 to 128 pixels was superimposed on the image and the number of boxes that contained any part of a marked patch was counted. The log of this number divided by the log of the size of a side of the box gives the mass fractal dimension. The value is obtained from the slope of the regression line of the two log values. The surface fractal dimension was determined in three different methods. First, the box-counting method described above was applied to only the outline of the stained patches; second, the length of the perimeter of stained patches was determined using "yardsticks" of different lengths, and the dimension was calculated as 1 – log of the number of yardsticks of a given length divided by the log of that length; and, third, the perimeter of the largest stained patch in each image was determined using discs of varying area since the perimeter length equals the area of dilated outline divided by 2 × the radius of the disc. The dimensions were determined as above. There was excellent agreement between these methods. Fractal fragmentation was calculated as –log of the number of complete marked patches larger than a given size divided by the log of that size.[31]

Considering the patch dynamics postulated as above, there are certain predictions that we can make about the fractal dimensions of the mosaic patch pattern. Since the patch is oscillating in time with the force to grow outward, balanced by the incursion force of the other cell type, the surface fractal dimension should reach a constant value independent of the proportion of the section that is of that patch cell type. The patch oscillations are dependent only on the iterative program of cellular division and therefore independent of proportion of cell type. As the cells are "unaware" of these differences, the oscillatory dynamic will occur despite a varying proportion of cell type, and surface fractal dimension remains independent of proportion.[31] We have previously demonstrated, though, that the patch sizes and the distribution of patch sizes is dependent on the proportion of the two cell types in the mosaic.[25,29]

On the other hand, the mass fractal dimension will vary directly and asymptotically with proportion. This occurs because as the section is filled with more and more of the geometrically complex patch pattern, the mass fractal dimension continues to increase until it reaches the maximum value of $D = 2$, which is the dimension of a plane and the upper limit. The curve rises steadily because the geometric complexity of the surface of the patches does not vary with proportion. The surface fractal dimension varies only with the patch dynamic, which remains constant despite changes in proportion. However, with increasing proportion, more of the section is steadily filled with geometrically complex surface, causing the mass fractal dimension to rise steadily rather than in some other nonuniform manner that might be expected if the surface fractal dimension did indeed change with proportion.

Finally, fractal fragmentation should occur if an iterative program of cellular division with random placement of daughter cells is responsible for organ parenchyma growth. Due only to adjacent placement of daughter cells, patches that were detached can aggregate while a single patch can fragment. These two processes lead to a size distribution of patches that exhibits fractal fragmentation. However, the fractal fragmentation should occur only at proportions that are within a specific range of values. At a very low proportion, there are too few patches to have any fractal fragmentation. As the proportion increases, more patches are formed and fractal fragmentation develops, but at some critical proportion all of the patches coalesce into one large single patch. It is at this point that any fragmentation of a patch would rapidly lead to interconnection with another. Fractal fragmentation would cease to exist at this proportion.

7.5 Mosaic pattern in chimeras

Biological studies have been undertaken to attempt to confirm these predictions of an iterative program of cellular division. By performing a partial hepatectomy and studying liver regeneration, the surface fractal dimension of the patches can be followed with time. Analysis showed that even despite clear signs of growth and cellular division the surface fractal dimension remained constant, which is predicted from the oscillatory patch dynamic.[30] By measuring many surface fractal dimensions with varying proportions of the marked cell type in the chimeric rat liver, it was shown that the surface fractal dimension does not change with proportion, again consistent with prediction. In the same study of chimeric rat liver, mass fractal dimension was also shown to vary smoothly with proportion approaching an asymptote of two as expected. Finally, the patches in the chimeric rat liver were shown to be fractally fragmented as postulated by the theory (Figure 8). However, these data in the chimeric rat liver do not exhibit the behavior of a loss of fractal fragmentation at very low and very high proportions as expected.[31] One possible explanation is that these studies were of sectioned tissue and therefore two-dimensional in nature while the patches are really three-dimensional. Moreover, it may be possible that the data examined may not be sufficiently unbalanced for this behavior to be exhibited, that is, an insufficient range of data were examined.

Nevertheless, it seems that a good deal of evidence exists supporting an iterative program of cellular division with random placement of daughter cells. It is important to note that not all growth models will give these same results. For example, a growth model in which cells grow primarily at the periphery so that cells surrounded by other cells are inhibited from growing (a form of contact inhibition) would not exhibit such behavior. In this case, growth occurs primarily at the periphery, so invasion of cells into neighboring patches is inhibited. Fractal fragmentation would not exist in this case, and the surface fractal dimension as well as the mass fractal dimension would vary with proportion. This would occur since only the edge of the patch has geometric complexity and the probability of observing an edge would vary with proportion.

It seems then that a fractal model of parenchyma expansion with an iterative program of cellular division and random placement of daughter cells is a plausible model for the development of rat liver mass. As has been mentioned before, the mosaic pattern is regulated so that different organs express different mosaic patterns. For example, the mosaic pattern of the adrenal gland is very different from that seen in the liver. The liver has patches which form "islands in a sea", while the patches in the adrenal gland radiate from the periphery into the center.[32] A possible explanation for this pattern would require only a slight modification in our fractal model suggested by computer simulations.[33] Rather than having random orientations of daughter cells after cell division, orientation of those daughter cells in a direction toward the center of the organ could produce the striped pattern seen in the adrenal gland. In fact, studies in yeast have shown that specific genes exist which specify the orientation at which buds form,[34] and analogous mammalian genes may also exist. Bias in daughter cell placement may be the critical element explaining the difference in the mosaic pattern of the adrenal gland and liver.[33,35,36]

7.6 Implications

While this fractal cell division model is only one possible way to explain the mosaic patterns, it has certain features that make it quite attractive. First, it easily explains how an extraordinarily complex pattern such as mosaic pattern can be generated by very simple rules. In fact, such rules could be encoded very easily into the genome, requiring few additional genes dedicated to this aspect of development. It also suggests that by slight modification of the cell division rules very different mosaic patterns are possible, suggesting that the differences in patterns between organs may be related by very small changes in genomic information which cause

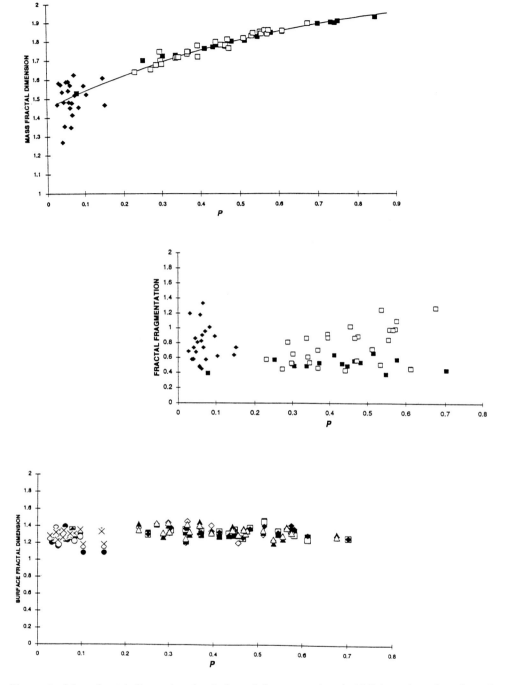

Figure 8 Mass fractal dimension (top), fractal fragmentation (middle), and surface fractal dimension (bottom), as defined in the text, determined from sections of the liver of rat chimeras prepared to reveal the parental lineage of the cells in the tissue. As the proportion of stained cells (PVG-RT1[a]) increases, the mass fractal dimension but not the surface fractal dimension increases asymptotically toward 2.

slightly different iterative programs of cellular division. In the light of the other models of pattern formation such as positional information and reaction-diffusion theory, the fractal model seems plausible from the standpoint of efficiency in genetic information. Since the mosaic patterns are regulated and conserved, it is likely that the forces that produce them are important to normal development. Obviously, the mosaic pattern itself is of no consequence to the organism since most individuals are not mosaic (although eutherian female mammals are mosaic as a consequence of X-chromosome inactivation).[37] Presumably the manner in which the cells divide has consequences in adrenal gland function, but it may not in the liver, while specific areas of cell death are critical to correct functional organization in the brain.

Another advantage of the fractal model is that it can be quite robust. Small errors in the placement of daughter cells will not be amplified into catastrophic errors. This is in sharp contrast to the reaction-diffusion model where even small convective currents could destroy the necessary pattern. Even in the mechanochemical theory, the misdirection of a motile cell could lead to misdirection of many of its progeny and companions, although these errors might be corrected by resetting the developmental potential of a misplaced cell to that appropriate for the particular microenvironment it finds itself in. It is probable, though, that a combination of these model's features is used at various times through development with iterative application of stereotypical cell division rules most important in the expansion of parenchyma mass during organ development.

7.7 Future studies

In order to generate further evidence for a fractal model of mosaic pattern formation, a number of studies could be performed. First, three-dimensional data should be collected, perhaps with ultra-high resolution magnetic resonance imaging (MRI) or with confocal microscopy. High resolution MRI would have the significant advantage of allowing for dynamic *in vivo* studies to be performed given appropriate cellular-bound MRI probes. Patch dynamics could actually be witnessed, and evidence for any cellular movement could be assessed in precise and accurate measurements.

Another avenue of further study would be to analyze cell divisions of both liver and adrenal cells. Measurements of any cellular orientation after cell division could then be assessed as a possible mechanism for differences in mosaic pattern between the liver and adrenal gland. The studies could be done in two and three dimensions using culture plates or matrigel, the latter to encourage three-dimensional growth to better simulate the natural environment of organ primordia. If the placement of daughter cells following division is a critical element of the normal development of an organ, it should be possible to identify the genes that regulate this element.

Finally, further mathematical study should be focused on the dynamics of patch formation toward a better understanding of the oscillations in the patch. Recent mathematical studies have shown that stable structures often exhibit chaotic behavior by settling in strange attractors. The patch may show extreme sensitivity to initial conditions in the dynamics that it shows but still form stable structures due to the nature of the attractor. Understanding such behavior would significantly raise our level of understanding of parenchyma expansion and provide clues to mechanisms important in early phases of visceral organogenesis.

Acknowledgment

The authors thank Stephen Iannaccone for preparation of the illustrations.

References

1. Iannaccone, P. M., in *Principals and Practice of Endocrinology*, Becker, K., Ed., J.B. Lippincott, Philadelphia, 1990, 880.
2. Costantini, F., Hogan, B., and Lacy, E., *Manipulating the Mouse Embryo*, Cold Spring Harbor Press, Cold Spring Harbor, NY, 1985.
3. Kaufman, M. H., *The Atlas of Mouse Development*, Academic Press, New York, 1992.
4. Markert, C. L. and Petters, R. M., *Science*, 202, 56, 1978.
5. Beddington, R. S., Smith, J. C., *Curr. Opinion Genet. Dev.*, 3, 655, 1993.
6. Gardner, R. L., *Philos. Trans. Royal Soc. Lond. [Biol.]*, 312, 163, 1985.
7. Edelman, G., *Topobiology*, Basic Books, New York, 1988.
8. Held, L. I., *Models for Embryonic Periodicity*, S. Karger, Basel, 1992.
9. Bodenstein, L., *Cell Differ.*, 19, 19, 1986.
10. Bodenstein, L., Sidman, R., *Dev. Biol.*, 121, 205, 1987.
11. Lewis, J. H. and Wolpert, L., *J. Theor. Biol.*, 62, 479, 1976.
12. Wolpert, L., *Development*, Suppl. 3, 1989.
13. Wolpert, L. and Hornbruch, A., *J. Embryol. Exp. Morphol.*, 63, 145, 1981.
14. Serbedzija, G. N., Bronner-Fraser, M., and Fraser, S. E., *Development*, 116, 297, 1992.
15. Harrison, L. G., *Kinetic Theory of Living Pattern*, Cambridge University Press, London, 1993.
16. Turing, A. M., *Phil. Trans. Royal Soc. Lond. (Part B)*, 237, 37, 1952.
17. Meinhardt, H., *Development*, Suppl. 169, 1989.
18. Oster, G. F., Murray, J. D., and Harris, A. K., *J. Embryol. Exp. Morphol.*, 78, 83, 1983.
19. Wolpert, L., Macpherson, I., and Todd, I., *Nature*, 223, 512, 1969.
20. Perelson, A. S., Maini, P. K., Murray, J. D., Hyman, J. M., and Oster, G. F., *J. Math. Biol.*, 24, 525, 1986.
21. Oster, G. F., Murray, J. D., and Maini, P. K., *J. Embryol. Exp. Morphol.*, 89, 93, 1985.
22. Murray, J. D. and Oster, G. F., *J. Math. Biol.*, 19, 265, 1984.
23. Weinberg, W. C., Howard, J. C., and Iannaccone, P. M., *Science*, 227, 524, 1985.
24. Iannaccone, P. M., Howard, J. C., Weinberg, W. C., Berkwits, L., and Deamant, F. D., in *Banbury Report 26: Developmental Toxicology: Mechanism and Risk*, McLachlan, J. M., Pratt, J. M., and Markert, C. L., Eds., Cold Spring Harbor Laboratory, NY, 73, 1987, 73.
25. Ng, Y. K. and Iannaccone, P. M., in *Current Topics in Developmental Biology*, vol. 27, Pedersen, R. A., Ed., Academic Press, New York, 1992, 235.
26. Weinberg, W. C., Ng, Y. K., and Iannaccone, P. M., in *The Role of Cell Types in Hepatocarcinogenesis*, Sirica, A. E., Ed., CRC Press, Boca Raton, FL, 1992, 29.
27. Papaioannou, V. E. and Dieterlen-Lievre, F., in *Chimeras in Developmental Biology*, Le Douarin, N. and McLaren, A., Eds., Academic Press, New York, 1984, 3.
28. Iannaccone, P. M., *FASEB J.*, 4, 1508, 1990.
29. Iannaccone, P. M., Weinberg, W. C., and Berkwits, L., *Development*, 99, 187, 1987.
30. Ng, Y. K. and Iannaccone, P. M., Fractal geometry of mosaic pattern demonstrates liver regeneration is a self-similar process, *Dev. Biol.*, 151, 419, 1992.
31. Khokha, M. K., Landini, G., and Iannaccone, P. M., Fractal geometry in rat chimeras demonstrates that repetitive cell division programs may generate liver parenchyma, *Dev. Biol.*, 165, 545, 1994.
32. Iannaccone, P. M. and Weinberg, W. C., *J. Exp. Zool.*, 243, 217, 1987.
33. Iannaccone, P. M., Lindsay, J., Berkwits, L., Lescinsky, G., DeFanti, T., and Lunde, A., in *Biomedical Modeling and Simulation*, Eisenfeld, J., Witten, J., and Levine, D. S., Eds., Elsevier, Amsterdam, 1992, 27–34.
34. Chant, J. and Herskowitz, I., *Cell*, 65, 1203, 1991.
35. Iannaccone, P. M., Joglar, J., Berkwits, L., Lindsay, J., and Lunde, A., *J. Theor. Biol.*, 141, 363, 1989.
36. Bodenstein, L., *Cell Differ.*, 19, 19, 1986.
37. Brown, S. D., Avner, P., Chapman, V. M., Hamvas, R. M., Herman, G. E., *Mammalian Genome.*, 1, S318, 1991.

chapter eight

Applications of Fractal Geometry in Pathology

Gabriel Landini

> *Nature forms patterns. Some are orderly in space but disorderly in time, others orderly in time but disorderly in space. Some patterns are fractal, exhibiting structures self-similar in scale. Others give rise to steady states or oscillating ones.*
>
> **James Gleick**

8.1 Current problems in histopathology

According to *Churchill's Illustrated Medical Dictionary*,[1] histopathology is "the study of the structural alterations of cells and tissues caused by disease". Although this defines histopathology, the purpose of such work is the interpretation of histological images plus clinical and laboratory data in order to reach a definite diagnosis, which in turn helps in the choice of appropriate treatment and may provide insight into the prognosis of the disease. Diagnosis is based on the assumption that any lesion can be categorized as an already known disease or process. Unfortunately, many of these categories are "open ended"; new relations between histological manifestations and clinical course of a disease or process are constantly drawn, and new markers of disease (such as cell proliferation markers or differentiation markers) are discovered which help to confirm, modify, or reject the established categories. In addition to these "ever changing" concepts, histopathological diagnosis suffers from a further technical problem. The diagnosis depends on the histopathologists' experience: it is not a simple procedure, but rather is a combination of subjective and expert knowledge of clinical data analysis, logic, image understanding, and pattern recognition. Time after time, this process has been shown to be poorly reproducible inter- and intra-observer and has prompted using the power of computers and statistics in order to increase sensitivity and accuracy, for example, of image characterization.

In cancer diagnosis, normality and frank malignancy are usually relatively easily discernible, but in "premalignant" conditions morphological changes are more subtle than in frank malignancy. Mathematical morphology, stereology, artificial intelligence, and digital image processing techniques have been applied with variable degrees of success to some problems in histopathology and have contributed to the creation of a relatively new branch of science known as analytical and quantitative microscopy. In general terms, images are captured by a video camera attached to a microscope and divided into small picture elements (or pixels), and the light intensity (grey scale) or color values are assigned a number (digitized). After this procedure, the images can be subjected to mathematical operations and statistical tests to extract information (such as number of objects, sizes, and optical density values) systematically and objectively. Despite the simplicity of this explanation, the main difference between the human and the machine approach is that while the observer uses extra-image knowledge (a metalanguage and a higher hierarchical knowledge) — i.e., clinical data, interpretation of the dynamic processes involved, the recognition of structural components: molecules, cells, fibers, tissues — computer-aided microscopy has to rely on relatively low-level information from the image by quantifying changes in color, grey-scale density, texture, shape, and size. The trade-off has the benefit of reproducibility, speed, and automation, but for the reasons outlined above, it is unlikely, at least in the near future, that fully automated expert imaging systems in histopathology will replace humans. Nevertheless, analytical and quantitative microscopy is a reality that is rapidly expanding. Regarding image analysis, some problems still remain in areas such as the characterization of irregular textures and shapes because the methodology has been based on classical geometry which cannot cope satisfactorily with geometric complexity. It is in this particular area, complexity assessment, that fractal geometry has a promising future.

8.2 More than 1 but less than 2

In classical Euclidean geometry, objects have integer dimensions: a point = 0 (zero dimension), a line = 1 (one dimension: length), a plane = 2 (two dimensions: length and width), a volume = 3. Euclidean geometry is useful for describing manmade objects or ideal shapes, but natural shapes are much more complex. As an example, there is no classical shape to describe the branching structure of a vascular tree or the irregular membrane of a lymphocyte. Interestingly, these and many other natural structures reveal more detail when the scale of observation is increased. The resultant increasing details themselves may make it very difficult to aid in the characterization of the scale of observation per se (and, hence, the object): a small branch of a tree with its sub-branches may look similar to the entire tree. As a consequence, the shape is similar at different scales, and in the absence of other reference marks, it would be difficult to know its actual size: the meaning of scale seems to be of relative value when describing a fractal object. This characteristic, called *self-similarity*, is one of the most important features of fractals. According to Mandelbrot,[2] who developed this new type of geometry, fractals are objects formed by sub-parts, each of which resembles the whole either exactly (self-exact fractals) or statistically (random fractals). This is also true in classical geometry; a line segment resembles an entire line, and a plane part is similar to a whole plane, but in this case it is called "classical" or "standard" scaling. To express the dimension of fractal objects, one usually needs a fractional (rather than integer) number: the fractal dimension. Thus, fractals can have 0.5, 1.33, or 2.87 dimensions, a concept somewhat confusing but which is illustrated by the so-called "Richardson effect"[3] in solving the question of "how long is the coast of Britain?". The quick answer is that it depends on the scale at which its length is measured: an ant will have to walk more coast than a human because at the ant's scale there are more details to walk through. Due to the self-similarity of coastlines, there is another way to look at the problem, as coastlines have a statistically predictable increase (that is not arbitrary) in length with increased observational scale. This increase obeys a power law of the type:

$$y = ax^\alpha \tag{1}$$

This power law means that if one measures the coastline length y for a large range of yardstick sizes x (and the coastline is self similar), then the plot of the logarithm of the measured coastline length vs. the logarithm of the yardstick size tends to a straight line with slope $\alpha = 1 - D$, where D is the fractal dimension. For a Euclidean object such as a straight line, the slope of log(length of the line) vs. log(yardstick size) for various yardstick sizes is 0 (and, hence, $D = 1$), since increasing or reducing the scale of observation does not reveal any change in detail. Fractal curves can have $D > 1$ (including fractional values) and the degree of irregularity or complexity of the curve relates directly to the size of D. However, the concept of fractal scaling is not only applicable to curves, but also can be extended to measures of mass and size distributions of objects.

Two types of fractals, as already mentioned, can be described based on the strict meaning of self-similarity. One group is composed of objects whose smaller scales replicate exactly their larger ones and are called "exact self-similar" objects of which the Koch curves in Figure 1 are classic examples. The second group is made up of "statistically self-similar" or "random" fractals, and they differ from exact self-similar fractals because the reproduced detail is not an exact copy but statistically the same, as shown in Figure 2 and also represented by photographs of coastlines at different altitudes.

The fractal dimension of an object is also a measure its "space filling" properties; the more space the object occupies, the higher the value of D and the more complicated the object. This concept of fractal dimension provides a formal measure of geometrical complexity which means that complex shapes can be characterized objectively.

Self-similar fractals (exact and random) are included in a larger group of "self-affine objects". Self-affinity is scaling by different amounts in different directions: a self-affine fractal must be rescaled by different amounts in different directions (a self-affine transform) to obtain a copy of the original figure, rather than increasing magnification alone (scaling all directions by the same amount). For example, natural landscape surfaces and profiles are self-affine. Self-similarity can be then considered to be a special case of self-affinity where the scaling ratios in different directions are the same. This has important consequences related to the methodology used to characterize fractals and is specifically described in Section 8.4.

8.3 What is fractal geometry good for?

Fractal geometry provides methods for characterizing complexity, and, consequently, in the context of image analysis it can be used to quantify morphologies that are considered random or irregular or which at present can only be qualitatively assessed. Fractal geometry also has provided insights into the mechanisms of pattern formation in phenomena such as viscous fingering, diffusion, fracture formation, and percolation.[4]

Can fractal geometry help in the quantification of irregular morphologies in pathological tissue samples? I believe that the answer is yes for certain problems in which morphological changes occur but are not easily quantified. Furthermore, because of the reproducibility, speed, and automation of computer methods, they are likely to become an important component of image analysis in pathology. There are two broad areas in which fractals are important in the medical sciences and, furthermore, to pathology: quantification of complex structures (not only

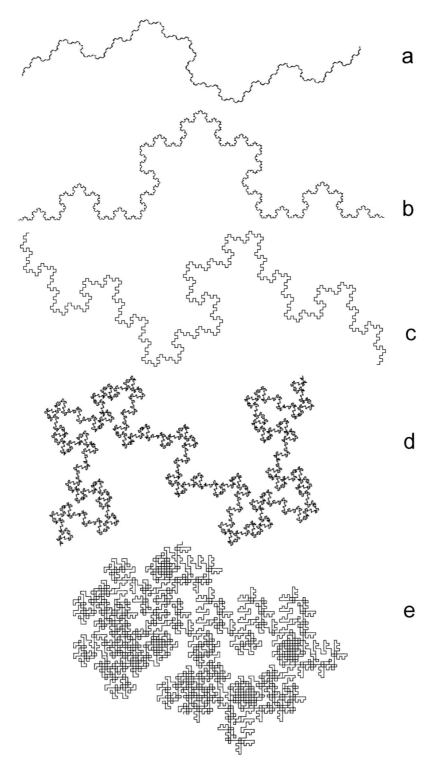

Figure 1 Examples of fractal Koch curves (self-exact or deterministic fractals) and theoretical self-similarity dimension: (a) $D = 1.1291$, (b) $D = 1.2618$, (c) $D = 1.3652$, (d) $D = 1.5859$, and (e) $D = 2.0849$. Note that these curves show a few iterations in the process of construction, although to be truly fractal the iterations should be performed an infinite number of times.

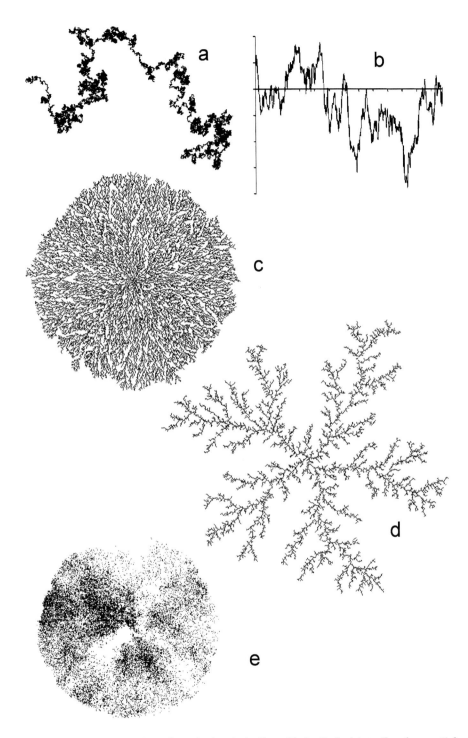

Figure 2 Examples of random fractals (statistically self similar): (a) trails of a particle in Brownian motion in the plane, $D = 2$; (b) one-dimensional Brownian motion (Brownian noise, self affine), $D = 1.5$; (c) a ballistic aggregation cluster in two dimensions; (d) a diffusion-limited aggregation (DLA) cluster in two dimensions, $D \sim 1.7$; and (e) a fractally fragmented system, fragmentation dimension = 1.05.

morphological aspects, but also processes in time) and modelling of biological, developmental, and pathological processes.

8.3.1 *Quantification and modelling of complex structures*

It is well known that pattern and shape recognition by the human brain is remarkable, but complex shapes and patterns are not easy to formalize or quantify. In histopathology this problem is linked with the "expertise" of the person making the diagnosis and is, therefore, thought to relate to the reproducibility and "quality" of diagnosis. The following are illustrated current problems in which fractal analysis has been applied successfully.

8.3.1.1 *Cell morphology*

In the diagnosis of oral cancer (squamous cell carcinoma), early identification of a "premalignant" stage allows expeditious surgical removal of the tissues and commonly preempts development of frank cancer. Histopathological diagnosis involves recognition of abnormal features in a tissue sample — for example, cell and nuclear shape. Apart from the problem of adequate sampling (is the biopsy representative of the main lesion?), the question of diagnosis reduces to one of whether the morphological features in the sample are characteristic of malignancy. This supposes a standard of malignancy features, which at present are regarded to include increased number of mitotic figures, increased nuclear/cytoplasmic ratio, prominent nucleoli, nuclear pleomorphism (increased variability of form and size), and patterns of invasion by the supposed tumor tissue. However, the assessment of such patterns is largely subjective and has been shown to be poorly reproducible, with high intra- and inter-observer variation as mentioned previously. Consequently, no weight or ranking can be systematically assigned to these variables in the diagnostic procedure at present, which would be beneficial in improving diagnosis.

In an attempt to standardize and quantify nuclear features, parameters such as departure from circularity, form factor, shape factor, and perbas (perimeter ratio before and after smoothing)[5] have been found to be statistically different in normal and malignant cells, although these parameters suffer from two drawbacks. First, they involve resolution-dependent measurements of irregular structures which exhibit the Richardson effect; for example, "form factor" and "shape" descriptors involve area (which remains finite for area measuring units $\varepsilon \to 0$) and perimeter (which increases for yardsticks $\varepsilon \to 0$). In Figure 3, the effects of resolution on the "form factor" parameter of a group of 66 cells digitized at different resolutions are shown. While some information can be extracted at a particular scale, information at all other scales is missed (assuming that the descriptor is scale independent). Second, most studies have assessed the projected outline of nuclei which are three dimensional, so the small detail of the nuclei margins is masked due to the non-negligible thickness of a histological section. As a consequence, nuclear shape measured by these techniques is not remarkably successful in terms of assessment of irregularity, and features such as *nuclear pleomorphism* remain subjectively assessed in histopathology at present.

We investigated the effect of resolution of the nuclear perimeter length in normal and malignant (squamous cell carcinoma) cells (using two-dimensional sections, Figure 4) and found that the Richardson effect is too marked to be neglected; furthermore, although irregular, nuclei do not have strictly fractal outlines. At very high resolutions, the nuclear membrane appears relatively smooth (hence nearly Euclidean) with a tendency to a finite maximum perimeter length, while at lower resolutions the irregularity is noticeable (Figure 5). We approached the quantitative analysis

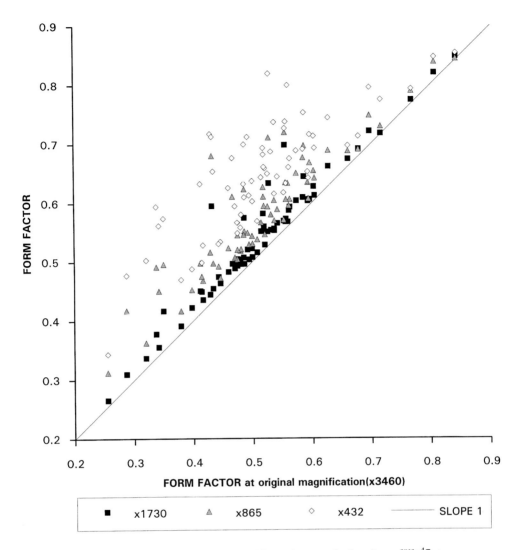

Figure 3 A graph showing the discrepancy of form factor calculated as $\frac{area \cdot 4\pi}{perimeter^2}$ in a group of 66 cell nuclei profiles obtained from TEM micrographs digitized at four different magnifications. A line of slope 1 represents the situation where there are no effects of resolution on the estimation of form factor, so form factor is not a useful parameter for determining the shape of complex objects using different magnifications.

of the irregularity of malignant and normal nuclei seen in transmission electron microscope images using an *asymptotic fractal* model (fully described below).[6-8] Using this model, the theoretical maximum boundary length (B_m) of nuclear section profiles and an asymptotic fractal dimension at low resolutions can be determined and those values used to characterize nuclear shape. We applied this asymptotic fractal model with a multivariate linear discriminant function, using the three parameters derived from the asymptotic fractal fitting — maximum boundary length asymptote (B_m), fractal asymptote (c), and a position constant (L) — as variables, to 672 normal and malignant cells (Figure 6). The nuclei were reclassified with 76.6% accuracy (84.8% of the normal and 67.5% of the tumor) using only the nuclear perimeter data[9] showing not only that an unbiased numerical estimate of the nuclear membrane irregularity was possible, but also that those estimates could quantify features effectively because of their reproducibility and automation. The results are very

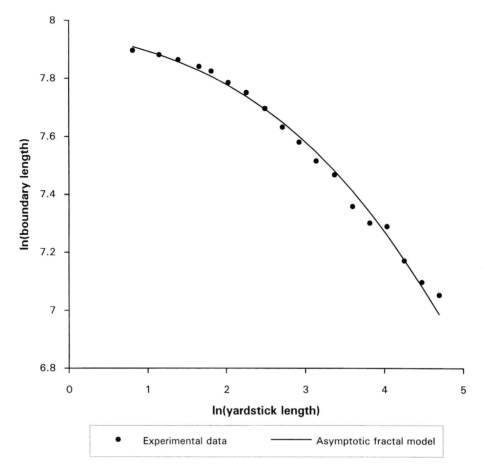

Figure 4 A graph showing the effects of resolution on the measurement of nuclear perimeter length from TEM micrograph of a tumour cell. Note that the logarithmic plot does not follow a "strict" fractal model (the data should tend to a straight line). The asymptotic fractal model takes into account the tendency toward a finite maximum boundary length at high resolutions and fractal behavior at low resolutions. The asymptotic fractal parameters for this curve are $B_m = 3049$, $L = 46.4$, and $c = 0.70$. B_m and L are in pixel units.

encouraging, and, although not perfect, one must bear in mind that diagnosis is not made at the electron microscopy level nor is it based on nuclear irregularity alone. Losa et al.[10,11] investigated the fractal dimension of the membrane of normal lymphocytes and acute lymphoblastic leukemia cells similarly using electron microscopy sections and found that the malignant cells undergoing proliferation had a smaller fractal dimension than normal lymphocytes.

8.3.1.2 Tumor shape

Tumor outlines are of special interest to histopathologists since they reveal the local growth behavior of the tumors. In general, benign tumors are described as "expansive" with smooth outlines, while malignant tumors have local aggressive features with invasion of the surrounding tissues and are described as "infiltrative". The infiltrative margins are usually irregular and fragmented (with detachment of islands from the tumor). For oral, esophageal, and laryngeal carcinomas, there have been attempts to quantify these patterns of invasion into categorical variables in an attempt to increase the accuracy of prognosis.[12-15] Some grading classifications have

chapter eight: Applications of Fractal Geometry in Pathology 213

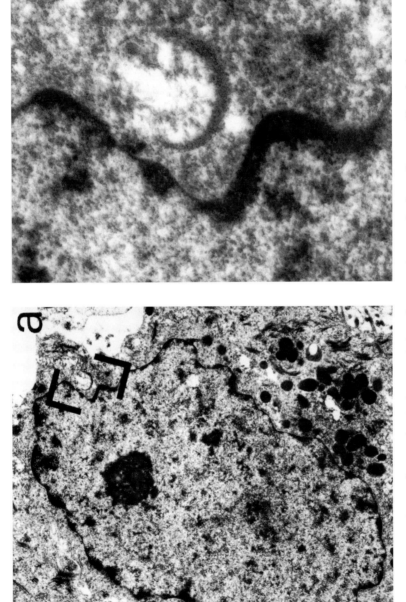

Figure 5 (a) A 9500× (original magnification) transmission electron micrograph of a malignant cell nucleus from a squamous cell carcinoma of the oral mucosa. (b) 95000× (original magnification) micrograph detail of the same cell. The combination of these micrographs shows that the low-scale irregularity of the nuclear membrane vanishes at this high resolution. (From Landini, G., and Rippin, J. W., *Fractals*, 1, 326, 1993. With permission).

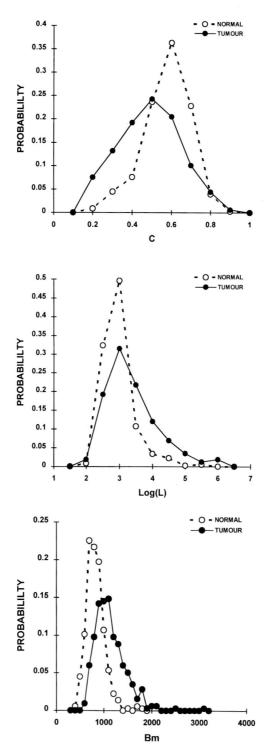

Figure 6 Graphs showing the distribution of the values of the three parameters of the asymptotic fractal formula c, $\log(L)$, and B_m in the normal and tumor groups showing the differences between normal and malignant cells of the oral mucosa. B_m and L are in pixel units (1 pixel = 35 nanometers). These distributions alone allow reclassification of 76.6% of 672 cells in the original diagnosed groups of normal and tumor patients. (From Landini, G., and Rippin, J. W., *Fractals*, 1, 326, 1993. With permission.)

been reported to be better prognostic indicators than others,[16] but, of course, malignancy is already implicit, so the issue is directed to prognosis (survival rates) and not diagnosis. Regarding premalignant conditions of the oral epithelium (such as epithelial dysplasia), there is no invasion of the lamina propria by epithelial cells, but the normal architecture of the epithelial tissue is affected (amongst other factors due to increased cell proliferation). In histological sections, this abnormal architecture is seen at the junction between the epithelium and lamina propria and is usually described as increased irregularity of the "epithelial-connective tissue interface" (ECTI). While in normal tissues the ECTI is smoothly undulating to flat, in premalignant lesions it becomes irregular; in carcinomas this is even more marked, including islands of epithelial cells that invade into the deep layers of the lamina propria. This increase in irregularity is well recognized but its diagnostic meaning is still relative, mainly because it is assessed in a subjective manner. We have investigated two approaches to the objective characterisation of ECTI irregularity.

8.3.1.2.1 Fractal approach. First we considered if the degree of irregularity in normal mucosa, premalignant lesions, and carcinomas ECTI could be expressed in terms of fractal dimension, and, if so, whether the fractal dimension could then be used as a quantitative parameter for diagnostic purposes.[17,18] The box-counting and yardstick methods (described in detail below) were applied to a set of 40 hematoxylin-eosin-stained histological sections of the floor of the mouth. Projected images of the junction between the basal epithelial cells and the adjacent lamina propria (the ECTI) were enlarged 23.7 times, hand traced, and then scanned into a personal computer achieving a final resolution of 1 pixel $\simeq$ 12 μm (Figure 7). The sections included 10 normal mucosae, 20 cases of epithelial dysplasia (including 10 keratoses with mild dysplasia and 10 with moderate-to-severe dysplasia), and 10 well differentiated squamous cell carcinomas. The range of self-similarity was small (just over one order of magnitude, from 40 to 885 μm for the box-counting and from 45 to 1134 μm for the yardstick method) due to the finite size of the cells which are the structural elements of the tissue. Nevertheless, using the fractal dimension of the ECTI alone, it was possible to differentiate statistically between normal/mild dysplasia (low D, not likely to become malignant), moderate/severe dysplasia (higher D, with higher chances of transforming into cancer), and squamous cell carcinoma (highest values of D) (Figure 8).

8.3.1.2.2 Multifractal approach. Although the results using the fractal dimension were encouraging, the problem remains to locate the structures of interest in the histological sections to be analysed. For example, the decision to perform the fractal analysis on a certain location of a sample (in this case, deciding where the lesion is) is an *a priori* decision by the histopathologist. This introduces a further problem which fractal analysis does not solve: an irregular lesion may be surrounded by normal tissues (less irregular) that will also contribute to the value of the fractal dimension of the ECTI. Thus, the value of the fractal dimension is generalized or an average because the ECTI does not necessarily have isotropic irregularity; some areas may be more irregular than others, which may indicate increased cell proliferation and either actual or potential tissue invasion. To address this problem of localization it was decided to use another approach (multifractal analysis) based on pointwise scaling properties (that is, local fractal dimensions) of the ECTI.[19] As the method is described in detail below, just an outline will be presented here. The ECTI obtained with light microscopy was represented by a 1 pixel-thick curve, and a computer program measured the total number of pixels locally connected in a box of increasing side size ε centered at a location x,y belonging to the ECTI. The scaling relation of

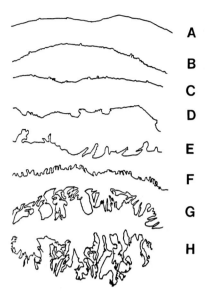

Figure 7 Eight profiles traced from projections of histological sections slides of epithelial-connective tissue interfaces as routinely diagnosed cases of normal tissues (A, B), mild epithelial dysplasia (C, D), moderate/severe epithelial dysplasia (E, F), and squamous cell carcinoma (G, H) of the floor of the mouth (all different individuals) showing the various degrees of irregularity. (From Landini, G., and Rippin, J. W., *Anal. Quant. Cytol. Histol.*, 15,144, 1993. With permission).

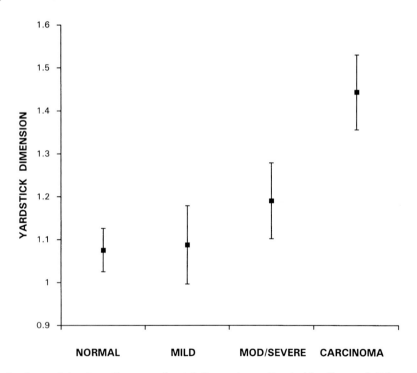

Figure 8 A graph to show the mean fractal dimension estimated by the yardstick method of 40 epithelial-connective tissue interfaces of the oral mucosa classified as four different types of lesions during routine diagnosis. The vertical bars indicate one standard deviation from the mean.

the number of locally connected pixels related to box size was found by the linear regression of the logarithm of $M(\varepsilon)$, the mass (number of pixels) for a box of size ε, on the logarithm of the size ε.[20] This relation, $M(\varepsilon) \propto F \varepsilon^{\alpha}$ involves an exponent α which corresponds to the local connected fractal dimension of the ECTI, so instead of a single number, one obtains a set of dimensions that represent the local irregularity for every possible position of the analysis. If the object was a locally filled area, then $\alpha = 2$ and the object was locally two-dimensional; if it was a straight line (one dimensional), then $\alpha = 1$; values in between characterized the local connected fractal dimension. The limits of the "local" attribute relate to the maximum box size ε, which was 31 pixels (372 µm). The statistical properties of the distribution of α values, namely average α, mode α, maximum α, and histogram entropy, were used as variables in a multivariate linear discriminant function analysis, a technique that reclassifies the cases from the original groups (normal, dysplasia, and carcinoma) (Figure 9). The discriminant analysis gave 85% correct classification: 100% of the normal and carcinoma cases, and 70% of the epithelial dysplasia cases. From the 20 epithelial dysplasias, 5 (4 mild dysplasia and 1 moderate/severe) were incorrectly classified as normal and 1 (moderate/severe) as carcinoma. As α estimates the local scaling for all the possible starting positions of the analysis, it allows the recognition of areas which have increased irregularity and allows for the new possibility for automatic detection (segmentation) based on the spatial distribution of α.[19]

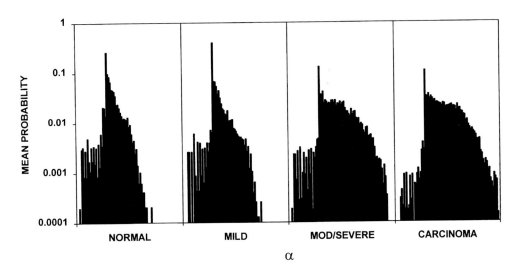

Figure 9 Graphs representing the mean probability (per type of lesion) of the local connected fractal dimension (α) seen at the epithelial-connective tissue interfaces of the oral mucosa in 40 cases (10 cases per lesion). The histogram bins range from 0.70 to 1.70 in steps of 0.01. Note the peaks at $\alpha = 1$ for all cases and the different distributions for high values of α.

8.3.1.3 Retinal vasculature analysis

The blood supply to the retina arises from the optic disc and is distributed on the retina in a very particular branch-like fashion, and several independent research groups have shown that patterns of retinal vasculature are fractal.[21-26] It is interesting to know not only that the hierarchical structure of retinal blood vessels can be defined in fractal terms, but also that it resembles a very particular type of fractal: diffusion limited aggregation (DLA)[27] clusters, which are models of stochastic growth in a diffusion field subjected to the Laplace diffusion equation. The mechanisms involved in vasculogenesis in general (the formation of the vessels) have been a matter of debate. Some authors have proposed deterministic models in which the sub-branching of vessels obeys certain deterministic or probabilistic rules[28-30] as intrinsic rules of growth, while other authors believe that mechanisms extrinsic to the vascular growth such as self-avoiding invasion percolation[31] and nonequilibrium diffusion processes, such as diffusion limited aggregation, are important and guide vascularization.[21,22,26] Apart from visual similarities between the patterns of retinal vessels (growing in a quasi two-dimensional space) and two-dimensional DLA computer simulations, there are geometrical similarities as well. Two-dimensional DLA models and the retinal vasculature have similar fractal dimensions ($D \sim 1.7$). It is possible that diffusion plays an important part in the formation of these patterns. The DLA-like patterns have been observed not only in the retina, but also in the pathological neovascularization of the cornea. Neovascularization of the cornea may occur following chemical or physical trauma and is known to involve diffusion of angiogenic factors, gases, cell mediators, and nutrients. Superficial and interstitial types of corneal neovascularisation have been simulated using a DLA model,[32] and, since then, it has been speculated that neovascularization advances towards the stimuli along a diffusion gradient and that this same process may also be occurring during the vascularization of the fetal retina.

The importance of these findings is related not only to the nature of the vascularization process, but also to the practical assessment of the status of the retina: the fractal dimension may be used for objective characterization of the retinal vessels in disease or during the monitoring of treatment. Retinal photography, can be performed either "red-free" (noninvasive method of data acquisition that consists of filtering color red from the retinal fundus) or by injecting fluorescein (a fluorescent dye) in the blood stream. In Figure 10, 1-pixel-thick tracings of 60° fluorescein angiograms of a normal individual and a case of central vein occlusion are shown. While the normal case shows a uniform pattern of vessels (except for the avascular macula area at the center), the pathological case has a lower number of vessels towards the bottom of the figure where the vein occlusion has occurred and a high density of small vessels in the top right area where the blood is flowing to compensate for the obstructed return. The angiograms appear to differ quite distinctly but surprisingly have practically the same fractal dimension (1.71 for the normal and 1.69 for the vein occlusion) (Figure 11a). Thus the fractal dimension alone was not capable of differentiating the patterns in normality and disease,[24] but another fractal parameter, lacunarity Λ (described later in detail), showed a difference between the two cases ($\Lambda = 1.94$ for the normal case and $\Lambda = 2.19$ for the vein occlusion). Figure 11b shows the distribution of local connected dimensions in the range of box sizes 25 μm to 975 μm. The probability distribution of local connected fractal dimensions (α) of the vein occlusion case shows a peak in the range of $\alpha 1 \sim 1.10$ not present in the normal case which corresponds with areas showing little branching, and a lower decay of the same probability distribution towards the high values of α which corresponds with the hyperemic area. This type of analysis may be particularly useful for diagnosing conditions such as diabetic proliferative retinopathy, in which proliferation of small vessels may change the architecture (and the local dimension) of the blood vessel tree.

8.3.1.4 Spread of herpes simplex virus

Herpes simplex virus (HSV) infection may occur in oral, ocular, and genital mucosae or on skin, producing ulcers. In the oral mucosa (herpetic gingivostomatitis), the ulcers are usually rounded to ovoid in shape and sometimes small ulcers coalesce to form a larger one. Herpes simplex virus lesions in the epithelium of the cornea (HS keratopathy) exhibit a characteristic arborescent or "dendritic" morphology.

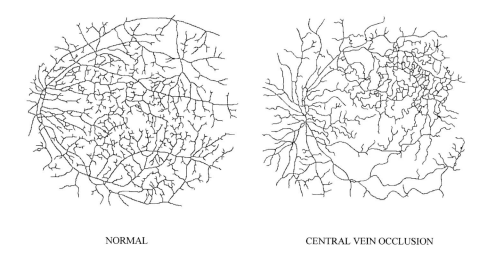

NORMAL   CENTRAL VEIN OCCLUSION

Figure 10 Two 1-pixel-thick digitized tracings of 60° retinal fundus fluorescein angiograms obtained from a normal patient and from a patient with a central retinal vein occlusion. The angiograms are centered on the macula area (avascular), and the optic disc is to the left. The vein occlusion is seen in the lower central area of the affected angiogram, and there is a compensatory hyperemia visible in the upper right area. (Original tracings courtesy of Drs. G. Misson and P. I. Murray, Birmingham and Midland Eye Hospital, England).

Most of the corneal lesions remain dendritic and localized and heal within 3 weeks with antiviral treatment. Occasionally, dendritic ulcers can enlarge progressively and develop a "geographic" or "ameboid" morphology, which is sometimes observed after the inappropriate use of topical corticosteroids. Although ameboid ulcers are still produced by the same HSV, they have a prolonged clinical course and are more difficult to treat.

We have recently shown that the shapes of HSV corneal ulcers have fractal outlines[33] and that there is an inverse relationship between D and the maximum diameter of the ulcer.[34] As the viruses do not have a metabolism of their own but use the host cell molecular machinery for replication, we suspected that the process involved in infection and ulcer formation (cell infection → viral replication → cell death → ulcer) could be regarded in algorithmic terms and that this mechanism depends on the host cells. Cellular automata (CA) seemed to be the most suitable model available to test this hypothesis,[35] because CA are mathematical models that contain a large number of simple identical elements (called "cells") which affect one another by means of local interactions following a set of simple fixed rules that are local but also govern the system dynamics. The CA model we adopted assumed a monolayer of corneal epithelium with two populations of epithelial cells having different degrees of permissivity to infection by HSV. The actual hypothesis was based on the presence or absence of a cell membrane receptor that facilitates HSV infection, so the simplest situation was programmed: cells were either permissive or resistant to infection (summary of CA in Table 1). The permissive and resistant cells were distributed randomly and could exist in three different states during the simulation: alive, infected, or dead (the dead cells formed the "ulcer"). The infection was triggered in a single cell and the evolution of the system (infection) depended only on the status of the neighbor cells ("spread by contiguity"). By varying the number of permissive cells in the tissue and the relative strength of the resistant cells (the number of neighboring cells that must be infected or dead to affect that resistant cell), it was possible to simulate a wide range of lesions, including dendritic,

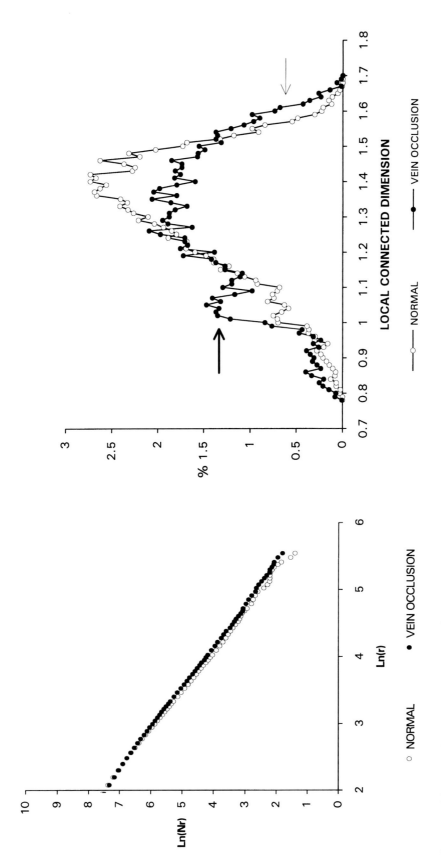

Figure 11 (a) Logarithmic plots used in the box-counting method to estimate the dimension of the digitized angiograms in Figure 10. Despite their different appearance, they assume practically the same value of D (1.71 and 1.69, respectively). (b) Using a multifractal analysis, however, the distributions of the local connected fractal dimensions show differences between the cases: a peak in the 1.00 to 1.10 region that corresponds to the area devoid of blood vessels (the vein occlusion area, thick arrow) and an increased number of areas with high dimension (hyperemic area, thin arrow) that occur through a reduction of the peak seen between 1.3 and 1.5, representing the normal range of values for local connected fractal dimension.

Table 1 Cellular Automaton Rules for Simulation of the Herpes Simplex Virus Spread

Rule 1
The tissue is a monolayer composed of two cell types: *permissive* (HSV infects them easily) and *resistant* (HSV does not infect them easily), distributed randomly in the tissue. The proportion of the permissive cells can be varied and it is recorded as:

$$p = \frac{\text{Number of permissive cells}}{\text{Total number of cells}}$$

Cell status can be *alive, infected*, or *dead*.

Rule 2
A *permissive* becomes infected if one or more neighbors are infected (viral infection).
A *resistant* becomes *infected* if n or more neighbors (n from 2 to 8) are infected (viral infection) or dead (lack of tissue support).

Rule 3
Infected cells become *dead* cells after one cycle of simulation.

---

ameboid, and round ulcers, which resembled the range of morphologies commonly seen in oral and corneal lesions (Figure 12). Additionally, when resistant cells were abundant the ulcers tended to be small and irregular; when they were scarce, then ulcers were larger, with smoother outlines, similar to the "dendritic" and "ameboid" appearances of corneal ulcers observed *in vivo*. For simulation of the shapes of corneal HSV corneal ulcers (that is dendritic or ameboid), resistant cells were infected when **5 or more** of the neighboring cells were infected or dead (called the "≥5" rule). We subsequently investigated this particular CA rule, performed 3600 simulations for different proportions of resistant and permissive cells, and estimated the fractal dimension of both the resultant simulated ulcers and that of corneal ulcers *in vivo* using the yardstick method. The fractal dimension of the ulcer outlines and its dependence on the ulcer size[34] from the CA simulations compare directly with data obtained *in vivo* (Figure 13).[35] This model suggested a possible mechanism for the transformation of dendritic to ameboid ulcers which is seen *in vivo* when corticosteroids are administered. Assuming corticosteroids depress the inflammatory response and transform "resistant" into "permissive" cells; this then allows the infection to percolate through the tissue. In essence, the viral infection can be regarded as a critical percolation phenomenon, and the critical threshold that allows the infection to spread (to make it unrestricted) is closely related to both the ulcer size and the fractal dimension of the outlines. This type of modelling, although simplistic, allows investigation of the relations between the elementary constituents of the system and the system attributes as a whole.

8.3.1.5 *Periodontal disease modelling*

Periodontal disease is a destructive inflammatory condition of the tissues that support the teeth which affects practically all dentate individuals with different degrees of severity. It can be present as early as at 3 years of age, and in its more severe presentations can cause tooth loss. There are many factors involved in periodontal disease including endogenous (host-dependent) and exogenous (nonhost-dependent) factors. In long-term epidemiological studies, it has been shown that, in general, periodontal breakdown (loss of alveolar bone) increases with age. Initially, it was thought that the breakdown advanced in a progressive linear fashion,[36,37] but recently intermittent periods of burst and remission[38-40] destruction have been proposed although the explanation of this burst and remission patterns remains elusive. A

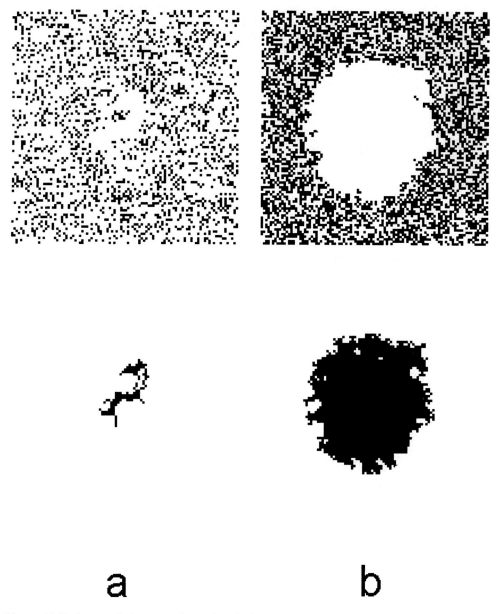

Figure 12 Background tissues and simulated ulcers produced by the CA rules from Table 1. (a) Dendritic ulcer produced by p = 0.35; top figure represents the background at the end of the simulation, with clear areas (resistant cells) and dark areas (permissive cells). (b) Ameboid ulcer produced by p = 0.75; top figure represents the background at the end of the simulation, with clear areas (resistant cells) and dark areas (permissive cells). For these two simulations, the minimum number of infected/dead neighbors necessary to infect/kill a resistant cell was set to 5.

computer simulation of periodontal breakdown based on a stochastic model has been proposed[41] but did not show a convincing interrelation between the causal factors and clinical observations of the disease. In view of this and the known multifactorial aetiology of periodontal disease, the author has investigated the results of integrating the multiple factors.[42] Events that increase and decrease the degree of periodontal breakdown probably occur by means of different mechanisms with

chapter eight: Applications of Fractal Geometry in Pathology

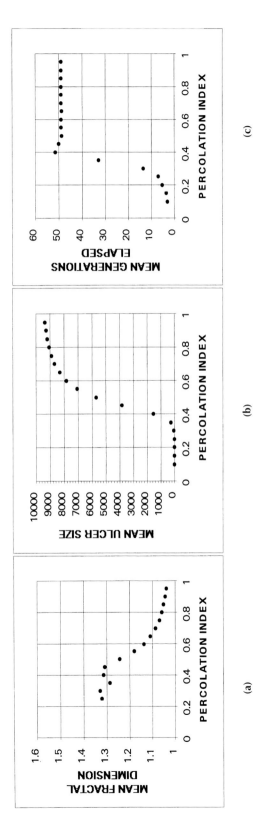

Figure 13 Three graphs to show some properties of the CA simulation of herpes simplex infection. (a) The fractal dimension of the ulcer outlines is reduced with increasing p (percolation index); (b) the size of the ulcers is increased with p; (c) the evolution of the infection suffers a phase transition at p = 0.4. This particular value of p is called the critical percolation threshold and relates (in the model) to the change in the fractal properties of the ulcer outlines and the abrupt increase in size of the ulcers. These characteristics resemble closely the observed infection of HSV *in vivo*.

varying durations and frequencies. Many of them have antagonistic or synergistic interactions — for example, oral hygiene combined with healing capacity and bruxism (tooth grinding) in the presence of dental plaque and calculus. These factors are also related to the immunological and hormonal states of individuals, and it reasonably could be assumed that a single factor might rarely correlate with the state of a periodontally affected site. Three points were considered in the design of the computer simulation of periodontal disease: (1) the model must be multifactorial; (2) the model output should be compatible with the epidemiological data of periodontal disease (an increase in destruction of periodontal tissues with time, advancing in bursts of destruction and remission); and (3) the model should emulate an accumulative process. To understand the dynamics of a multifactorial disease, one must be able to integrate the interactions of the multiple factors. An ensemble of multiple random variables operating at different frequencies has been proposed as the origin of the ubiquitous 1/f noise (a very widespread fractal fluctuation which occurs in nature),[43] and this ensemble of multiple variables resembles the accepted multifactorial origin of periodontal disease.[42]

An algorithm to produce "pseudo" 1/f noise[44] was used to simulate periodontal destruction. Starting with a value of 0 periodontal breakdown, the amount of breakdown (bone loss) was increased at each subsequent step in the simulation, according to the value of the pseudo 1/f generator output. When the output value was higher than a threshold 0, it was added to the periodontal loss; whereas, if the output values were 0 or negative, the destruction would stop. The latter were considered as instances when the pathogenic factors were reduced (or the reactive-healing processes were enhanced) and the periodontal breakdown stopped until a new value higher than 0 was generated. The hypothetical periodontium, therefore, showed periods of remission in the destructive disease process, recognized as segments with slope 0 (Figure 14) which ultimately made the disease accumulative. Changing the influence of one variable changed the overall behaviour of the simulated disease. This is the only multifactorial model proposed so far that explains why bursts and remission episodes occur and why those bursts and remissions can occur at a large range of scales. The model also questions the accuracy and adequacy of the clinical methods used in the assessment of periodontal breakdown. As significant bone changes occur at levels smaller than the most accurate clinical measurement are capable of determining,[45] the bursts will be detected only if their sizes are equal to or larger than the resolving power of the method used for diagnosis, or if the accumulated effects are of that size.

8.3.2 Why fractal?

The previous examples show how fractal geometry can be used to quantify changes in geometrically complex patterns, but it is also interesting to speculate why fractal patterns are common in nature in the first place. Two main reasons for the existence of fractal morphologies in biological systems appear to be possible. First, fundamental methods of constructing fractals in the computer usually involve very simple recursive and iterative formulae, which in some ways resemble biological growth processes: values for the specific variables are used as inputs, the formula is solved, and the result is then used as the variable in the next cycle, so the output of the operation is redirected to the input (recursion) and repeated (iteration). Complex structures can be (and usually are) produced as a result of these two mechanisms and some of these even resemble biological forms[46] (Figure 2); usually the formulas involved are very simple. The entire resulting fractal is contained (coded) in a very small set of instructions. The information stored is very small which benefits the

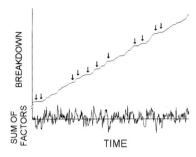

Figure 14 Diagramatic representation of a simulation of periodontal breakdown in untreated periodontal disease by the integration of four independent factors operating at different frequencies. As the periodontal breakdown is a cumulative process, it was assumed that the interactions of the factors could produce destruction if the integrated result was larger than a theoretical threshold beyond which the destruction occurred (a "burst of destruction"). When the sum of the factors does not reach this threshold, the destruction stops and the breakdown shows a "remission period". Some remission periods are shown with vertical arrows.

information system (in this case, the amount of DNA used to code the structure). Likewise, DNA coiling in chromosomes has been described as hierarchical structure from the alpha-helix to the chromatid[47] and consequently is seen as a recursive procedure for packing and unpacking DNA efficiently. Blood vessels have an optimized shape to reach and perfuse the tissues they supply, the lungs are optimized for gas exchange with blood, and exocrine glands have branching ductal architecture which tends to maximize the volume of secreting parenchyma per unit volume of gland. More insights into fractal pattern formation may be obtained through Lindenmayer systems (L-systems),[48] which are recursive grammars originally developed as models of plant growth. The instructions to build up an L-system are extremely short yet produce very detailed, plant-like results: certain micro algae are known to follow growth strategies similar to those modelled by L-systems. Interestingly, specific genes that control the direction of division in yeasts have already been identified,[49,50] which means yeasts can reprogram their growth behavior through gene expression, giving further support for the existence of L-system-like mechanisms. Deterministic fractal mechanisms then could be seen as advantageous in evolutionary terms, because of the capabilities of system optimization, robustness, coding efficiency, and simplicity. Weibel[51,52] has proposed that fractal geometry is a design principle for biological organisms.

A second reason for the presence of fractal patterns in biological systems may be related to randomness and stochastic phenomena. Here, morphology is not related to deterministic processes or morphology for which the organism has been coded, but to other mechanisms that result in fractal morphologies. For example, stochastic growth processes such as diffusion limited aggregation (DLA), gelation, sedimentation, viscous fingering, and invasion percolation produce fractal patterns. In relation to biological systems, replicating cellular systems with random location of daughter cells produce specific types of fractal pattern.[53-55] As shown previously, the development of blood vessels can be modelled as both a deterministic and a stochastic process. Herpes simplex virus ulcers are fractal, although this is not a direct consequence of a particular strategy developed by the virus, but rather as a result of percolation in disordered media, and the change in nuclear shape of malignant cells is the result of alterations of structural proteins. These examples show that the fractal patterns generated depend on underlying physical processes rather than strategies embedded in the genome.

8.4 How to estimate dimension

This section outlines the procedures involved in the estimation of fractal dimensions in order to facilitate their implementation using a computer or image analyzer, and the rigorous proofs of the methods are left to their respective references for brevity. The assumption for these methods is that "the object" is a set of square pixels (forming a boundary or an aggregate) in a binary image (image pixels = 1, background = 0). It also should be noted that not all the methods measure the same attributes of an object. While some are designed to estimate the irregularity of a boundary or "coastline" (usually called the "surface dimension"), others relate to the entire filled object ("mass dimension") or size distribution of particles ("fragmentation dimension").

8.4.1 Self-similarity dimension

This dimension is obtained by relating the number of parts or pieces that compose the object to their size reduction factor.[3] The method is used to calculate the self-similarity dimension of such self-exact fractals as the Koch curves (Figure 1), Sierpinski gasket, and Vicsek fractal. The relation of the number of parts to the reduction factor can be expressed as:

$$D_{self\text{-}similarity} = \frac{\log(a)}{\log\left(\frac{1}{s}\right)} \quad (2)$$

where a is the number of pieces and s is the scaling factor. The value of D increases for increasingly complex curves. For nonintersecting fractal curves embedded in a plane, D ranges between 1 and 2. Note that in the case of self-intersecting curves, D can exceed the value of 2 as in Figure 1e, but the box dimension (defined below) cannot exceed 2.

8.4.2 Length-resolution methods

The following methods are graphical implementations based on the method applied to describe the "Richardson effect" mentioned earlier[2,3] and in which Equation (1) can be used to calculate the increase in length of a digitized boundary as the resolution of the measurements is increased. Equation (1) takes the form:

$$L(\varepsilon) = F\varepsilon^{1-D} \quad (3)$$

where $L(\varepsilon)$ is the length of the boundary measured with a yardstick of size ε, F is a prefactor (constant), and D is the fractal dimension of the boundary. As the boundary to be analyzed is represented in a discrete square matrix of picture elements (pixels), the screen resolution poses a finite scale to the measurements and therefore a high resolution imaging system is preferable.

8.4.2.1 Yardstick method

This method is also known as the "ruler", "divider", "polygonation", or "structured walk" method. Using the technique,[3,6] the length $L(\varepsilon)$ of the boundary to be analyzed is measured by "walking" along the boundary and using yardsticks or rulers of different length ε. Although the technique can be performed by hand and a pair of dividers, it is in the graphic implementation in a computer that speed, automation, and, therefore, the practical nature become apparent. The ruler lengths are usually measured in pixels and may vary from 1 to a maximum which is smaller than the maximum diameter of the object. An automated tracking subroutine checks the distance from a starting point sequentially along the boundary until it finds the best approximation to the defined ruler length. It must be noted that because of the squareness of pixels representing the image, the ruler length values are approximations and therefore must be corrected to mean values. For instance, the possible distances along a boundary (eight-neighbor connected) for a ruler $\varepsilon = 2$ are 2, $\sqrt{5}$, and $\sqrt{8}$. A useful test for finding the best approximated distance for a particular ruler length is

$$\varepsilon - 0.5 \leq d < \varepsilon + 1 \qquad (4)$$

where d is the distance under evaluation from the starting point to the point under consideration. If d falls within the range of $\varepsilon - 0.5$ and $\varepsilon + 1$, then one ruler length is counted and d added to the boundary length. The end of the ruler is now the starting point for the next measurement and the procedure continues until the whole outline has been measured. If the last segment is smaller than the ruler (which is usually the case), then two approaches can be taken: either to discard this last segment, which results systematically in an underestimation of $L(\varepsilon)$, or to add the length of the last segment to the total boundary estimate.[6] The sum of all the ds (plus the remainder, if considered) is the length of the boundary $L(\bar{\varepsilon})$, and this number divided by the number of rulers counted (plus the fraction of the ruler corresponding to the remainder) gives the average ruler length $\bar{\varepsilon}$. Finally, the value of $D_{yardstick}$ is calculated from the formula:

$$D_{yardstick} = 1 - \frac{\log(L(\bar{\varepsilon}))}{\log(\bar{\varepsilon})} \qquad (5)$$

One minus the slope of the linear regression of $\log(L(\bar{\varepsilon}))$ on $\log(\bar{\varepsilon})$ is thus $D_{yardstick}$. The range of ε where the log-log plot is linear gives the range of scales when the object can be considered fractal, and it is necessary that this range is of at least one order of magnitude (this is also a requirement for all the methods described below).

The principle of the yardstick method can also be implemented by varying observational magnification and using a fixed yardstick length;[56] however, the boundary to be measured must be easily resolved at all magnifications used.

8.4.2.1.1 *Extension to surface analysis.* The same principles used for measuring the length of a curve with different yardstick lengths can be applied to the problem of measuring the area of a complex surface to estimate D. The triangulation method[57] is based precisely on this principle: the surface is covered with triangular grids of different sizes (triangle vertices are determined by the height of the surface), and the total normalized surface area $A(\varepsilon)$ is measured for different triangle side sizes ε.

A similar approach has been used by Caldwell et al.[58] using the surface-area relation to estimate the dimension of X-ray images to characterize parenchymal texture in mammograms (the optical density of the image is the "height" of the landscape) at different pixel sizes ε (by averaging the image in areas of size ε). These methods are appropriate for self-similar surfaces but not for self-affine ones[59] because of the different ratio of scaling in the x-, y-, and z-directions of the surface. Self-affine scaling cases require that other methods (such as Fourier analysis described below) be used.

8.4.2.2 Interpolated yardstick variant

A variant of the yardstick method uses the "exact" ruler length, by interpolating the n and $n-1$ points in the boundary (n being the point at which we exceed the ruler length ε) to find the perfect match to ε.[60] This version involves intensive computation (coordinate interpolation), and it has been reported to be as accurate as the noninterpolated method because, for the latter, large errors only occur when searching an appropriate length fit at very small ε.

8.4.2.3 Censored intercept method

This method devised by Flook[61] is based on Buffon's needle problem[62] (the probability of a randomly tossed needle to land on any of a set of parallel lines separated by a distance equal to the length of two needles) and the stereological relationship:

$$B_A = \frac{\pi}{2} I_L \qquad (6)$$

where B_A is called the boundary length density and I_L is the number of intersections of the boundary with the test lines. One uses different boundary estimates from intercepts which subtend a chord greater than a specified value, and the minimum chord length used when counting the intercepts is the method's "resolution". This method has been applied to particle analysis and is suitable for image analyser instruments (they handle intercepts very efficiently) and situations in which other methods would be more laborious to implement.

8.4.2.4 Coordinate skipping method

In this variant of the yardstick method devised by Schwarz and Exner,[63] instead of varying the ruler size, one varies the number of coordinate points m to skip while "walking" along the boundary. If the number of skipped points is m, then the automatic tracking routine measures the distance between the n and $n + m$ points in the boundary. The sum of all the segments' lengths is the length of the boundary $L(\bar{\varepsilon})$ which, divided by the number of segments counted, gives the average ruler length $\bar{\varepsilon}$, and, the value of D_{skip} is estimated using Equation (5). This method is extremely fast since the fitting of a segment to a ruler size does not have to be found, only the calculation of the distance between two points separated by m sequential points. However, the deviations in $L(\bar{\varepsilon})$ are non-negligible (especially for large εs) since m covers a shorter segment in convoluted curves than in smooth ones.

8.4.2.5 Equally spaced test-lines method

This variant is an intermediate version of the "censored intercept" and the "coordinate skipping" methods. The boundary of an object is divided by parallel lines spaced at ε pixels, and the boundary is reduced to a polygon with vertices at the points of intersection of the boundary with the test lines where $L(\varepsilon)$ is the perimeter of such polygon. The dividing procedure is repeated for various spacings, and D is calculated using Equation (5).

8.4.2.6 Dilation method

This technique uses the approach of Cantor and Minkowsky to "tame" nondifferentiable curves.[2,64] The length $L(\varepsilon)$ of the outline is measured after progressive dilation with discs of known radii ε producing a "sausage" (i.e., each point in the original curve is replaced by a disc of radius ε). As the curve is dilated, it is smoothed and loses details smaller than ε, and by measuring the area of the dilated outline $A(\varepsilon)$ (of the "sausage"), its length $L(\varepsilon)$ can be determined as:

$$L(\varepsilon) = \frac{A(\varepsilon)}{2\varepsilon} \qquad (7)$$

and $D_{dilation}$ is

$$D_{dilation} = 1 - \frac{\log(L(\varepsilon))}{\log(\varepsilon)} \qquad (8)$$

which can be calculated as one minus the slope of the regression of $\log(L(\varepsilon))$ on $\log(\varepsilon)$. As most image analyzers have image dilation and erosion functions based on specific structural elements, the implementation is straightforward. Flook has proposed corrections when using octagonal rather than circular structural elements (since the octagon is the nearest approximation to a circle in a square matrix):

$$\varepsilon_{eq} = \sqrt{\frac{A_w}{\pi}} \qquad (9)$$

where A_w is the area of one picture point dilated w steps. For open boundaries, the same author found that a further correction was necessary. As a curve is dilated, it also becomes "longer" due to the dilation at the extremes, and this results in an overestimation of $A(\varepsilon)$ and, hence, $L(\varepsilon)$ in Equation (7). If the open boundary is a straight line (two free ends), then the overestimation of area for dilating with discs of radius ε is

$$A'(\varepsilon) = \frac{\pi \varepsilon^2}{2} \qquad (10)$$

for each free end, and for the length of the entire line based on Equation (7):

$$L'(\varepsilon) = \frac{\pi \varepsilon}{4} \qquad (11)$$

For fractal curves of dimension D, the correction for each open ending is

$$L'(\varepsilon) = K\varepsilon \left(\frac{\pi}{4}\right)^{1-D} \qquad (12)$$

where K is a constant that has been found experimentally to take values close to D (for a curve under study D is unknown and K has to be approximated using an iterative procedure).[64]

Graphical implementation of this method provides the opportunity for some optimizations. For example, instead of measuring the area after each dilation, if the coordinate of the pixels forming the original boundary are stored in memory, then the dilations are carried out successively, starting with the maximum ε. A dilation of the original boundary with a small disc will cover only parts already dilated by a larger disc. By labelling each dilation with a different color (which corresponds to a given ε) and performing a cumulative histogram of the image, the number of pixels — the area $A(\varepsilon)$ — for each color (ε) is obtained.

8.4.3 Box-counting method

The box-counting method[2,65] consists of a grid with boxes of size ε superimposed onto the image, the number of boxes that contain any part of the figure is recorded as $N(\varepsilon)$. The procedure is repeated for different sizes of ε, and the box dimension D_{box} is

$$D_{box} = -\frac{\log(N(\varepsilon))}{\log(\varepsilon)} \qquad (13)$$

D is calculated from the negative value of the slope of the linear regression of $\log(N(\varepsilon))$ on $\log(\varepsilon)$. As the starting position of the grid is arbitrary, it is preferable to repeat all the measurements for many initial positions of the grid and use an average of the number of boxes $\overline{N}(\varepsilon)$ in Equation (13). One advantage of this method is that it can be extended easily to higher dimensions (e.g., in three dimensions, the boxes are "cubes") or lower dimensions (when estimating Cantor-type sets on the unit interval, the boxes become "intervals").[66] A further advantage of the technique is that it allows measurements of other types of objects that cannot be analyzed with the "yardstick" approach — for example, complicated networks (blood vessels, trees, neuron branchings) or cluster aggregates. Even though in such cases there is no explicit "boundary" or "path" to follow, there may still be self-similar structures and the scaling of the "mass" (number of particles or pixels related to size) of the object can be analyzed. In some instances, these objects can be characterized as either "surface fractals" or "mass fractals", while in other cases objects can be characterized both by the "boundary" or surface dimension and by its mass. Objects that can be characterized by more than one type of fractal dimension are the surface and mass dimensions of herpes simplex virus ulcers in the cornea[33,34] and the mass and perimeter dimensions of percolation clusters.[4]

8.4.4 Area-perimeter relation

The method based on the area-perimeter relation, also called "slit island analysis",[67,68] is useful for determination of the dimension of outlines of objects of various sizes. In contrast to the length-resolution methods, this technique uses measurement at a single resolution of objects which (supposedly) have the same D. The areas A and perimeters lengths L of the objects ("islands") are measured at a single resolution and plotted in logarithmic scale. The dimension D of the coastlines of the islands is estimated from the expression:

$$A \propto L^n \qquad (14)$$

where n is the slope of the linear regression of $\log(A)$ on $\log(L)$ and

$$D = \frac{2}{n} \qquad (15)$$

This method has been used mostly in materials science for analysis of the geometry and dimension of the fracture surfaces.

8.4.5 Mass-radius relation

The mass-radius relation[3] is useful for estimating the dimension of cluster-like objects (networks, blood vessels, aggregation clusters). It consists of selecting an origin point in the object (usually the center of mass) and counting the number of particles (mass = pixels) that make up the object at a radius r from the origin. For a two-dimensional Euclidean object (a plane), the mass-radius relation is

$$M(r) \propto r^2 \quad (16)$$

For example, the area of a plane under discs of increasing size is proportional to the square of the radius of the measuring disc. The exponent is therefore the dimension (in effect, a square is two-dimensional), but the mass of a fractal object embedded in two dimensions changes with a fractional exponent:

$$M(r) \propto r^D \quad (17)$$

and the fractal dimension $D_{mass\text{-}radius}$ is obtained from:

$$D_{mass\ radius} = \frac{\log(M(r))}{\log(r)} \quad (18)$$

and computed as the slope of the linear regression of $\log(M(r))$ on $\log(r)$.

8.4.5.1 Radius of gyration

Similar to the mass-radius relation, the "radius of gyration" R_g of an object also can be used to estimate D.[69] In growth processes (such as bacterial colonies or electrodeposition clusters), in which the cluster size (in number of particles N) can be monitored as a function of time, R_g is calculated as:

$$R_g = \sqrt{\frac{1}{N}\sum_{i=1}^{N} r_i^2} \quad (19)$$

where N is the number of particles in the cluster at a given time, and r_i is the distance from the ith particle to the center of mass of the cluster. If R_g can be calculated for several N and the object is self similar, then the following relation allows computation of the fractal dimension:

$$N \sim R_g^D \quad (20)$$

The slope of the linear regression of $\log(N)$ on $\log(R_g)$ is D.[69]

8.4.5.2 Pair correlation function

Also related to the mass-radius relation is the pair correlation function, $C(r)$,[4] which is the probability of finding a mass particle at a distance r from an existing particle at x. For a fractal it decays according to the power law:

$$C(r) \propto r^{-\alpha} \qquad (21)$$

where α = (embedding dimension – fractal dimension), and the pair correlation function is computed from:

$$C(r) = \frac{1}{N} \sum_r |p(x)p(x+r)| \qquad (22)$$

where N = number of pixels, and p is a function of the probability of finding an occupied pixel between x and $x + r$ distances from the starting point; p is 1 for occupied sites and 0 for unoccupied sites. For an object embedded in two dimensions (the image is two dimensional), the slope of the regression of log($C(r)$) on log(r) is then ($D - 2$) where D is the fractal dimension:

$$D_{pair\ correlation} = 2 + \frac{\log(C(r))}{\log(r)} \qquad (23)$$

While the mass-radius relation usually is used as an estimator from a single starting position (as previously mentioned, most often the center of mass or symmetry), the $C(r)$ is an average over the object from different origins. The values of r usually range from a few pixels to values smaller than the size of the whole object. Special care is necessary when calculating $C(r)$ from origins near the borders of the object in order to avoid the empty areas outside the object. When these procedures are implemented graphically in computers and image analyzers (using square pixels) there are two sources of error which must be taken into account.[70] The first is associated with underestimation of area of the "circle scan" in a square matrix; the second arises from the overestimation of areas at small radii (circles of small radii have more pixels than the area of a theoretical circle). These errors may be avoided by using the sandbox method,[71] which uses a square box of side ε rather than a circle of radius r.

One advantage of mass-radius and related methods is that they give "pointwise" measurements,[72] which may be of interest in situations such as quantification of structures with special symmetries, data sets with different local fractal dimensions,[20] or multifractals.[66] The mass-radius relation, radius of gyration, and pair correlation function techniques can be extended easily to three dimensions by using "balls" of radius r (or cubes of side ε).

8.4.6 Fragmentation dimension

This type of fractal dimension is derived from Korčak's empirical law[73] and is not necessarily a measure of boundary complexity, but instead is a size distribution dimension. The relation between number (distribution) and size was first discussed by Korcak in relation to the number of islands and their areas in archipelagos. The relation was empirically obtained as:

$$N(A > a) = F a^{-B} \qquad (24)$$

in which the number of islands N of area A larger than a value a scales as that area to an exponent $-B$ (the fractal fragmentation dimension) and a constant F. This type of distribution relation has been found in many natural systems (crater sizes,[74] cellular replicating models,[75] fractures of materials[76]). If the system under investigation is composed of "islands" or particles on a background, the area of all the islands in the image are estimated and then sorted in ascending order. Starting with the smallest island size a, the number of islands of size larger than a, $N(A > a)$, is determined and this procedure is repeated for all sizes (a large range of a). Since

$$B = -\frac{\log(N(A > a))}{\log(a)} \quad (25)$$

B is the negative of the slope of the linear regression of $N(A > a)$ on $\log(a)$. Although Korčak's claimed that B was equal to 0.5 for real archipelagos, Mandelbrot[3] found it to be always larger than 0.5. Furthermore, when the system is composed of fractal-shaped elements (such as real islands), the relation can be extended to:

$$B = \frac{D}{2} \quad (26)$$

where D is the fractal dimension of the coastlines. However, this relation between fractal dimension of coastlines and fractal fragmentation dimension does not always hold (a set of fractally distributed Euclidean objects is possible), and in these cases, the fractal fragmentation dimension becomes a measure of the scaling distribution of sizes (fragmentation) in the system.

8.4.7 Other methods

Most of the methods described above make the implicit assumption that the space in which the supposedly fractal objects are embedded is isotropic. In the image, this means that the choice of an x coordinate is equivalent to the choice of the y coordinate (up or down and left or right are relative). This is true in many cases; for example, the outline of an island observed from the zenith has no special direction in the x-y directions, but real landscape silhouettes are not isotropic because the gravitational force operates in the vertical direction. In these cases, the x and y directions mean different things and scale differently. The same situation occurs with events in time, such as certain time records, waves, and noises.[3,77] Rescaling the x-axis (time) by a fixed value requires the y-axis (amplitude) to be scaled by a different value if the image is to remain invariant. This quality is called *self-affinity*. As a consequence of unequal scaling ratios (called *affine transformation*), some of the procedures above (such as the box-counting and yardstick methods) will give D estimates which depend on the choice of scales for the coordinate system. One simple method of investigating these unequal ratios of self-affine curves has been devised by Matsushita and Ouchi.[78] The following techniques allow the analysis of signals, time series, and profiles and, if they are fractal, also allow estimation of their dimensions.

8.4.7.1 Fourier analysis
The procedure consists of analyzing the signal (a time record, a surface cross-section, etc.) in the frequency domain using Fourier analysis (usually the fast Fourier transform, or FFT),[79] which describes the original data as the sum of a set of sine and cosine waves of different frequencies. The power spectrum $S(f)$ or power spectral density is then calculated from the mean sum of squares of the sine and cosine

amplitudes for each frequency f. For self-affine signals, there is invariance in the spectral power related to frequency[79] that follows the relation:

$$S(f) \propto \frac{1}{f^\beta} \qquad (27)$$

where β can be estimated from the slope of the linear regression of $\log(S(f))$ on $\log(f)$. Furthermore, β relates to the fractal dimension:

$$D_{spectral} = E + \frac{3-\beta}{2} \qquad (28)$$

where E is the Euclidean space in which the signal is embedded ($E = 1$ for time series, 2 for surfaces). Note that for signals with spectral densities $1 < \beta < 3$ (fractional Brownian motion models in one dimension[80]) and D ranges from 1 to 2. For series with β outside this range, it may be appropriate to use integration or differentiation of the series[79] and to characterize those fluctuations instead. For a fractal series with power spectrum β, the corresponding differentiation yields a series with β reduced by 2, while integration increases β by 2. Therefore, fractional Brownian noises with $-1 < \beta < 1$ correspond to the differentiation of fractional Brownian motion series.[79]

The spectral technique can be used to analyze surfaces, in which case a two-dimensional FFT is performed and the average radial power spectrum calculated. Another possibility for surface analysis is to obtain cross-sections of the surface (thus reducing the dimension by 1) and to calculate the spectral density of the profiles as if they were a time series; 1 is then added to the value of D as obtained in Equation (28). These procedures are appropriate for imaging purposes (where the height z of the image is the intensity of the pixel at location x,y), but for real surfaces, there might be "hangouts" or "tunnels" in the surface which will not be recognized and therefore will not be recorded. Furthermore, some preprocessing of the series is usually necessary (such as the elimination of trends in the data and subsequent "windowing") so that the series begins and ends with zero.[81]

8.4.7.2 Root-mean-square fluctuation

A procedure for characterizing correlations of time records is the root-mean-square (RMS) fluctuation.[82] For a quantity which varies with time steps, the fluctuation $f(l)$ of such quantity after l steps is the sum of the individual increments $u(i)$ from time $i = 1$ to l, and the difference d_l of $f(l)$ over a distance l:

$$d_l = f(l_0 + l) - f(l_0) \qquad (29)$$

The RMS fluctuations $F(l)$ of d_l are calculated as the square root of the difference between the average of the square and the square of the average:

$$F(l) = \sqrt{\overline{d_l^2} - (\overline{d_l})^2} \qquad (30)$$

averaging over all the possible locations of l_0. If the analysis is repeated for various lengths of l and the fluctuation has no characteristic length (it is fractal), then F is expected to scale as:

$$F(l) \sim l^\alpha \qquad (31)$$

where α is also related to the spectral density exponent β of the increments u by the relation:

$$\alpha = \frac{\beta + 1}{2} \qquad (32)$$

and to the fractal dimension D of the series:

$$D - 2 - \alpha \qquad (33)$$

for $0 \leq \alpha \leq 1$. When $\alpha = 0.5$, the data has no long-range correlations (the phenomenon is a random event); for $\alpha > 0.5$, the data is increasingly positively correlated (if the data trend is now going up, it will tend to go up in the future), while for $\alpha < 0.5$ it has negative correlations (if the data trend is now going up, it will tend to go down in the future or vice versa). For example, for a fractional Brownian motion process in which the underlying fluctuation u is uncorrelated (Gaussian white noise with $\beta = 0$), $\alpha = 0.5$ and the series (which is an integration of the fluctuations u) has $D = 1.5$.

8.4.7.3 Hurst's rescaled range analysis

Using Hurst's rescaled range analysis (R/S), a fractal time series can be characterized by the Hurst exponent H which relates to the fractal dimension of the time series by:

$$D_H = 2 - H \qquad (34)$$

for $0 \leq H \leq 1$. Note that for fractional Brownian motion models,[80] Equations (33) and (34) are equivalent, but here H is an estimator of the stationary process that (by integration) is responsible for the time series. To estimate H, a time period (the lag τ) shorter than the total record is chosen, then the time series is divided in nonoverlapping parts of length τ. For each lag, the increments I between every two adjacent times are calculated (the fluctuations or change in the signal) and its average A computed, after which a new set of increments (I-A) are calculated. The range of the new increments $R(\tau)$ is defined as the maximum minus the minimum values and $S(\tau)$ is the standard deviation. The procedure is then repeated for various sizes of τ. It was found empirically that for certain natural phenomena:[77, 83, 84]

$$\frac{R(\tau)}{S(\tau)} \sim \tau^H \qquad (35)$$

where H can be estimated by the linear regression of $\log(R(\tau)/S(\tau))$ on $\log(\tau)$. For a random record, $H = 0.5$, while for most natural phenomena $H > 0.5$. Mandelbrot relates H to the "persistence" of the data (for $H > 0.5$, the data is called *persistent*, while for $H < 0.5$, it is *antipersistent*). At least two variants of R/S analysis exist for estimating H called G Hurst and F Hurst, but according to Wallis and Matalas,[85] R/S analysis is a biased estimator of H (overestimating for $H < 0.7$ and underestimating for $H > 0.7$). The bias is reduced for increasingly large data sets.

8.4.7.4 Semivariogram method

The semivariogram[86] $\gamma(\tau)$ of a time series or a profile sample (the profile may represent a "section" through an image to characterize texture) is calculated as:

$$\gamma(\tau) = \frac{1}{2N(\tau)} \sum_{i=1}^{N(\tau)} (z(i) - z(i+\tau))^2 \qquad (36)$$

where $N(\tau)$ is the number of pair measurements separated by a distance τ, and $z(i)$ is the elevation at the point i. For a fractal sample,[80] the semivariogram is expected to scale within a large range of lags τ where:

$$\gamma(\tau) \sim a\, \tau^{2H} \qquad (37)$$

where a is a constant and H is the Hurst exponent estimated from half the slope of the linear regression of $\log(\gamma(\tau))$ on $\log(\tau)$. Thus, D can be computed using Equation (34).

8.4.7.5 Other proposed methods for estimation of dimension

Further methods (some of them slight variations of the ones mentioned above) have been described in the literature in specific applications such as the relative dispersion analysis, correlation analysis,[87] maximum likelihood estimator,[88] maximum entropy method, and roughness-length method.[81]

8.4.8 Fractal harmonics

Clark[89] devised an approach to describe macroscopic shape attributes in particle analysis called fractal (or polygonal) harmonics. These harmonics relate to the yardstick method, in which, after completing the measurement of a boundary, one arrives either at the starting point (perfect fit) or there is a remainder. Instead of stopping after one cycle of measurements, the procedure is repeated again and again from the ending points of the previous cycle. The author found that for most shapes at certain yardstick lengths, the walk on the boundary finally "walks on its own footsteps" forming a regular polygon with n vertices. The "persistence" of the nth fractal harmonic has been used to characterize particle shape at large scales (near the size of the object) together with fractal dimension to characterize the boundary complexity. No applications of fractal harmonics have been reported so far in medical imaging.

8.4.9 Multifractals and local dimensions

It was stated at the beginning of this section that the object to be analyzed is a binary image. In many instances, one can "extract" the object of interest (or, in image analysis terms, "segment" the object) by, for example, thresholding the image based on the optical density histogram. In these cases, a pixel "belongs" to the object if its optical density is larger than or equal to the thresholding value (or vice versa); however, in other instances the data cannot be easily defined as a set. An example of this is a grey-scale function in an image, where it is not an object that is being characterized but the grey-scale function itself (texture characterization). Fractal geometry can deal with situations where there are no binary "sets" but "measures" (in this particular example, the optical density can be considered a measure[90]), and this area is the study of multifractals.[91] Regarding morphology, the concept of fractal geometry is playing an important role in the understanding and characterization of shape complexity, but sometimes, even though objects can be segmented as a set, they still are not adequately quantified by a single D because different parts may have different attributes — for example the outline of a focus of malignant cells surrounded by normal tissues. In such cases, the estimation of D with the methods described above is global, and it may or may not reflect deviations of D in subparts of the object. One method of investigating multifractality is related to box counting, but instead of counting the number of boxes that contain any part of the figure, the probability $P(m,\varepsilon)$ of finding m points in a box of side ε is estimated over different origins of the boxes and for different box sizes ε.[79] The q moments (the value of the qth power of a variable, where q is the integer indexing the set of moments) are defined as:

$$M^q(\varepsilon) = \sum_{m=1}^{N} m^q P(m,\varepsilon) \tag{38}$$

for $q \neq 0$, while for $q = 0$, the configuration entropy is used:

$$S(\varepsilon) = \sum_{m=1}^{N} \log m P(m,\varepsilon) \tag{39}$$

The scaling exponents D_q are

$$D_q = \frac{1}{q} \frac{\log(M^q(\varepsilon))}{\log(\varepsilon)} \tag{40}$$

for $q \neq 0$, while for $q = 0$ it is

$$D_0 = \frac{1}{q} \frac{S(\varepsilon)}{\log(\varepsilon)} \tag{41}$$

For a uniform fractal, the same D_q is expected for all the moments. When D_q takes different values, the set is considered multifractal.[10,79]

Using another multifractal approach, subsets of the original object can be quantified and D expressed in terms of its distribution for a particular subset scale size. D becomes a local dimension (D_{local}) relative to an origin located at a particular point in the set and a scale length l. To implement the local dimension analysis, one can use a procedure such as the mass-radius relation outlined above, for small-scale lengths centered at all possible origins in the set. The D_{local} then becomes an estimate around that origin, and instead of a single D the object is characterized by a spectrum of local Ds. The concept has been extended further to the local dimension of the connected pixels to the origin.[20] This approach allows the effects of neighboring structures that are not locally connected to the set to be neglected (for example, the dimension of a straight line will not be affected by a neighboring two-dimensional object if it is not locally connected to it, thus preserving $D_{local} = 1$ for the line).

8.4.10 Lacunarity

The fractal dimension is a number that characterizes an object's geometry, but this may not be enough to differentiate between two fractal objects; they may share the same D but still look very different. Mandelbrot[2] has proposed two additional parameters to characterize the "texture" of fractals: lacunarity (*lacuna* is Latin for gap) and succolarity (from *subcolare*, "to almost flow through"). There are no formal definitions of these two parameters but there are some alternative interpretations of lacunarity. For the mass-radius relation, the proportionality in Equation (17) also involves a prefactor F which corresponds to the y intercept in the linear regression of $\log(M(r))$ on $\log(r)$. If we estimate D_{local} for all possible origins in a fractal, we also can investigate the discrepancy in the expected value of $\bar{F}$ (the mean F). Mandelbrot proposes the second order expression:

$$\Lambda = \overline{\left(\frac{F}{\bar{F}} - 1\right)^2} \tag{42}$$

where the horizontal lines mean average. Note that in the case where F and $\bar{F}$ are equal, $\Lambda = 0$. Obert[92] has investigated lacunarity in deterministic fractals using the mass-radius relation, giving numerical examples and estimates of D and Λ.

Another alternative definition of lacunarity[79] relates to a variation of the box-counting method mentioned earlier, involving the probability $P(m,\varepsilon)$ of finding m points in a box of side ε. For a fixed size ε, $\Lambda(\varepsilon)$ is defined as:

$$\Lambda(\varepsilon) = \frac{\langle M^2(\varepsilon)\rangle - \langle M(\varepsilon)\rangle^2}{\langle M(\varepsilon)\rangle^2} \tag{43}$$

where the statistical moments $M(\varepsilon)$ and $M^2(\varepsilon)$ are calculated using Equation (38):

$$M(\varepsilon) = \sum_{m=1}^{N} mP(m,\varepsilon) \tag{44}$$

and

$$M^2(\varepsilon) = \sum_{m=1}^{N} m^2 P(m,\varepsilon) \tag{45}$$

These methods have been used together with an estimation of D to describe and segment images of natural textures.[93] Another possible measure of lacunarity recently has been proposed by Mandelbrot[94] based on the filling rate of ε-neighborhoods (related to the dilation method described above).

8.4.11 Implementation details

It may be useful to enumerate some common issues that arise during implementation of the methods described above.

8.4.11.1 Image digitization/binarization

Transferring real images to a computer via a camera or scanner usually involves some kind of thresholding procedure to isolate elements of the picture. If the dimension of the thresholded object is to be estimated, the thresholding procedure must not introduce spurious detail in the image. This is particularly important when the object contrast with the background is not very good. As there might be complex relations in the grey-scale function of the image, it is a good idea to investigate various thresholds and to determine the relation of thresholding levels with estimates of D.

8.4.11.2 Dimension reduction, projection, and zerosets

It is possible to reduce the dimension of a geometric object by one (whether Euclidean or fractal) when it intersects a plane. Thus, a sphere is reduced to a disc, a disc to a line, and a line to a point. By intersecting a fractal embedded in three dimensions, a two-dimensional section is obtained. The fractal dimension of the whole object can be obtained theoretically as the fractal dimension of the object in the section +1. Interestingly, for a self-affine fractal, such as a fractional Brownian motion, it is possible to reduce it to a self-similar set of points of dimension $D-1$ — or, $D = 1-H$, modifying Equation (34) — by intersecting the motion with the time axis and producing a set of disconnected points called the *zeroset*.[79] For the same reason, landscapes (self-affine surfaces) produce islands with self-similar boundaries (in this case the "water line" acts as a plane section).

In general, it is assumed that the projection of a fractal onto a plane maintains its dimension unchanged only if D is smaller than the dimension of the plane (the same applies to higher dimensions). This is an important issue in image analysis because image capture (unless it is the image of a section) projects a three-dimensional structure into two dimensions. According to Vicsek,[71] there are only heuristic arguments supporting the preservation of D under projection and considerable deviations may occur when D is slightly smaller that the plane. Therefore, for fractal objects with $D > 2$ (such as a branching structure in three dimensions), the analysis of the mere projection onto two dimensions is not a valid procedure for estimating D.

8.4.12 How fractal are biological fractals?

In a strict sense, one can very appropriately ask whether biological structures are *really* fractal. Although fractal geometry has been called "the geometry of Nature", there is some controversy over applying the concept of fractals to biological structures because of the existence of limited ranges of self similarity. Surely blood-vessel trees have some kind of self-similarity (vessel branching is repeated at several scales), and when analytical tools are applied, the patterns reveal fractal dimension. However, this is usually within a finite range, while ideal fractals have self-similarity *ad infinitum*. One finite limit is the lowest resolved scale, that is, the size of the object itself (the characteristic size of the object); the other, the elementary constituents of the structure (using the example of blood vessels, it is the smallest capillary vessels after which no further branching occurs). It may be more appropriate to say that there is a fractal structure within a range of scales; the same applies to real coastlines, but in this case the range of fractality is extremely large. The lower limits for fractality usually depend on the fundamental structural elements of the object under study. If we are considering a tree, below the smallest branch scale there is no more "tree structure" but only a part of it. This limited self-similar domain has been used to question if natural shapes are fractals; while fractals in physics show scaling properties over large orders of magnitude, in biological systems this range is rarely above three orders of magnitude. However, we may argue that in this sense, fractal geometry is an idealization much like Euclidean geometry: strictly speaking, how many ideal Euclidean shapes exist in nature?

The problem of limited self-similarity is not trivial because those limits need to be defined by some means. When actually performing the fitting of data in a log-log plot (for example, perimeter length vs. yardstick length), usually a linear regression by the least-squares method is used to find the slope that relates to the exponent of the power law and to D. Although there is no formal method of deciding which ranges to include in the estimation of the slope, there are several alternative procedures.

8.4.12.1 Arbitrary range

When the constitutive elements of an object are known (such as cell size in a tissue or a bacterium in a colony) one can expect that below the cell size the fractal structure of the tissue will vanish (there can, of course, be further scaling structure at the cell scale, but not as a tissue) and so these sizes can be used as cutoffs. When the limits are not clear, the problem can be approached as proposed by Nonnenmacher,[95] using an automatic procedure based on the usual least-squares fit that searches for the widest range of scales within which the standard deviation of the estimated slope does not exceed a given (although still arbitrary) limit.

8.4.12.2 Local slope

When the scaling relation in the log-log plot is not linear, a useful procedure is to estimate the changes in local slopes throughout the plot.[96] The set of local slopes has been used as a "fractal signature" of grey-scale texture images by Peleg et al.[97] Instead of deriving a power-law characteristic of a fractal, the authors measured the distance between signatures and were able to classify textures as to whether they were fractal or not.

When the log-log plots used to obtain the scaling relation show different slopes within different ranges, this may indicate different scaling properties.[76] One can calculate these scale-bounded slopes and express a different scaling relationship at each scale range.[26]

8.4.12.3 Equally spaced data points

The log-log plots of the variables usually produce "clumping" of data, and this can affect the least-squares fitting procedure. This is particularly noticeable in power spectrum estimates since all the frequencies are present (there will be much more data points at higher frequencies in the log scale). A useful procedure to overcome this situation is to reduce the data points into log-spaced "bins" (and calculate mean values for those bins in the x and y coordinates) which will allow the least squares to be equally weighted throughout the plot.[98]

8.4.12.4 Extended fractal models

As mentioned before, many natural objects show limited ranges of self similarity. Beyond those limits, the geometry tends to be Euclidean (and produces a concavity in the log-log plots). It has been proposed that all data (instead of some ranges) be used to characterize the fractal behavior plus any transition to Euclidean geometry. After all, in a log-log plot characterizing boundary length vs. yardstick length, all the points are real data that describe the geometry of the object; by not considering the points that do not fit the power law, we neglect geometric data of the object at those scales.

8.4.12.4.1 Asymptotic fractal model.

This model developed by Rigaut[6-8] proposes that the relation of boundary length to resolution in most biological shapes naturally has two asymptotes: one is a fractal asymptote (c) at low resolutions (same meaning as in the yardstick method) and the other an Euclidean asymptote (that is, a finite maximum boundary length (B_m)) at high resolutions. The asymptotic fractal formula takes the form:

$$B(\varepsilon) = \frac{B_m}{1 + \left(\frac{\varepsilon}{L}\right)^c} \quad (46)$$

where $B(\varepsilon)$ is the boundary length measured with a yardstick of size ε, B_m is the asymptotic maximum boundary for $\varepsilon \to 0$ (Euclidean asymptote), L is a constant of position of the crossover, and c is the asymptotic fractal dimension minus the boundary's topological dimension (that is, $D - Dt$) for $\varepsilon \to \infty$. The estimation of the values of B_m, L, and c can be achieved by two iterative procedures borrowed from enzyme kinetics, namely those of Hill-Wyman and Lineweaver-Burk. The Hill-Wyman method involves searching for the highest coefficient of determination (R^2) of the linear regression of $\ln((B_m - B(\varepsilon))/B(\varepsilon))$ on $\ln(\varepsilon)$. The iteration is on B_m, (starting with B_m as the estimate for the smallest yardstick size $B(\varepsilon) + 1$, to avoid $\ln(0)$), using either an optimization searching subroutine or a monotone increase of B_m. When the highest R^2 is found, the linear regression line with formula $y = a + bx$ is used to estimate the parameters c and L:

$$c = b \quad (47)$$

and

$$L = e^{(-a/b)} \quad (48)$$

where e is the base of natural logarithms (~2.718281...). Interestingly, when the object is fractal, no solution is possible, since $B_m \to \infty$ for $r \to 0$ and then R^2 increases very slowly. In such a case, an arbitrary value of B_m (for example, $B_m > B_1^* 10$) can be used to terminate the procedure, and the object is considered as fractal rather than asymptotic fractal. The second procedure is known as the Lineweaver-Burk plot, which searches for the highest coefficient of determination (R^2) of the linear regression of $1/B(\varepsilon)$ on ε^c, iterating on c. When the highest R^2 is found, the linear regression line with formula $y = a + bx$ is used to estimate the parameters B_m and L:

$$B_m = \frac{1}{a} \quad (49)$$

and

$$L = e^{\frac{\ln(a/b)}{c}} \quad (50)$$

where e is the base of natural logarithm and $\ln$ indicates the natural logarithm. An extensive chapter on this subject can be found in Reference 8.

8.4.12.4.2 Log-logistic approach. Sernetz et al.[99-101] have proposed a model of limited self-similarity based on the logistic function which assumes a sigmoid-shaped log-log plot with two horizontal (Euclidean) asymptotes at B_{min} and B_{max}. The model has been used to measure porosity of organic tissue (cartilage, cornea).[102]

Nonnenmacher[95] and Struzik and Dooijes[103] have investigated other formulae to fit log-log plots with an upper Euclidean asymptote.

Acknowledgment

The author wishes to thank Dr. J. W. Rippin, Oral Pathology Unit, and Dr. R. M. Shelton, Biomaterials Unit, School of Dentistry, The University of Birmingham, for many fruitful discussions during the preparation of the manuscript.

References

1. *Churchill's Illustrated Medical Dictionary*, Churchill Livingstone, London, 1989.
2. Mandelbrot, B. B., *The Fractal Geometry of Nature*, W.H. Freeman, New York, 1982.
3. Mandelbrot, B. B., How long is the coast of Britain? Statistical self-similarity and fractional dimension, *Science*, 156, 636, 1967.
4. Bunde, A., and Havlin, S., *Fractals and Disordered Systems*, Springer-Verlag, New York, 1991.
5. Payne, C. M., Bjore, C. G., Cromey, D. W., and Roland, F., A comparative mathematical evaluation of contour irregularity using form factor and PERBAS, a new analytical shape factor, *Anal. Quant. Cytol. Histol.*, 11, 341, 1989.
6. Rigaut, J. P., An empirical formulation relating boundary lengths to resolution in specimens showing "non-ideally fractal" dimensions, *J. Microsc.*, 133, 41, 1984.
7. Rigaut, J. P., Fractals, semi-fractals et biomètrie, in *Fractals, Dimensions Non Entières et Applications*, Cherbit, G., Ed., Masson, Paris, 1987.
8. Rigaut, J. P., Fractals in biological image analysis and vision, in *Gli Oggetti Frattali in Astrofisica, Biologia, Fisica e Matematica*, vol. 3, Losa, G., Merlini, D., and Moresi, R., Eds., CERFIM, Locarno, 1991.
9. Landini, G., and Rippin, J. W., An "asymptotic fractal" approach to the morphology of malignant cell nuclei, *Fractals*, 1, 326, 1993.
10. Losa, G. A., Baumann, G., and Nonnenmacher, T. F., The fractal dimension of pericellular membrane from lymphocytes and lymphoblastic leukemic cells, *Acta Stereol.*, 11 (suppl.), 1992, 335.
11. Losa, G., Fractal properties of pericellular membrane from lymphocytes and leukemic cells, in *Fractals in Biology and Medicine*, Nonnenmacher, T. F., Losa, G. A., and Weibel, E. R., Eds., Birkhäuser-Verlag, Basel, 1994, 190.
12. Anneroth, G., Batsakis, J., and Luna, M., Review of the literature and a recommended system of malignancy grading in oral squamous cell carcinomas, *Scand. J. Dent. Res.*, 95, 229, 1987.
13. Willen, R., Nathanson, A., Moberger, G., and Anneroth, G., Squamous cell carcinoma of the gingiva. Histological classification and grading of malignancy, *Acta Otolaryngol.*, 79, 146, 1975.
14. Rubio, C. A., Liu, F.-S., and Zhao, H.-Z., Histological classification of intraepithelial neoplasias and microinvasive squamous carcinoma of the esophagus, *Am. J. Surg. Pathol.*, 13, 685, 1989.
15. Jakobsson, P. A., Eneroth, C. M., Killander, D., Moberger, G., and Martensson, B., Histologic classification and grading of malignancy in carcinoma of the larynx, *Acta Radiol. Ther. Phys. Biol.*, 12, 1, 1973
16. Bryne, M., Koppang, H. S., Lilleng, R., Stene, T., Bang, G., and Dabelsteen, E., New malignancy grading is a better prognostic indicator than Broder's grading in oral squamous cell carcinomas, *J. Oral Pathol. Med.*, 18, 432, 1989.

17. Landini, G., and Rippin, J. W., Fractal dimensions of the epithelial-connective tissue interfaces in premalignant and malignant epithelial lesions of the floor of the mouth, *Analyt. Quant. Cytol. Histol.*, 15, 144, 1993.
18. Landini, G., and Rippin, J. W., Fractal dimension as a characterisation parameter of premalignant and malignant epithelial lesions of the floor of the mouth, in *Fractals in Biology and Medicine*, Nonnenmacher, T. F., Losa, G. A., and Weibel, E. R., Eds., Birkhäuser-Verlag, Basel, 1994, 315.
19. Landini, G., and Rippin, J. W., Multifractal analysis of the epithelial-connective junction in oral cancer, *Proc. Intl. Assoc. Oral Pathol.*, York, July 18-22, 1994. Also in: Landini, G., and Rippin, J.W., Tumor shape and local connected fractal dimension analysis in oral cancer and pre-cancer, in *Fractal Reviews in the Natural and Applied Sciences*, Novak, M.M., Ed., Chapman & Hall, 1995.
20. Voss, R. F., and Wyatt, J. C. Y., Multifractals and the local connected fractal dimension: classification of early Chinese landscape paintings, in *Applications of Fractals and Chaos*, Crilly, T., Earnshaw, R. A., and Jones, H., Eds., Springer-Verlag, Berlin, 1993, 171.
21. Family, F., Masters, B. R., and Platt, D. E., Fractal pattern formation in human retinal vessels, *Physica D*, 38, 98, 1989.
22. Masters, B. R., Family, F., and Platt, D. E., Fractal analysis of human retinal vessels, *Biophys. J.*, 55, 575a, 1989.
23. Mainster, M. A., The fractal properties of retinal vessels: embryological and clinical implications, *Eye*, 4, 235, 1990.
24. Misson, G., Landini, G., and Murray, P., Fractals and ophthalmology, *Lancet*, 339, 872, 1992.
25. Daxer, A., Fractals and retinal vessels, *Lancet*, 339, 618, 1992.
26. Landini, G., Misson, G., and Murray, P. I., Fractal analysis of the normal human retinal fluorescein angiogram, *Curr. Eye Res.*, 12, 23, 1993.
27. Witten, T. A., and Sander, L. M., Diffusion-limited aggregation, *Phys. Rev. B*, 27, 5686, 1983.
28. Kiani, M. F., and Hudetz, A. G., Computer simulation of growth of anastomosing microvascular networks, *J. Theor. Biol.*, 150, 547, 1991.
29. Bittner, H. R., Modelling of fractal vessel systems, in *Fractals in the Fundamental and Applied Sciences*, Peitgen, H.-O., Henriques, J. M., and Penedo, L. F., Eds., Elsevier, Amsterdam, 1991, 59.
30. Bassingthwaighte, J. B., Fractal vascular growth patterns, *Acta Stereol.*, 11 (suppl. 1), 305, 1992.
31. Daxer, A., The fractal geometry of proliferative diabetic retinopathy: implications for the diagnosis and the process of retinal vasculogenesis, *Curr. Eye Res.*, 12, 1103, 1993.
32. Landini, G., Misson, G., Simulation of corneal neovascularization by inverted diffusion limited aggregation, *Invest. Ophthalmol. Vis. Sci.*, 34, 1872, 1993.
33. Landini, G., Misson, G., and Murray, P. I., Fractal properties of herpes simplex dendritic keratitis, *Cornea*, 11, 510, 1992.
34. Misson, G. P., Landini, G., and Murray, P. I., Size-dependent variation in fractal dimensions of herpes simplex epithelial keratitis, *Curr. Eye Res.*, 12, 957, 1993.
35. Landini, G., Misson, G., and Murray, P. I., Fractal characterisation and computer modelling of herpes simplex virus spread in the human corneal epithelium, in *Fractals in the Natural and Applied Sciences*, Novak, M. M., Ed., North-Holland, Amsterdam, 1994, 241.
36. Löe, H., Anerud, A., Boysen, H., and Smith, M., The natural history of periodontal disease in man, *J. Periodontol.*, 49, 607, 1978.
37. Becker, W., Berg, L., and Becker, B. E., Untreated periodontal disease: a longitudinal study, *J. Periodontol.*, 50, 234, 1979.
38. Socransky, S. S., Haffajee, A. D., Goodson, J. M., and Lindhe, J., New concepts of destructive periodontal disease, *J. Clin. Periodontol.*, 11, 21, 1984.
39. Goodson, J. M. Clinical measurements of periodontitis, *J. Clin. Periodontol.*, 13, 446, 1986.
40. Haffajee, A. D., and Socransky, S. S., Attachment level changes in destructive periodontal diseases, *J. Clin. Periodontol.*, 13, 461, 1986.

41. Manji, F., and Nagelkerke, N., A stochastic model for periodontal breakdown, *J. Periodont. Res.*, 24, 279, 1989.
42. Landini, G., A fractal model for periodontal breakdown in periodontal disease, *J. Periodont. Res.*, 26, 176, 1991.
43. West, B. J., and Shlesinger, M. F., On the ubiquity of 1/f noise, *Intl. J. Mod. Phys. B*, 3, 795, 1989.
44. Gardner, M., White and brown music, fractal curves, and one-over-f noise, *Sci. Am.*, 238, 16, 1978.
45. Landini, G., Videodensitometrical study of the alveolar bone crest in periodontal disease, *J. Periodontol.*, 62, 528, 1991.
46. Levin, M., A computer model of field-directed morphogenesis. Part I. Julia sets, *Comput. Appl. Biosci.*, 10, 85, 1994.
47. Takahashi M., A fractal model of chromosomes and chromosomal DNA replication, *J. Theoret. Biol.*, 141, 117, 1989.
48. Prusinkiewicz, P., and Lindenmayer, A., *The Algorithmic Beauty of Plants*, Springer-Verlag, New York, 1990.
49. Chant, J., and Herskowitz, I., Genetic control of bud site selection in yeast by a set of gene products that constitute a morphogenetic pathway, *Cell*, 65, 1203, 1991.
50. Chant, J., Corrado, K., Pringle, J. R., and Herskowitz, I., Yeast bud5, encoding a putative GDP-GTP exchange factor, is necessary for bud site selection and interacts with bud formation gene bem1, *Cell*, 65, 1213, 1991.
51. Weibel, E. R., Fractal geometry: a design principle for living organisms, *Am. J. Physiol.*, 261, 1991.
52. Weibel, E. R., Design of biological organisms and fractal geometry, in *Fractals in Biology and Medicine*, Nonnenmacher, T. F., Losa, G. A., and Weibel, E. R., Eds., Birkhäuser-Verlag, Basel, 1994, 68.
53. Iannaccone, P. M., Fractal geometry in mosaic organs: a new interpretation of mosaic pattern, *FASEB J.*, 4, 1508, 1990.
54. Khokha, M. K., Landini, G., and Iannaccone, P. M., Iterated cell division programs may generate the fractal geometry of the mosaic pattern in chimeric rat liver, *Proc. Intl. Conf. Complex Geometry of Nature*, Budapest, August 1993.
55. Iannaccone, P. M., Khokha, M., and Landini, G., Can iterating division rules explain the rapid expansion of parenchyma in liver organogenesis? A fractal analysis in rat chimeras, *Proc. First World Congress on Computational Medicine, Public Health and Biotechnology*, Austin, April, 1994.
56. Paumgartner, D., Losa, G., and Weibel, E. R., Resolution effect on the stereological estimation of surface and volume and its interpretation in terms of fractal dimensions, *J. Microsc.*, 121, 51, 1981.
57. Denley, D. R., Practical applications of scanning tunneling microscopy, *Ultramicroscopy*, 33, 83, 1990.
58. Caldwell, C. B., Stapleton, S. J., Holdsworth, D. W., Jong, R. A., Weiser, W. J., Cooke, G., and Yaffe, M. J., Characterisation of mammographic parenchymal pattern by fractal dimension, *Phys. Med. Biol.*, 35, 235, 1990.
59. Mandelbrot, B. B., Self-affine fractal sets, II: Length and surface dimensions, in *Fractals in Physics*, Pietronero, L., and Tosatti, E., Eds., Elsevier, Amsterdam, 1986, 17.
60. Clark, N. N., Three techniques for implementing digital fractal analysis of particle-shape, *Powder Technol.*, 46, 45, 1986.
61. Flook, A. G., Fractal dimensions: their evaluation and their significance in stereological measurements, *Proc. Third European Symp. Stereology*, 2nd Part, *Acta Stereol.*, 1, 79, 1982.
62. McCracken, D. D., The Monte Carlo method, *Sci. Am.*, 192, 90, 1955.
63. Schwarz, H., and Exner, H. E., The implementation of the concept of fractal dimension on a semi-automatic image analyser, *Powder Technol.*, 27, 207, 1980.
64. Flook, A. G., The use of the dilation logic on the Quantimet to achieve fractal dimension characterisation of textured and structured profiles, *Powder Technol.*, 21, 295, 1978.
65. Barnsley, M., *Fractals Everywhere*, Academic Press, San Diego, 1988.

66. Peitgen, H.-O., Jürgens, H., and Saupe, D., *Chaos and Fractals, New Frontiers of Science*, Springer-Verlag, New York, 1992.
67. Mandelbrot, B. B., Passoja, D. E., and Paullay, A. J., Fractal character of fracture surfaces of metals, *Nature*, 308, 721, 1984.
68. Lovejoy, S., Area-perimeter relation for rain and cloud areas, *Science*, 216, 185, 1982.
69. Matsuyama, T., and Matsushita, M., Self-similar colony morphogenesis by gram-negative rods as the experimental model of fractal growth by a cell population, *Appl. Environ. Microbiol.*, 58, 1227, 1992.
70. Landini, G., and Rippin, J. W., Notes on the implementation of the mass-radius method of fractal dimension estimation, *Comp. Appl. Biosci.*, 9, 547, 1993.
71. Vicsek, T., *Fractal Growth Phenomena*, World Scientific, Singapore, 1989.
72. Theiler, J., Estimating fractal dimension, *J. Optical Soc. Am. A*, 7, 1055, 1990.
73. Korcak, J., Deux types fondamentaux de distribution statistique, *Bulletin de l'Institut International de Statistique*, 3, 294, 1938.
74. Takayasu, H., *Fractals in the Physical Sciences*, Manchester University Press, Manchester, 1990.
75. Landini, G., Rippin, J. W., Fractal fragmentation in replicative systems, *Fractals*, 1, 239, 1993.
76. Kaye, B. H., *A Random Walk Through Fractal Dimensions*, VCH, Weinheim, 1989.
77. Mandelbrot, B. B., and Wallis, J. R., Some long-run properties of geophysical records, *Water Resour. Res.*, 5, 321, 1969.
78. Matsushita, M., and Ouchi, S., On the self-affinity of various curves, *Physica D*, 38, 264, 1989.
79. Voss, R. F., Fractals in nature: from characterization to simulation, in *The Science of Fractal Images*, Peitgen, H.-O., and Saupe, D., Eds., Springer-Verlag, New York, 1988, 21.
80. Mandelbrot, B. B., and Van Ness, J. W., Fractional Brownian motions, fractional noises and applications, *SIAM Rev.*, 10, 422, 1968.
81. Gallant, J. C., Moore, I. D., Hutchinson, M. F., and Gessler, P., Estimating fractal dimension of profiles: a comparison of methods, *Math. Geol.*, 26, 455, 1994.
82. Peng, C.-K., Buldyrev, S. V., Goldberger, A. L., Havlin, S., Sciortino, F., Simons, M., and Stanley, H. E., Long-range correlations in nucleotide sequences, *Nature*, 356, 168, 1992.
83. Mandelbrot, B. B., and Wallis, J. R., Robustness of the rescaled range R/S in the measurement of the noncyclic long run statistical dependence, *Water Resour. Res.*, 5, 967, 1969.
84. Plotnick, R. E., and Prestegaard, K. L., Fractal and multifractal models and methods in stratigraphy, in *Fractals in Petroleum Geology and Earth Processes*, Barton, C. C., and Lapointe, P. R., Eds., Plenum Press, New York, 1995.
85. Wallis, J. R., and Matalas, N. C., Small sample properties of H- and K-estimators of the Hurst coefficient h, *Water Resour. Res.*, 6, 1583, 1970.
86. Olive, M. A., and Webster, R., Semi-variograms for modelling the spatial pattern of landform and soil properties, *Earth Surf. Processes Landforms*, 11, 491, 1986.
87. Schepers, H. E., van Beek, J. H. G. M., and Bassingthwaighte, J. B., Four methods to estimate the fractal dimension from self-affine signals, *IEEE Eng. Med. Biol.*, June, 57, 1992.
88. Fortin, C., Kumaresan, R., Ohley, W., and Hoefer, S., Fractal dimension in the analysis of medical images, *IEEE Eng. Med. Biol.*, June, 65, 1992.
89. Clark, N. N., A new scheme for particle-shape characterization based on fractal harmonics and fractal dimensions, *Powder Technol.*, 51, 243, 1987.
90. Levy-Vehel, J., and Berroir, J. P., Image analysis through multifractal description, in *Fractals in the Natural and Applied Sciences*, Novak, M. M., Ed., North-Holland, Amsterdam, 1994, 261.
91. Evertsz, C. J. G., and Mandelbrot, B. B., Multifractal measures, in *Chaos and Fractals, New Frontiers of Science*, Peitgen, H.-O., Jürgens, H., and Saupe, D., Eds., Springer-Verlag, New York, 1992, 921.
92. Obert, M., Numerical estimates of the fractal dimension D and the lacunarity L by the mass radius relation, *Fractals*, 1, 711, 1993.

93. Keller, J. M., Chen, S., and Crownover, R. M., Texture description and segmentation through fractal geometry, *Comp. Vision, Graphics Image Process.*, 45, 150, 1989.
94. Mandelbrot, B. B., A fractal's lacunarity and how it can be tuned and measured, in *Fractals in Biology and Medicine*, Nonnenmacher, T. F., Losa, G. A., and Weibel, E. R., Eds., Birkhäuser-Verlag, Basel, 1994, 8.
95. Nonnenmacher, T. F., Spatial and temporal fractal patterns in cell and molecular biology, in *Fractals in Biology and Medicine*, Nonnenmacher, T. F., Losa, G. A., and Weibel, E. R., Eds., Birkhäuser-Verlag, Basel, 1994, 22.
96. Sernetz, M., Wubbeke, J., Wlczek, P., Three-dimensional image analysis and fractal characterization of kidney arterial vessels, *Physica A*, 191, 13, 1992.
97. Peleg, S., Naor, J., Hartley, R., and Avnir, D., Multiple resolution texture analysis and classification, *IEEE Trans. Pattern Anal. Machine Intelligence*, 6, 518, 1984.
98. Voss, R. F., The fractal dimension of percolation cluster hulls, *J. Phys. A*, 17, L373, 1984.
99. Sernetz, M., Bittner, H. R., Willems, H., and Baumhoer, C., Chromatography, in *The Fractal Approach to Heterogeneous Chemistry*, Avnir, D., Ed., John Wiley & Sons, Chichester, 1989, 61.
100. Bittner, H. R., and Sernetz, M., Self-similarity within limits: description with the log-logistic function, in *Fractals in the Fundamental and Applied Sciences*, Peitgen, H.-O., Henriques, J. M., and Penedo, L. F., Eds., Elsevier, Amsterdam, 1991, 47.
101. Bittner, H. R., Limited self-similarity, in *Fractal Geometry & Computer Graphics*, Encarnação, J. L., Peitgen, H.-O., Sabas, G., and Engleth, G., Eds., Springer-Verlag, Berlin, 1992, 213.
102. Sernetz, M., Bittner, H. R., Bach, P., Glittenberg, B., Fractal characterization of the porosity of organic tissue by interferometry, in *Characterization of Porous Solids II*, Rodríguez-Reinoso, F., Sing, K. S. W., Rouquerol, J., Eds., Elsevier, Amsterdam, 1991, 141.
103. Struzik, Z. R., and Dooijes, E. H., Towards fractal metrology, in *Fractals in the Natural and Applied Sciences*, Novak, M. M., Ed., North-Holland, Amsterdam, 1994, 417.

Section V
Organs

chapter nine

Fractals and the Heart

Ary L. Goldberger, Chung-Kang Peng, Jeffrey Hausdorff, Joseph Mietus, Shlomo Havlin, and H. Eugene Stanley

> *Yes, for all these years, we have been living with fractal arteries, not far from fractal river systems, draining fractal mountainscapes under fractal clouds, toward fractal coastlines.*
>
> M. Schroeder

9.1 Introduction

This chapter is intended to provide an overview of selected applications of fractals to the structure and dynamics of the heart. Self-similarity appears to be a fundamental property of the healthy function of the human cardiovascular system at both microscopic and macroscopic levels of observation. Furthermore, quantifying the spatial and temporal breakdown of fractal organization promises to provide important new diagnostic and prognostic measures for clinicians.

We begin, therefore, with a brief review of fractal-like anatomic structures in the cardiovascular system. Next, we describe in detail one specific application of fractal analyses to the understanding of cardiac dynamics — namely, beat-to-beat regulation of heart rate. We show how the fractal concept provides a unified framework for quantifying the complex fluctuations and long-range, power-law correlations of the heartbeat observed in healthy subjects. Then, we show how disruption of the regulatory mechanisms in individuals with potentially lethal cardiovascular syndromes, such as congestive heart failure, can lead to alteration of this fractal behavior.

The importance of fractals in cardiac physiology relates intimately to their connection with long-range power-law correlations.[1-3] This topic is discussed in more detail in Chapter 2, which describes long-range correlations between nucleotides in noncoding DNA sequences. Traditionally, correlations in physiology are usually assumed to be short-range in nature, decaying in an exponential fashion. Processes which display short-range correlations are usually simulated using Markov chain models. In contrast, a variety of different physical processes near so-called critical points display long-range correlations which decay, not exponentially, but more slowly as a power law.[4] From a physiological perspective, such long-range correlations might confer some adaptive benefits, since they indicate the presence of a type of "memory" effect. In a dynamical system with long-range correlation properties, the current value of some variable, e.g., heart rate, will be influenced not just by very recent events, but also by much more remote ones. Not all fractal systems generate nontrivial, long-range power correlations;[2]* however, it appears that fractal mechanisms are necessary long-range (spatial or temporal) correlations. Such long-range (fractal) correlations appear to be an important organizing principle for physiological structures and processes that lack a characteristic (single) scale of length or time. What is the evidence for fractal organization of the human cardiovascular system?

9.2 Fractal geometry of the heart

A number of cardiopulmonary structures have a fractal-like appearance (Table 1).[1,5-9] Examples of self-similar architechures include the arterial and venous trees (Figure 1) and the branching of certain cardiac muscle bundles (Figure 2), as well as the ramifying tracheo-bronchial tree and the His-Purkinje network (Figure 3). The latter provides a highly efficient mechanism for rapidly distributing the electrical impulse to the ventricles.[10] Recently, there has been interest in modeling the electrogenesis of certain electrocardiographic (EKG) waveforms using a fractal-like conduction system, as well as for studying alterations in the frequency content of the normal EKG due to changes in His-Purkinje geometry or in myocardial conduction properties. For example, Abboud et al.[11] have shown that abnormally slow conduction in myocardial cells activated by such a fractal network can lead to "late potentials" or selective attenuation of higher frequency content of the QRS complex, simulating changes observed in certain pathologic states such as myocardial infarction.

Table 1 Examples of Fractals in Cardiovascular Physiology

I. Structural
   A. Vascular: arterial and venous trees
   B. Muscular:
      1. Hierarchical organization of muscle bundles
      2. Branching of certain intracardiac muscles
   C. Electrical: His-Purkinje network
   D. Connective tissue: chordae tendineae; aortic valve leaflets

II. Dynamical ($1/f$ spectra)
   A. Regulation of healthy heartbeat fluctuations
   B. Regulation of healthy beat-to-beat blood pressure fluctuations

From an information theory viewpoint, these self-similar cardiopulmonary architectures all serve a common physiologic function: rapid and efficient transport over a complex, spatially distributed system. In the case of the electrical conduction system, the quantity being transported is the electrical stimulus that regulates the timing of cardiac contraction. In the case of the vasculature, fractal branchings provide a rich, redundant network for distribution of O_2 and nutrients and for the collection of CO_2 and other waste products of metabolism. The relationship between fractal branching of the vasculature and the *heterogeneous* pattern of blood-flow distribution in the myocardium has been described by Bassingthwaighte and colleagues.[9,12] The fractal tracheo-bronchial tree provides an enormous surface area for exchange of gases at the vascular-alveolar interface, coupling pulmonary and cardiac function.

---

* For example, the classical random walk model of Brownian motion is a fractal process with only trivial long-range correlations (see Section 9.3.1 below).

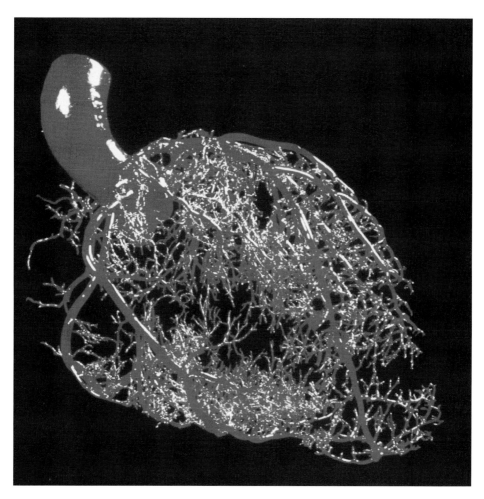

Figure 1 Cast of the human coronary artery tree. This rich, redundant network supplies blood to the heart muscle with enormous efficiency and reserve. (Courtesy of Ciba-Geigy.)

In a less apparent way, fractal geometry also underlies important aspects of cardiac mechanical function. For example, the mitral and tricuspid valves are anchored to the underlying papillary muscles by a branching canopy of connective tissue strands called the *chordae tendineae*.[5] Peskin and McQueen[13] have shown how fractal organization of connective tissue in the aortic valve leaflets relates to the distribution of mechanical forces. The branching pattern of certain muscle bundles within the heart is also fractal-like (see Figure 2). In a more general way, the overall organization of muscle tissues can be viewed as a fractal hierarchy of bundles within bundles within bundles, etc. (Figure 4). Thus, the largest muscle units are composed of bundles composed of fibers composed of fibrils comprised of the myofilaments, actin and myosin.

9.3 Human heartbeat dynamics

An important extension of the fractal concept was the recognition that it could be applied not just to irregular geometric forms that lack a characteristic length scale, but also to complex processes that lack a single scale of time (Table 1). Fractal processes generate fluctuations on multiple time scales;[14] furthermore, such variability is statistically self similar. A crude, qualitative appreciation for the self-similar

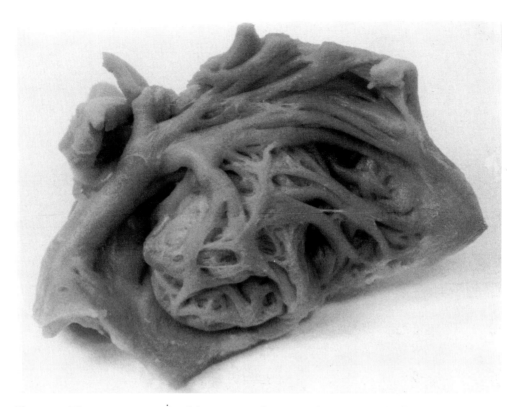

Figure 2 The pectinate muscle of the canine right atrium shows a complex branching pattern. (Adapted from Goldberger, A. L., and West, B. J., *Yale J. Biol. Med.*, 60, 421, 1987.)

nature of fractal processes can be obtained by plotting the time series in question at different "magnifications", i.e., different temporal resolutions. For example, Figure 5 shows the time series of the heartbeat from a healthy subject plotted on three different scales. What is notable is that all three graphs show that the normal heartbeat has an irregular ("wrinkly") appearance. Furthermore, the irregularity on different time scales is not visually distinguishable. The roughness of the "time-scape" (analogous to a fractal landscape) appears to have a self-similar property.[2,3] Quantifying this fractal scaling and its relationship to long-range correlations and power-law distributions is the topic of the remainder of this chapter.

Clinicians have traditionally described the normal activity of the heart as "regular sinus rhythm", but, in fact, cardiac interbeat intervals normally fluctuate in a complex, apparently erratic manner[6,15] as illustrated in Figures 5 and 6. This highly irregular behavior has recently motivated us to apply time series analyses that derive from statistical physics, especially methods for the study of critical phenomena where fluctuations at all length (time) scales occur.[2,3] These studies show that under healthy conditions, interbeat interval time series exhibit long-range power-law correlations reminiscent of physical systems near a critical point.[4] Furthermore, as described below, certain disease states may be accompanied by a breakdown in this scale-invariant (fractal) correlation property.[2,3]

Our analysis begins with the beat-to-beat heart rate fluctuations of digitized electrocardiograms recorded with an ambulatory (Holter) monitor. The time series obtained by plotting the sequential intervals between beat i and beat $i + 1$, denoted by $B(i)$, typically reveals a complex type of variability (Figures 6a and b). The

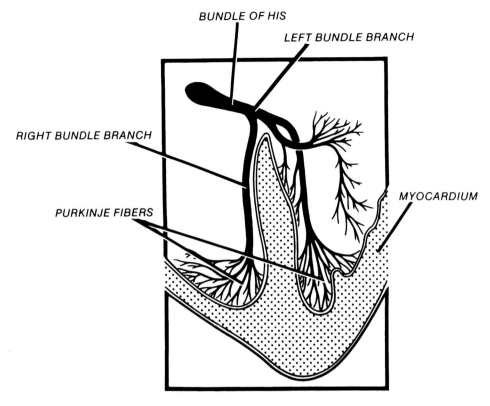

Figure 3 Schematic diagram of the His-Purkinje network, a tree-like structure which rapidly transmits the electrical impulses from the atria to the ventricles. (Adapted from Goldberger, A. L., Bhargava, V., West, B. J., and Mandell, A. J., *Biophys. J.*, 48, 525, 1985.)

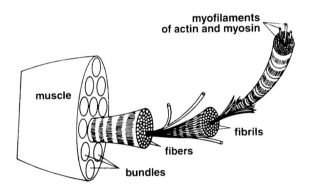

Figure 4 Cardiac muscle can be viewed as a self-similar collection of contractile "cables" within "cables" over multiple scales from bundles to fibers to fibrils to myofilaments. (Adapted from *The Dictionary of Science*, Lafferty, P. and Rowe, J., Simon and Schuster, New York, 1994.)

mechanism underlying these fluctuations appears to be related primarily to countervailing neuroautonomic inputs. Parasympathetic stimulation decreases the firing rate of pacemaker cells in the heart's sinus node. Sympathetic stimulation has the opposite effect. The nonlinear interaction (competition) between the two branches of the autonomic (involuntary) nervous system is the postulated mechanism for the type of erratic heart rate variability recorded in healthy subjects.[6,17]

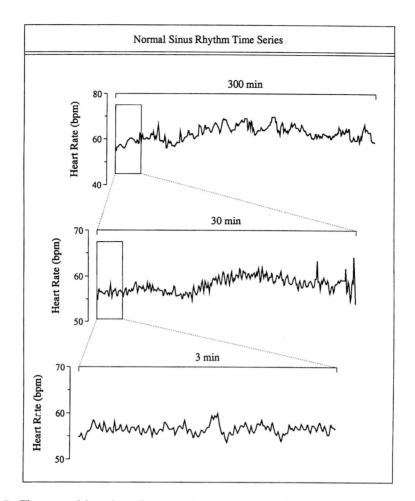

Figure 5 The normal heartbeat fluctuates in an apparently irregular fashion; furthermore, these fluctuations, when examined at different temporal resolutions, show a similar type of "rough" surface or landscape. See also Figure 6. (Adapted from Goldberger, A. L., Rigney, D. R., and West, B. J., *Sci. Am.*, 262, 42, 1990.)

An immediate problem facing researchers applying time series analysis to interbeat interval data is that the heartbeat time series is often highly nonstationary.* A number of approaches can be taken to reduce these nonstationary effects. We will discuss two such methods — interbeat increment analysis and detrended fluctuation analysis — and their physiological implications separately in the following sections.**

9.3.1 Increment of interbeat interval

A standard treatment in time series analysis is to remove nonstationarities by taking the first difference of the original time series.[2] In our case, studying the interbeat interval measurement, $I(i) \equiv B(i + 1) - B(i)$, is physiologically relevant since one important mechanism for regulating heart rate is the baroreflex, which keeps the arterial blood pressure in a proper range. This reflex operates via arterial pressure sensors, the function of which is to collect information about the blood pressure and

---

* Traditional physiological analyses have tended to ignore these nettlesome nonstationarities or to assume that they do not exist.
** Other approaches to studying nonstationary processes may be helpful, including the wavelet transform.

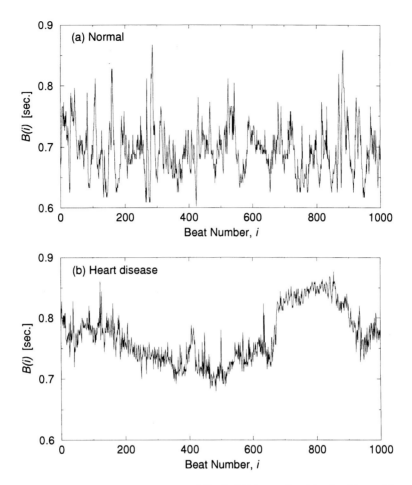

Figure 6 The interbeat interval time series $B(i)$ of 1000 beats for (a) a healthy subject and (b) a patient with severe cardiac disease (dilated cardiomyopathy). The healthy heartbeat time series shows more complex fluctuations (see Figure 5) compared to the diseased heart rate fluctuation pattern.

possibly its rate of change, which, in turn, is related to the increment of heart rate. This feedback information can then be used to regulate the pressure in the arteries by adjusting heart rate and vascular tone.

Unlike the original time series $B(i)$, it turns out that $I(i)$ is stationary, i.e., the average and the variance are independent of the sampling position. We have verified this stationarity property experimentally by measuring the stability of the average and variance in a moving window. An advantage of the stationarity property of $I(i)$ is that, unlike the original interbeat intervals, a comparison of histograms between data from healthy and diseased subjects becomes feasible.

We find that $I(i)$ for the two time series in Figures 6a and b have virtually identical histograms and can be well described by a Lévy stable distribution (see Figure 7):

$$P(I, \psi, \gamma) = \frac{1}{\pi} \int_0^\infty \exp(-\gamma q^\psi) \cos(qI)\, dq,$$

with $\psi = 1.7$ and $\gamma > 0$.[18] Since the histograms of the increments are the same for both normal and diseased conditions, the different fluctuation patterns observed in health and disease in this example must relate to the ordering of these increments, i.e., to

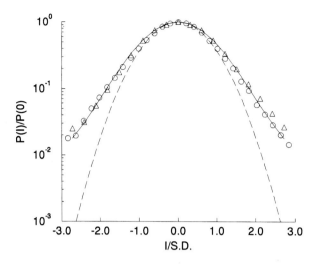

Figure 7 The histogram of $I(i)$ for the healthy (circles) and diseased (triangles) subjects shown in Figure 6. $P(I)$ is the probability of finding an interbeat increment in the range $(I - \Delta I/2, I + \Delta I/2)$. To facilitate comparison, we divide the variable I by the standard deviation (S.D.) of the increment data and rescale P by $P(0)$. In Lévy stable distributions, ψ is related to the power-law exponent describing the distribution for large values of the variable, while the width of the distribution is characterized by γ. Since we have rescaled I by the width, ψ is the only relevant parameter. Both histograms are indistinguishable and are well fit by a Lévy stable distribution with $\psi = 1.7$ (solid line). The dashed line is a Gaussian distribution, which is a special case of a Lévy stable distribution with $\psi = 2$. Although the second moment diverges for a Lévy stable distribution, for a finite sample the second moment remains finite. Similar fits were obtained for 8 of the 10 normal subjects and all 10 subjects with heart disease we studied. The slow decay of Lévy stable distributions for large increment values may be of adaptive importance and may relate to the plasticity of response of the cardiovascular system.

the correlations between the length of successive increments produced by the underlying dynamics of the heartbeat.

To investigate the dynamical differences we observed visually in Figure 6, it is helpful to study further the correlation properties of the time series. Since $I(i)$ is stationary, we can apply standard spectral analysis techniques. Figures 8a and b show the power spectra $S_I(f)$, calculated as the square of the Fourier transform amplitudes for $I(i)$, derived from the same data sets used in Figure 6. The fact that the log-log plot of $S_I(f)$ vs. f is linear implies

$$S_I(f) \sim f^\beta$$

Furthermore, β can serve as an indicator of the presence and type of correlations:

1. If $\beta = 0$, there is no correlation in the time series $I(i)$ ("white noise").
2. If $-1 < \beta < 0$, then $I(i)$ is correlated such that positive values of I are likely to be close (in time) to each other, and the same is true for negative I values.
3. If $0 < \beta < 1$, then $I(i)$ is also correlated; however, the values of I are organized such that positive and negative values are more likely to alternate in time ("anticorrelation").[19]

For the diseased data set, we observe a flat spectrum ($\beta \simeq 0$) in the low frequency region (Figure 8b), confirming that $I(i)$ are not correlated over long time scales (low frequencies); therefore, $I(i)$, the first derivative of $B(i)$, can be interpreted as being

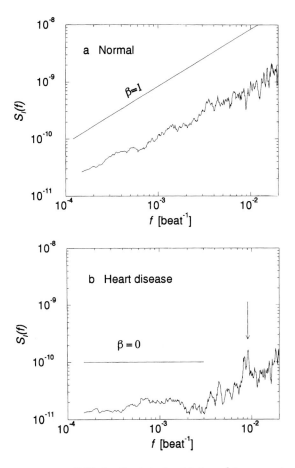

Figure 8 The power spectrum $S_I(f)$ for the interbeat interval increment sequences over approximately 24 hours for the same subjects in Figure 6. (a) Data from a healthy adult. The best-fit line for the low frequency region has a slope $\beta = 0.93$. The heart-rate spectrum is plotted as a function of "inverse beat number" (beat^{-1}) rather than frequency (time^{-1}) to obviate the need to interpolate data points. The spectral data are smoothed by averaging over 50 values. (b) Data from a patient with severe heart failure. The best-fit line has slope 0.14 for the low frequency region, $f^- < f_c = 0.005$ beat^{-1}. The appearance of a pathologic, characteristic time scale is associated with a spectral peak (arrow) at about 10^{-2} beat^{-1} (corresponding to Cheyne-Stokes respiration, an abnormal type of periodic breathing commonly associated with low cardiac output in heart failure). (From Peng, C.-K., Mietus, J., Hausdorff, J. M., Havlin, S., Stanley, H. E., and Goldberger, A. L., *Phys. Rev. Lett.*, 70, 1343, 1993. With permission.)

analogous to the velocity of a random walker, which is uncorrelated on long time scales, while the values of $B(i)$ — corresponding to the position of the random walker — are correlated. However, this correlation is of a trivial nature since it is simply due to the summation of uncorrelated random variables.

In contrast, for the data set from the healthy subject (Figure 8a), we obtain ($\beta \simeq 1$), indicating nontrivial long-range correlations in $B(i)$ — these correlations are not the consequence of summation over random variables or artifacts of nonstationarity. Furthermore, the "anticorrelation" properties of $I(i)$ indicated by the positive β value are consistent with a nonlinear feedback system that "kicks" the heart rate away from extremes. This tendency, however, does operate not only on a beat-to-beat basis (local effect) but also over a wide range of time scales.

9.3.2 Interbeat interval time series: detrended fluctuation analysis

In this section we discuss a second, complementary approach for analyzing the original interbeat interval time series.[2,20] An important question is whether the observed heterogeneous heart rate time series structure arises trivially from changes in environmental conditions having little to do with the intrinsic dynamics of the heart rate itself. Alternatively, these heart rate fluctuations may arise from a complex nonlinear dynamical system rather than being an epiphenomenon of environmental stimuli.

From a practical point of view, if the fluctuations driven by uncorrelated stimuli can be decomposed from intrinsic fluctuations generated by the dynamical system, then these two classes of fluctuations may be shown to have very different correlation properties. If that is the case, then a plausible consideration is that only the fluctuations arising from the dynamics of the complex, multiple-component system should show long-range correlations. Other responses should give rise to a different type of fluctuations (although highly nonstationary) having characteristic time scales (i.e., frequencies related to the stimuli). This type of "noise", although physiologically important, can be treated as a "trend" and distinguished from the more subtle fluctuations that may reveal intrinsic correlation properties of the dynamics. To this end, we introduced a modified root-mean-square analysis of a random walk, termed *detrended fluctuation analysis* (DFA),[20-23]* to the analysis of physiological data. The advantages of DFA over conventional methods (e.g., spectral analysis and Hurst analysis) are that it permits the detection of long-range correlations embedded in a seemingly nonstationary time series and also avoids the spurious detection of long-range correlations that are an artifact of nonstationarities. This method has been validated on control time series that consist of long-range correlations with the superposition of a nonstationary external trend.[22] The DFA method has also been applied successfully to detect long-range correlations in highly heterogeneous DNA sequences[21-23] and other complex physiological signals.[20,24,25]

To illustrate the DFA algorithm, we use the interbeat time series shown in Figure 6a as an example. Briefly, the interbeat interval time series (of total length N) is first integrated: $y(k) = \sum_{i=1}^{k} [B(i) - B_{ave}]$, where B_{ave} is the average interbeat interval. Next, the integrated time series is divided into boxes of equal length, n. In each box of length n, a least-squares line is fit to the data, representing the trend in that box (Figure 9). The y coordinate of the straight line segments is denoted by $y_n(k)$. We then detrend the integrated time series, $y(k)$, by subtracting the local trend, $y_n(k)$, in each box. The root-mean-square fluctuation of this integrated and detrended time series is calculated by

$$F(n) = \sqrt{\frac{1}{N} \sum_{k=1}^{N} [y(k) - y_n(k)]^2}.$$

This computation is repeated over all time scales (box sizes) to provide a relationship between $F(n)$, the average fluctuation as a function of box size, and the box size n (i.e., the number of beats in a box which is the size of the window of observation). Typically, $F(n)$ will increase with box size n. A linear relationship on a double log graph indicates the presence of scaling. Under such conditions, the fluctuations can be characterized by a scaling exponent α, the slope of the line relating $\log F(n)$

---

* Computer software of DFA algorithm is available upon request; contact C.-K. Peng (e-mail: peng@chaos.BIH.Harvard.Edu).

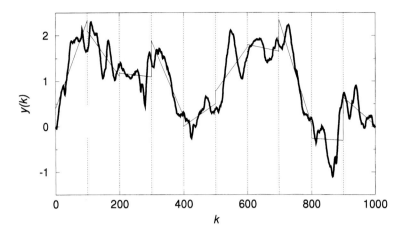

Figure 9 The integrated time series: $y(k) = \sum_{i=1}^{k} (B(i) - B_{ave})$, where $B(i)$ is the interbeat interval shown in Figure 6a. The vertical dotted lines indicate box of size $n = 100$; the solid straight line segments represent the "trend" estimated in each box by a linear least-squares fit. (From Peng, C.-K., Havlin, S., Stanley, H. E., and Goldberger, A. L., *Chaos*, 5, 82, 1995. With permission.)

to log n. Consider first a process where the value at one interbeat interval is completely uncorrelated from any previous values, i.e., white noise. This can be achieved by using a time series for which the order of the points has been shuffled (a so-called "surrogate" data set). For this type of uncorrelated data, the integrated value, $y(k)$, corresponds to a random walk, and therefore $\alpha = 0.5$.[22,26] If there are only short-term correlations, the initial slope may be different from 0.5, but α will approach 0.5 for large window sizes. An $\alpha > 0.5$ and ≤ 1.0 indicates persistent long-range power-law correlations such that a large (compared to the average) interbeat interval is more likely to be followed by a large interval and vice versa. In contrast, $0 < \alpha < 0.5$ indicates a different type of power-law correlation such that large and small value of the time series are more likely to alternate.[2,19] A special case of $\alpha = 1$ corresponds to $1/f$ noise.[16,27] For $\alpha \geq 1$, correlations exist but cease to be of a power-law form; $\alpha = 1.5$ indicates "brown noise", the integration of white noise. The α exponent also can be viewed as an index that describes the "roughness" of the original time series:[1,28] the larger the value of α, the smoother the time series. In this context, $1/f$ noise can be interpreted as a compromise between the complete unpredictability of white noise (very rough "landscape") and the relatively smooth landscape of Brownian noise.

Figure 10 compares the DFA analysis of representative 24-hour interbeat interval time series of a healthy subject (○) and a patient with congestive heart failure (△). Notice that for large time scales (asymptotic behavior), the healthy subject interbeat interval time series shows almost perfect power-law scaling over two decades ($20 \leq n \leq 10000$) with $\alpha = 1$ (i.e., $1/f$ noise), while for the pathologic data set $\alpha \approx 1.3$ (closer to Brownian noise). This result is consistent with our findings that on the interbeat interval increment there is a significant difference in the long-range scaling behavior between healthy and diseased states (Figure 8).[2,3]

9.4 Normal vs. pathologic time series

Can fractal analytic techniques be used to enhance the accuracy of diagnostic and prognostic tools in bedside clinical medicine? To test further for the possible clinical applicability of the DFA method, we analyzed cardiac interbeat data from two

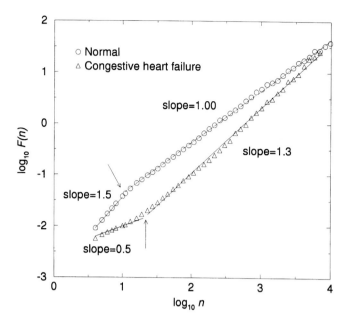

Figure 10 Plot of log $F(n)$ vs. log n (see description of DFA computation in text) for two very long interbeat interval time series (approximately 24 hours). The circles are from a healthy subject while the triangles are from a subject with congestive heart failure. Arrows indicate "crossover" points in scaling. (From Peng, C.-K., Havlin, S., Stanley, H. E., and Goldberger, A. L., *Chaos*, 5, 82, 1995. With permission.)

different groups of subjects reported in our previous work:[2] 12 healthy adults without clinical evidence of heart disease (age range, 29 to 64 years; mean, 44) and 15 adults with severe heart failure (age range, 22 to 71 years; mean, 56).* Data from each subject consist of approximately 24 hours of ECG recording. Data from patients with heart failure due to severe left ventricular dysfunction are likely to be particularly informative in analyzing correlations under pathologic conditions since these individuals have abnormalities in both the sympathetic and parasympathetic control mechanisms[17,29] that regulate beat-to-beat variability. Previous studies have demonstrated marked changes in short-range heart rate dynamics in heart failure compared to healthy function, including the emergence of intermittent relatively low frequency (~1 cycle/minute) heart rate oscillations associated with the well recognized syndrome of periodic (Cheyne-Stokes) respiration, an abnormal breathing pattern often associated with low cardiac output (see Figure 8).[29]

We have observed the following scaling exponents** (for time scales $10^2 \sim 10^4$ beats) for the group of healthy cardiac interbeat interval time series (mean value ± S.D.): $\alpha = 1.00 \pm 0.11$. This result is consistent with previous reports of $1/f$ fluctuations in healthy heart rate[5,30,31] and blood pressure.[32] The pathologic group showed a significant ($p < 0.01$ by Student's *t*-test) deviation of the long-range correlation

---

* ECG recordings of Holter monitor tapes were processed both manually and in a fully automated manner using our computerized beat recognition algorithm. Abnormal beats were deleted from each data set. The deletion had practically no effect on the DFA analysis since less than 1% of total beats were removed. Patients in the heart failure group were receiving conventional medical therapy prior to receiving an investigational cardiotonic drug; see Reference 20.

** Typical regression fit shows excellent linearity on double log graph (indicated by correlation coefficient $r > 0.97$) for both groups. However, usually data from healthy subjects show even better linearity on log-log plots than data from subjects with heart disease. This observation also supports the notion of a degradation of long-range scaling behavior with disease.

exponent from normal. For the group of heart failure subjects, we found that α = 1.24 ± 0.22. Of interest, some of the heart failure subjects show an α exponent very close to 1.5 (Brownian noise), indicating random walk-like fluctuations, also consistent with our previous findings in this group. The group-averaged exponent α is less than 1.5 for the heart failure patients, suggesting that pathologic dynamics may operate only transiently in the random-walk regime or may only approach this extreme state as a limiting case. We obtained similar results when we divided the time series into three consecutive subsets (of approximately 8 hours each) and repeated the above analysis; therefore, our findings are not simply attributable to different levels of daily activities.

9.4.1 Crossover phenomena

Although this asymptotic scaling exponent may serve as a useful index for selected diagnostic purposes, a drawback is that very long data sets are required (at least 24 hours) for statistically robust results. For practical purposes, clinical investigators are often interested in the possibility of using substantially shorter time series. In this regard, we note that for short time scales, there is an apparent crossover exhibited for the scaling behavior of both data sets (arrows in Figure 10). For the healthy subject, the α exponent estimated from very small n (<10 beats) is larger than that calculated from large n (>10 beats). This is probably due to the fact that on very short time scales (a few beats to ten beats), the physiologic interbeat interval fluctuation is dominated by the relatively smooth heartbeat oscillation associated with respiration, thus giving rise to a large α value. For longer scales, the interbeat fluctuation, reflecting the intrinsic dynamics of a complex system, approaches that of $1/f$ behavior as previously noted. In contrast, the pathologic data set shows a very different crossover pattern (Figure 10). For very short time scales, the fluctuation is quite random (close to white noise, α ≈ 0.5). As the time scale becomes larger, the fluctuation becomes smoother (asymptotically approaching Brownian noise, α ≈ 1.5). These findings are consistent with the interbeat-increment analysis described above which indicated altered correlation properties under pathologic conditions.[2,3]

9.4.2 Clinical applications: preliminary results

The observation of a differential crossover pattern for healthy vs. pathologic data motivated us to extract two parameters from each data set by fitting the scaling exponent α over two different time scales: one short, the other long. To be more precise, for each data set we calculated an exponent α_1 by making a least-squares fit of log $F(n)$ vs. log n for $4 \leq n \leq 16$. Similarly, an exponent α_2 was obtained from $16 \leq n \leq 64$. Since these two exponents are not extracted from the asymptotic region, relatively short data sets are sufficient, thereby making this technique applicable to "real world" clinical data.

We applied this quantitative fluctuation analysis to the two different groups of subjects mentioned above to measure the two scaling exponents α_1 and α_2. All data set records were divided into multiple subsets (each with N = 8192 beats ~ 2 hours) and the two exponents were calculated for each subset. For healthy subjects, we find the following exponents for the cardiac interbeat interval time series: α_1 = 1.201 ± 0.178 and α_2 = 0.998 ± 0.124. For the group of congestive heart failure subjects, we find that α_1 = 0.803 ± 0.259 and α_2 = 1.125 ± 0.216, both significantly ($p < 0.0001$ for both α_1 and α_2) different from normal. Furthermore, we show in Figure 11 that fairly good discrimination between these two groups can be achieved by using these two scaling exponents. We note that not all subjects in our preliminary study show an obvious crossover in their scaling behavior. Only 8 out of 12 healthy subjects exhibited this crossover, while 11 out of 15 pathologic subjects exhibited a "reverse" crossover. However, the two scaling exponents (α_1 and α_2) measured from relatively short data sets still can be potentially useful indicators to distinguish normal from pathologic time series.

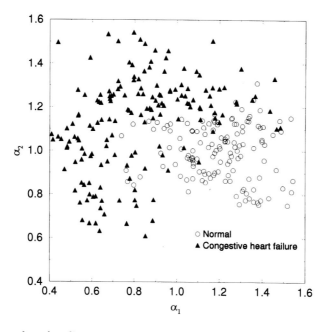

Figure 11 Scatter plot of scaling exponents α_1 vs. α_2 for the healthy subjects (○) and subjects with congestive heart failure (▲). The αs were calculated from interbeat interval data sets of length 8192 beats. Longer data set records were divided into multiple data sets (each with 8192 beats). Note good separation between healthy and heart disease subjects, with clustering of points in two distinct "clouds". (From Peng, C.-K., Havlin, S., Stanley, H. E., and Goldberger, A. L., *Chaos*, 5, 82, 1995. With permission.)

To test the effect of data length on these calculations, we repeated the same DFA measurements for longer data sets ($N = 16,384$) and also for shorter data sets ($N = 4096$). As expected, the results for shorter data sets are less reliable (more overlap between two groups) due to anticipated statistical error related to finite sample size.[33] On the other hand, longer data sets result in little improvement for the distinction between groups; therefore, the data length of 8192 seems to be a statistically reasonable choice.*

Furthermore, we note that data from normal interbeat interval time series are tightly clustered, suggesting that there may exist a "universal" scaling behavior for physiologic interbeat time series. In contrast, the pathologic data show more variation, a finding which may be related to different clinical conditions and varying severity of the pathologic states.

---

* We also tested these calculations by varying the fitting range of α_2. We find that the results are very similar when we measure α_2 from 16 to 128 beats; however, moving the upper fitting range for α_2 leads to less obvious separation from normal subjects. This is partly due to the fact that, for finite length data sets, the calculation error of $F(n)$ increases with n.[33] Therefore, scaling exponents obtained over larger values of n will have greater uncertainty.

The potential practical applications of DFA analysis to the assessment of patient survival using Holter monitor data was supported in a preliminary report of a prospective population-based study.[34] We found that the DFA analysis confirmed an alteration of long-range correlations in subjects with heart failure vs. age- and sex-matched controls. Furthermore, the DFA analysis appeared to add prognostic information about mortality not extractable from traditional methods of heart rate variability analysis.

9.5 Heart rate dynamics: conclusions

Our finding of nontrivial, long-range correlations in healthy heart rate dynamics is consistent with the observation of long-range correlations in other biological systems that do not have a characteristic scale of time or length.[1,5,6] Such fractal behavior may be adaptive for at least two reasons:[2]

1. The long-range correlations serve as an organizing principle for highly complex, nonlinear processes that generate fluctuations on a wide range of time scales.
2. The lack of a characteristic scale helps prevent excessive mode-locking that would restrict the functional responsiveness of the organism.

Support for these related conjectures is provided by observations from severe diseased states such as heart failure, where the breakdown of long-range correlations is often accompanied by the emergence of a dominant frequency mode (e.g., the Cheyne-Stokes frequency). Analogous transitions to highly periodic regimes have been observed in a wide range of other disease states including certain malignancies, sudden cardiac death, epilepsy, and fetal distress syndromes.[5,6]

The complete breakdown of normal long-range (fractal) correlations in any physiological system could, theoretically, lead to three possible classes of pathologic dynamics schematized in Figure 12: (1) a random walk (brown noise), (2) highly periodic behavior, or (3) completely uncorrelated behavior (white noise). Cases (1) and (2) both indicate only "trivial" long-range correlations of the types described above in severe heart failure. Case (3) may correspond to certain cardiac arrhythmias such as fibrillation. More subtle or intermittent degradation of long-range correlation properties may provide an early warning of incipient pathology. Finally, we note that the long-range correlations present in the healthy heartbeat indicate that the neuroautonomic control mechanisms actually drive the system away from a single steady state. Therefore, the classical theory of homeostasis, according to which stable physiological processes seek to maintain "constancy",[35] should be extended to account for this dynamical, far from equilibrium, behavior.

In summary, we have tried to show how the concept of fractals may be useful in understanding the spatial and temporal "structure" of the human heart (Table 1). We discussed how fractal analysis can be used to detect the presence of long-range correlations in one important physiological time series: namely, cardiac interbeat interval fluctuations. We find that this approach is capable of identifying crossover behavior due to differences in scaling over short vs. long time scales. This finding is of interest from a basic physiologic viewpoint since it motivates new modeling approaches to account for the control mechanisms regulating cardiac dynamics on different time scales. From a practical point of view, quantification of these scaling exponents may have potential applications for bedside and ambulatory monitoring, in particular for identifying high-risk patients.

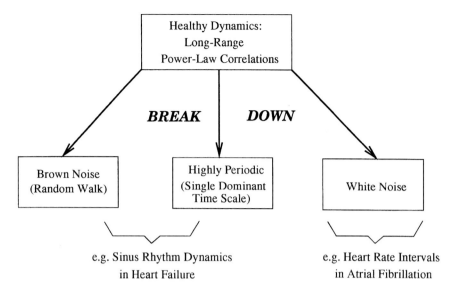

Figure 12 The breakdown of long-range power-law correlations may lead to any of three dynamical states: (1) a random walk ("brown noise") as observed in low frequency heart rate fluctuations in certain cases of severe heart failure; (2) highly periodic oscillations, also observed in Cheyne-Stokes pathophysiology in heart failure; (3) completely uncorrelated behavior ("white noise"), exemplified by the heart rate during atrial fibrillation. (From Peng, C.-K., Buldyrev, S. V., Hausdorff, J. M., Havlin, S., Mietus, J. E., Simons, M., Stanley, H. E., and Goldberger, A. L., in *Fractals in Biology and Medicine*, Nonnenmacher, T. F., Losa, G. A., and Weibel, E. R., Eds., Birkhaüser-Verlag, Basel, 1994. With permission.)

Acknowledgments

We are grateful to S. V. Buldyrev, G. B. Moody, and D. R. Rigney for valuable discussions. Partial support was provided to Chung-Kang Peng by an NIH/NIMH NRSA postdoctoral fellowship, to Jeffrey Hausdorff by the NIA, to H. Eugene Stanley and Shlomo Havlin by the NSF, and to Ary L. Goldberger by the G. Harold and Leila Y. Mathers Charitable Foundation, NIDA, and NASA. This review is mainly based on References 2, 3, and 20.

References

1. Bunde, A., Havlin, S., Eds., *Fractals in Science*, Springer-Verlag, Berlin, 1994.
2. Peng, C.-K., Mietus, J., Hausdorff, J. M., Havlin, S., Stanley, H. E., and Goldberger, A. L., Long-range anti-correlations and non-Gaussian behavior of the heartbeat, *Phys. Rev. Lett.*, 70, 1343, 1993.
3. Peng, C.-K., Buldyrev, S. V., Hausdorff, J. M., Havlin, S., Mietus, J. E., Simons, M., Stanley, H. E., and Goldberger, A. L., Fractal landscapes in physiology and medicine: long-range correlations in DNA sequences and heart rate intervals, in *Fractals in Biology and Medicine*, Nonnenmacher, T. F., Losa, G. A., and Weibel, E. R, Eds., Birkhaüser-Verlag, Basel, 1994.
4. Stanley, H. E., *Introduction to Phase Transitions and Critical Phenomena*, Oxford University Press, London, 1971.
5. Goldberger, A. L., and West, B. J., Fractals in physiology and medicine, *Yale J. Biol. Med.*, 60, 421, 1987.
6. Goldberger, A. L., Rigney D. R., and West B. J., Chaos and fractals in human physiology, *Sci. Am.*, 262, 42, 1990.

7. Sernetz, M., Gelléri, B., and Hofmann, J., The organism as bioreactor: interpretation of the reduction law of metabolism in terms of heterogeneous catalysis and fractal structure, *J. Theor. Biol.*, 117, 209, 1985.
8. Weibel, E. R., Fractal geometry: a design principle for living organisms, *Am. J. Physiol. (Lung Cell. Mol. Physiol.)*, 5, L361, 1991
9. Bassingthwaighte, J. B., Liebovitch, L. S., and West, B. J., *Fractal Physiology*, Oxford University Press, New York, 1994.
10. Goldberger, A. L., Bhargava, V., West, B. J., and Mandell, A. J., On a mechanism of cardiac electrical stability: the fractal hypothesis, *Biophys. J.*, 48, 525, 1985.
11. Abboud, S., Berenfeld, O., and Sadeh, D., Simulation of high-resolution QRS complex using a ventricular model with a fractal conduction system. Effects of ischemia on high-frequency QRS potentials, *Circ. Res.*, 68, 1751, 1991.
12. Bassingthwaighte, J. B., King, R. B., and Roger, S. A., Fractal nature of regional myocardial blood flow heterogeneity, *Circ. Res.*, 65, 575, 1989.
13. Peskin, C. S., and McQueen, D. M., Mechanical equilibrium determines the fractal fiber architecture of aortic valve leaflets, *Am. J. Physiol.*, 266, (*Heart Circ. Physiol.*, 35), H319, 1994.
14. Shlesinger, M. F., Fractal time and $1/f$ noise in complex systems, *Ann. N.Y. Acad. Sci.*, 504, 214, 1987.
15. Kitney, R. I., and Rompelman, O., *The Study of Heart-Rate Variability*, Oxford University Press, New York, 1980.
16. Bak, P., Tang, C., Wiesenfeld, K., Self-organized criticality: an explanation of $1/f$ noise, *Phys. Rev. Lett.*, 59, 381, 1987.
17. Levy, M. N., Sympathetic-parasympathetic interactions in the heart, *Circ. Res.*, 29, 437, 1971.
18. Lévy, P., *Théorie de l'Addition des Variables Aléatoires*, Gauthier-Villars, Paris, 1937.
19. Havlin, S., Selinger, R. B., Schwartz, M., Stanley, H. E., and Bunde, A., Random multiplicative processes and transport in structures with correlated spatial disorder, *Phys. Rev. Lett.*, 61, 1438, 1988.
20. Peng, C.-K., Havlin, S., Stanley, H. E., and Goldberger, A. L., Quantification of scaling exponents and crossover phenomena in nonstationary heartbeat time series, *Chaos*, 5, 82, 1995.
21. Buldyrev, S. V., Goldberger, A. L., Havlin, S., Peng, C.-K., Stanley, H. E., and Simons, M., Fractal landscapes and molecular evolution: modeling the myosin heavy chain gene family, *Biophys. J.*, 65, 2675, 1993.
22. Peng, C.-K., Buldyrev, S. V., Havlin, S., Simons, M., Stanley, H. E., and Goldberger, A. L., On the mosaic organization of DNA sequences, *Phys. Rev. E*, 49, 1691, 1994.
23. Ossadnik, S. M., Buldyrev, S. V., Goldberger, A. L., Havlin, S., Mantegna, R. N., Peng, C.-K., Simons, M., and Stanley, H. E., Correlation approach to identify coding regions in DNA sequences, *Biophys. J.*, 67, 64, 1994.
24. Hausdorff, J. M., Peng, C.-K., Ladin, Z., Wei, J. Y., and Goldberger, A. L., Is walking a random walk? Evidence for long-range correlations in the stride interval of human gait, *J. Appl. Physiol.*, 78, 349, 1995.
25. Hausdorff, J. M., Purdon, P., Peng, C.-K., Ladin, Z., Wei, J. Y., and Goldberger, A. L., Fractal scaling of human gait: stability of long-range correlations in stride interval fluctuations, *J. Appl. Physiol.*, in press.
26. Montroll, E. W., and Shlesinger, M. F., On the wonderful world of random walks, in *Nonequilibrium Phenomena. II. From Stochastics to Hydrodynamics*, Lebowitz, J. L., and Montroll, E. W., Eds., North-Holland, Amsterdam, 1984.
27. Press, W. H., Flicker noise in astronomy and elsewhere, *Comments Astrophys.*, 7, 103, 1978.
28. Peng, C.-K., Buldyrev, S., Goldberger, A. L., Havlin, S., Sciortino, F., Simons, M., and Stanley, H. E., Long-range correlations in nucleotide sequences, *Nature*, 356, 168, 1992.
29. Goldberger, A. L., Rigney, D. R., Mietus, J., Antman, E. M., and Greenwald, S., Nonlinear dynamics in sudden cardiac death syndrome: heart rate oscillations and bifurcations, *Experientia*, 44, 983, 1988.

30. Kobayashi, M., and Musha, T., $1/f$ fluctuation of heartbeat period, *IEEE Trans. Biomed. Eng.*, 29, 456, 1982.
31. Butler, G. C., Yamamoto, Y., Xing, H. C., Northey, D. R., and Hughson, R. L., Heart rate variability and fractal dimension during orthostatic challenge, *J. Appl. Physiol.*, 75, 2602. 1993
32. Marsh, D. J., Osborn, J. L., and Cowley, Jr., A. W., $1/f$ fluctuations in arterial pressure and the regulation of renal blood flow in dogs, *Am. J. Physiol.*, 258 (*Renal Fluid Electrolyte Physiol.*, 27), F1394, 1990.
33. Peng, C.-K., Buldyrev, S. V., Goldberger, A. L., Havlin, S., Simons, M., and Stanley, H. E., Finite size effects on long-range correlations: implications for analyzing DNA sequences, *Phys. Rev. E*, 47, 3730, 1993.
34. Ho, K. K. L., Moody, G. B., Peng, C.-K., Mietus, J. E., Larson, M. G., Goldberger, A.L., and Levy, D., Assessing survival in heart failure cases and controls using nonlinear and conventional indices of heart rate dynamics: the Framingham heart study, *Circulation*, 90, I-330, 1994.
35. Cannon, W. B., Organization for physiological homeostasis, *Physiol. Rev.*, 9, 399, 1929.

chapter ten

Fractal Probability Density and EEG/ERP Time Series

Bruce J. West, Marcos N. Novaes, and Voyko Kavcic

Life is a fractal in Hilbert space.

R. Rucker

10.1 Introduction

We examine the feasibility of applying newly developed data processing techniques to electroencephalogram/event-related potentials (EEG/ERP) time series to assess the capability of objectively determining a student's mastery of classroom material and how that mastery may influence future performance in the field. The student that has completed the requisite course work, the novice who has finished a vigorous regimen of simulator training, or the expert cross-training in a new area could all benefit from a diagnostic tuned to the activity of the brain associated with memory and the correlation of that memory with performance. As systems become more complex and students are required to process ever increasing amounts of data and to transfer that memory of facts and figures to performance, the need for such a diagnostic becomes more and more clear. Physiological measurements of the electrical activity of the brain may well provide the predictive information necessary for a sensitive measure of the student's command of the desired material.

The biophysical mechanisms dominating the generation and propagation of electromagnetic signals in the brain are not understood; therefore, it is necessary to have techniques for the analysis of spontaneous brain activity, as well as event-related potentials, that will allow us to extract meaningful information from the ongoing brain wave activity without this detailed mechanistic understanding. The data processing methods that we use have been developed over the past decade in the emerging area of nonlinear dynamical systems theory. These techniques have been applied with varying degrees of success to interpret the dynamical content of a number of measured irregular time series — for example, to construct relatively simple mathematical descriptions of oscillating chemical reactions, the dynamics of epidemics, the beating of the heart, interspike interval distributions in single neurons, and epileptic seizures in EEG records (see, for example, West[1] for a review of these and other applications). Herein we review some of the successes in applying these processing methods to EEG/ERP time series, refining and further developing these techniques, and extending their application to draw certain tentative conclusions regarding learning and performance.

In this chapter, we develop a technique that allows us to calculate a time-dependent indicator of the degree of learning that is taking place in the performance of a task. Indeed, much of our recent research has been directed towards the development of this new technique, increasing the efficiency of the computer programs we have developed, and applying the procedure to our own EEG/ERP data. In this way we could independently test and verify the published results in the literature, which we have done with new and provocative results. The above indicator of learning is called the *fractal dimension* and is a measure of the degree of irregularity in EEG/ERP time series.

The EEG/ERP data we use herein comes from a database developed at the University of North Texas to assess the differences in neurophysiological measures among attention deficit hyperactivity disorder (ADHD) in normal 8- to 10-year-old boys and to assess the effects of methylphenidate on the performance and etechophysiological measures of ADHD boys (see Miller et al.[2]). The specific data used in the present project consisted of EEG recordings of one control group subject collected while the subject performed the first task of a letter pair test. A description of the methodology used in the Miller et al.[2] study is given in the text.

10.1.1 Measures of EEG/ERP time series

The activity of brain waves is quite similar to a wide variety of other natural phenomena that exhibit irregular and apparently unpredictable or random behavior. Examples that immediately come to mind are changes in the weather over a few days' time; the height of the next wave breaking on the beach as you sit in the hot sun, shivering from a cold wind blowing down your neck; and the infuriating intermittency of the time interval between drips from the bathroom faucet just after you crawl into bed at night. In some cases, such as the weather, the phenomenon always appears to be random, but in other cases, such as the dripping faucet, sometimes the dripping is periodic and other times each drip appears to be independent of the preceding one, thereby forming an irregular sequence. The formal property that these phenomena and brain wave time series have in common is nonlinearity and the constrained randomness that is generated by deterministic nonlinear dynamical equations, which is to say *chaos*.

We shall discuss chaos and other such technical terms in the appropriate sections, but for now we shall use a working definition of this type of irregularity in order to distinguish it from *noise*. The difference between noise and chaos is that the former arises from a system being in contact with a large complex environment and the latter from the intrinsic dynamical properties of a nonlinear system. Thus, although both noise and chaos give rise to irregular time series, the fluctuations in chaotic time series contain information about the phenomenon of interest, whereas noise tells us nothing about the system. Because chaos is generated by deterministic dynamical equations, one might assume that a sufficiently clever processing technique would enable one to reconstruct the equations of motion directly from the time series. This can in fact be done for "simple" chaotic systems, but not in general.

What scientists have been able to do, rather than reconstruct the equations of motion for chaotic systems, is to reconstruct the solutions to those equations. These solutions lie on what is called an *attractor*. We use such a reconstruction technique, to be described later, on EEG/ERP time series data to help us understand the several states of variability that are so apparent in the human brain. In addition, a number of measures are used to determine the geometrical structure of the attractor underlying the brain wave activity. The measure that is used above all others in the above variety of fields is the fractal dimension.

The fractal concept developed in recent years by Mandelbrot[3a,b] is primarily geometrical. A fractal structure is not smooth and homogeneous, but rather, when examined with stronger and stronger magnifying lenses, reveals greater and greater levels of detail. Many objects in nature, including trees, coral formations, cumulus clouds, and coastlines are fractal. In addition, mammalian lungs and hearts, as well as many other anatomic structures, possess fractal properties.[4,5] A fractal object is said to have a noninteger dimension, or a fractal dimension. For example, the surface of a cloud is greater than two but less than three, meaning that the cloud's surface is highly irregular. The fractal dimension gives a measure of the degree of irregularity of the surface (process). A fractal dimension near two means the surface is merely that of a smooth two-dimensional plane. As the dimension increases, the surface becomes more and more wispy, with the surface penetrating deeper and deeper into the volume of the cloud.

What does this have to do with the chaos noted above and to brain wave data? When the manifold (attractor) on which the solutions to the equations of motion evolve has an integer dimension, the motion of the system (time series) is regular or at most quasi-periodic. However, when the manifold has a fractal dimension, then the time series generated by the equations of motion is chaotic. The irregular behavior in the time series is a consequence of the infinitely complex geometrical structure of the attractor. Thus when the experimental data is used to reconstruct the attractor, if the attractor is found to have a noninteger dimension, then there is a good chance that the irregular behavior observed in the EEG/ERP signal is the result of chaos and not noise. Therefore, these fluctuations may contain useful information about the operation of the brain and should not be smoothed through filtering as is done in more traditional analyses.

One of the more dramatic results that has been obtained in recent years is the precipitous drop in the fractal dimension of the EEG time series when an individual undergoes an epileptic seizure. The brain's attractor is found to have a dimensionality on the order of four or five in deep sleep and to have the much lower dimensionality of approximately two in the epileptic state. This sudden drop in dimensionality was successfully captured in the Freeman model of the olfactory system of a rat[6] in which he calculated the EEG time series for a rat undergoing a seizure.[1] The normal erratic behavior of the brain characterized by the relatively large fractal dimension is markedly reduced with the onset of the coherent activity of an epileptic seizure with a corresponding halving of the fractal dimension.

It has been clearly established that the degree of complexity of the EEG time series measured by the fractal dimension is correlated with the cognitive activity of the subject. Another measure of the correlation between brain wave activity and the cognitive task is the event related potential. This characteristic of neural response to the environment has been proposed as an indicator of mental workload, in part because it is a noninvasive measure of cognitive information processing. We shall describe how the ERP has been used to measure the attentiveness of a subject to a prescribed task and how that attention can be divided among several tasks. The particular ERP measure described herein gives a quantitative indication of the mental workload associated with the tasks. We explicitly calculate the fractal dimension associated with ERPs and are able to determine the dimension of certain other states of the brain.

10.1.2 Fractal dimensions and learning

It is known that learning results in long-term changes in the brain which may possibly be assessed using modern neuroscience techniques. One measure that has been shown to reflect changes in the brain structure and function is the ERP. With this measure, individuals are exposed to a stimulus that is presumed to be related to the function being measured, and the brain electrical and magnetic activity is recorded just prior to, and following, the stimulus. The shape of the waveform that is recorded is affected by the perceptual, cognitive, and motor processes associated with the task. We have identified and developed a specific ERP processing technique that indicates the change in fractal dimension of EEG/ERP time series that occurs during learning. In other words, we have developed a reliable and efficient algorithm for determining the fractal dimension of fluctuating time series data. Thus, the fractal dimension of the background EEG signal is determined in nearly real time, and the change in dimension that occurs with an ERP is also determined. In this way, we have a quantitative measure of the attention being paid to a particular task: the greater the fractal dimension, the greater the attention and subsequently the more learning that is taking place.

There is a clear progression of the dimension magnitude from quiet, awake, and eyes closed to quiet, awake, and eyes closed using verbal memory. In addition to this distinct ordering of the mental task performed and the magnitude of the dimension, there is a decrease in the variance of the dimension as the state of the brain changes from no task to one involving cognitive activity. The trend in these data supports the hypothesis that the dimension of the EEG time series is closely tied to the cognitive activity of the brain. Further, the ERP results verify that the brain wave activity provides an objective measure of attention to a task and the learning of that task.

Gregson, Campbell, and Gates[7] directly demonstrate that the fractal dimension of the EEG time series increases with the complexity of task load in visual scanning tasks. They find that the dimension increases as the task becomes more complicated and support the interpretation that interrelates brain dynamics to implied cognitive processes. Nan and Jinghua[8] find that the fractal dimension increases during the performance of a task and decreases back to the background EEG level afterwards. They further find that the two hemispheres of the brain in the resting states are relatively symmetric in their dimensions and become antisymmetric during an active mental task. We have obtained similar results independently using our newly developed processing technique.

10.1.3 Fractal dimensions and performance

One purpose of the present research is to identify and develop specific EEG/ERP processing techniques that might be useful in assessing learning and performance. We have developed a technique to quantify learning by means of the variation in the fractal dimension of the time series data; this new procedure is outlined in the Technical Overview section.

Hoyert[9] applied nonlinear dynamics to predicting the behavior of pigeons responding to a fixed-interval schedule of reinforcement. He concludes that even the "noisiest" behavior might be the product of purely deterministic mechanisms: "Should low-dimensional motion prove to be ubiquitous, there is cause for optimism. This suggests that even the most complex behavior might stem from very simple deterministic systems. This is a much more helpful alternative than the idea that unpredictable, variable behavior is the product of a very large number of complexly interacting variables whose effects must be understood independently and in combination."

As Siegler[10] points out, "If cognitive variability does indeed facilitate learning, it would be adaptive if such variability were most pronounced when learning, rather than efficient performance, is most important." This does, in fact, appear to be the case; expertise brings with it decreasingly variable performance.

Molnár and Skinner[11] point out that any cognitive effort should increase the value of the fractal dimension, as was noted above; however, they find that the same stimulus did *not* evoke an increase in the dimension when it was familiar. Rapp et al.[12,13] reported a lower dimension value for the EEG epochs following an attended target compared to when the stimulus was ignored. Similar results were obtained by Molnár and Skinner during their "odd-ball" experiment. They point out that the significance of the dimension decrease when the signal is a "target stimulus" is not yet understood. They go on to say, "A dimensional increase, evoked by either a 'novel stimulus' or 'cognitive effort' suggests that the system has become more complex; its opposite may mean that the system has become *less* complex. But why fewer degrees of freedom or less complexity would be necessary to process the same signal when it has meaning compared to when it is ignored, as in the present data, or why greater degrees of freedom or more complexity would be necessary to process the signal when it is unfamiliar and/or require effort is simply not yet understood."

In the present context, that of learning and performance, it is possible to speculate that, as a task is repeated and becomes familiar, a control process may be initiated that is similar in nature to that predicted by Ott et al.[14] These authors determined that the stability properties of a chaotic attractor could be enhanced by means of a feedback of the output signal into the input. Using the fact that a chaotic attractor has an infinity of unstable periodic orbits embedded in it, they showed how, with the proper feedback, one could stabilize the system on a particular periodic orbit. The importance of this result is that only very small variations in the parameter(s) controlling the system are necessary. It is like balancing a broom on the end of your finger; only small movements of the hand are required to keep the broom balanced. But without this feedback, the unstable broom would fall.

It is possible that the process of familiarizing oneself with a stimulus is actually the formation of a feedback loop that is pattern specific. In this way, only very little information would be required to trigger the pattern associated with that particular stimulus. This may in fact be the actual process of learning. We point out that this is only a speculation at the present time, but one which is attractive.

10.1.4 Technical overview

The processing methods previously alluded to include the attractor reconstruction technique (ART) of Takens,[15] the Grassberger-Procaccia algorithm (GPA)[16a,b] for determining estimates of the fractal dimension of the reconstructed attractor, and a substantial number of variations on the ART theme. In Section 10.2 we critique these various methods after describing them in detail and conclude that the GPA approach to characterizing the EEG/ERP data is much too uncertain to be useful for the real-time determination of the state of the brain. We stress that here because the GPA has become the method of choice over the past decade for estimating the fractal dimension of experimental time series; therefore, it is important to emphasize that this entire procedure has been superseded herein. Another technique that has been frequently used is the singular value decomposition method. This method is used on noisy data to reduce the dimensionality of the processed time series to only that associated with the signal. This method has been criticized because it does not give an unambiguous separation between signal and noise. This is also discussed in Section 10.2 and is ruled out as a viable way of processing EEG/ERP time series.

The three technical problems we addressed in applying the new processing technique to EEG/ERP data for the purpose of assessing learning and performance are (1) limitations on the data sets, (2) restricted sampling rates, and (3) stationarity of the time series. A technique which focuses on the probability distribution rather than the correlation function as used in the GPA has been implemented in Section 10.3 and clearly demonstrates its superiority over all other approaches based on the ART and the modified GPA in particular. We note here that the correlation function used in the GPA is merely a way to estimate the probability density and thereby the fractal dimension of the underlying process. This is discussed in detail in Section 10.2.4.

We have found a fractal dimension for an EEG time series collected at the University of North Texas and were able to estimate the confidence interval for the estimated fractal dimension using the fractal probability density. Attractive features of the fractal probability distribution approach are that the limitations on the data sets are no longer an obstruction to determining the fractal dimension, because the histogram for the probability density can be constructed using relatively few data points and therefore a short time series; the restriction on the sampling rates does not limit the algorithm, because the distribution function includes information about the small-scale and large-scale properties of the time series; and, finally, there is an estimate of the error made in estimating the dimension that is not possible to obtain in the standard implementations of ART. The problem of stationarity remains, as does the specification of a lag time, but we believe that we may have a procedure for handling the latter difficulty in the EEG/ERP context.

In addition to using the fractal probability density we can replace the notion of an embedding dimension used in ART by the various channels of the EEG. In this way we replace the scalar time series of a single channel by a vector time series using multiple channels. Various authors have used this technique as a way of reconstructing attractors without introducing a lag time and have successfully mapped out the spatial dynamics of the brain. We refer to this approach as MART and review it in Section 10.2. In Section 3, we combine this method with the fractal probability approach and thereby bypass *all* the difficulties associated with the traditional attractor reconstruction techniques except for stationarity. Even the problem of stationarity is partially handled by considering short time series. In this way, the fractal dimension can be made a function of time and we can track the evolution of the EEG through a sequence of dimensions. This does in fact give a better characterization of the EEG time series than the single numbers investigators have attempted to obtain in the past.

We conclude that we have solved, at least in part, the major problems associated with the methods of determining the fractal dimension of EEG/ERP time series under various task loadings using actual data. To be specific: (1) it is no longer necessary to determine a lag time for EEG/ERP time series, (2) it is no longer necessary to postulate an embedding dimension for the phase space, (3) one can now use limited (short) data sets efficiently, and (4) stationarity of the time series is not a problem if the data sets are sufficiently short.

10.2 Attractor reconstruction technique: a critique

It has been well over a century since it was discovered that the mammalian brain generates a small but measurable electrical signal. The electroencephalograms of small animals were measured by Caton in 1875 and in man by Berger in 1925. It had been thought by the mathematician N. Wiener, among others, that generalized harmonic analysis would provide the mathematical tools necessary to penetrate the mysterious relations between the EEG time series and the functioning of the brain. The progress along this path has been slow, however, and the understanding and interpretation of EEGs remains quite elusive. After 112 years, one can only determine intermittent correlations between the activity of the brain and that found in EEG records. There is *no* taxonomy of EEG patterns which delineates the correspondence between those patterns and brain activity. The clinical interpretation of EEG records is made by a complex process of visual pattern recognition and association on the part of the clinician, and significantly less often through the use of Fourier transforms. To some degree, the latter technique is less useful than it might be because most EEG centers are not equipped with the computers necessary for detailed analysis of the time series.

The electroencephalographic signal is obtained from a number of standard contact configurations of electrodes attached by conductive paste to the scalp. The actual signal is in the microvolt range and must be amplified several orders of magnitude before it is recorded. Layne et al.[17] emphasize that the EEG is a weak signal in a sea of noise so that the importance of skilled electrode placement and inspection for artifacts of the recording protocol cannot be overestimated.[18] Note that pronounced artifacts often originate from slight movements of the electrodes and from contraction of muscles below the electrodes.

The long-standing use of frequency decomposition in the analysis of EEG time series has provided ample opportunity to attribute significance to a number of frequency intervals in the EEG power spectrum. The power associated with the EEG signal is essentially the mean-square voltage at a particular frequency. The power is distributed over the frequency interval 0.5 to 100 Hz, with most of it concentrated in the interval 1 to 30 Hz. This range is further subdivided into four subintervals, for historical rather than clinical reasons: the delta, 1 to 3 Hz; the theta, 4 to 7 Hz; the alpha, 8 to 14 Hz; and the beta for frequencies above 14 Hz. Certain of these frequencies dominate in different states of awareness. A typical EEG signal looks like a random time series, with contributions from throughout the spectrum appearing with random phases. This apparent aperiodic signal changes throughout the day and changes clinically with sleep, i.e., its high frequency random content appears to attenuate with sleep, leaving an alpha rhythm dominating the EEG signal. The erratic behavior of the signal is so robust that it persists, as pointed out by Freeman,[6] through all but the most drastic situations, including near-lethal levels of anesthesia, several minutes of asphyxia, or the complete surgical isolation of a slab of cortex. The random aspect of the signal is more than apparent; in particular, the olfactory EEG has a Gaussian amplitude histogram, a rapidly attenuating autocorrelation function, and a broad spectrum that resembles "$1/f$ noise".[19]

The EEG/ERP time series are erratic functions of time that might not be so mysterious after one examines the brain's conduction system such as depicted in Figure 1. The question is whether the fluctuations are due to chaos or are the result of noise. If the time series is chaotic, then the fluctuations contain valuable information about the internal structure of the brain. If the erratic behavior is the result of noise, then the fluctuations are masking the information of interest. A technique was devised slightly over a decade ago to determine if a given time series was generated by a set of deterministic nonlinear dynamical equations or not. If the dynamics is generated by such a set of equations then the evolution of such a system unfolds on a manifold called an *attractor*. The term attractor is used because the manifold attracts to itself configurations of the dynamical variable which are initially off the manifold. In this way the attractor determines the asymptotic properties of the time series.

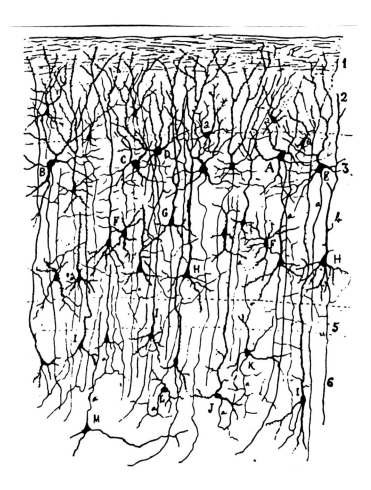

Figure 1 The complex ramified structure of typical nerve cells in the cerebral cortex is depicted. (Sketch by Cajal, 1888.)

The method alluded to above is the attractor reconstruction technique (ART) that we mentioned previously. If the time series is chaotic, then the attractor often has a fractal dimension, and this is the measure that most scientists have estimated in processing time series data, whether from the physical, social, or life sciences. The observation that the fractal dimension is a monotonic function of levels of brain activity suggests that we can relate the fractal dimension to learning.

These early results on processing EEG time series suffer from a number of technical difficulties; for example, the number of data points needed to sample the attractor is of the order of ten to the dimension of the attractor. Thus, 1000 data points are needed to treat a three-dimensional attractor, but 10,000 points are needed for a four-dimensional attractor. The latter just means that at a minimum four variables are required to describe the dynamics underlying the time series. The more data points required, the longer the time series needed; however, we do not expect the brain to remain in one state for an extended period of time so that the time series is not stationary. Therefore, the shorter the data set necessary for characterizing the attractor the better. After describing the limitations of the traditional ART methods, we outline a new processing procedure due to Judd[20a,b] based on the probability density empirically determined by time series data. We modify and extend the theory

10.2.1 The ART of Takens

The processes in which we are interested are those that can in principle be described by the Q-dimensional vector $X(t) = \{X_1(t), X_2(t), ..., X_Q(t)\}$ and whose evolution through time is determined by the set of differential equations

$$\frac{dX(t)}{dt} = F(X, t). \quad (1)$$

A Hamiltonian system, for example, has N-degrees of freedom ($Q = 2N$), where X_1, ..., X_N denote the generalized coordinates, and X_{N+1}, ..., X_{2N} denote the canonical generalized momenta; however, we are not limited to Hamiltonian systems in our discussion. In general, the system of interest may be dissipative rather than conservative; we only require that the evolution be confined to a finite-dimensional attractor in the phase space. Recall that the phase space of a dynamical system is defined by the set of variables describing the system dynamics, e.g., the rate Equation (1) yields solutions defined in a phase space with axes ($x_1, x_2, ..., x_{2N}$). A given point is a 2N-tuple, and this point uniquely defines the complete state of the system. As the evolution unfolds, this point traces out a curve, and this curve is the trajectory of the system through time. The uniqueness property of the solutions to Equation (1) requires that no two trajectories intersect in phase space, since this would imply a fundamental ambiguity in the solutions at the point of intersection. In this approach we hypothesize the existence of a limiting set, called an *attractor*, to which these trajectories are drawn asymptotically.

In general, one does not know the form of the dynamical equations describing the evolution of the biological system of interest nor whether such equations in fact exist. The experimentalist more often than not does not know what the important variables are, so that even the number of variables Q is not well defined. What is available is one or more different measurements that can be performed, such as the cardiac time series given by the electrocardiogram to characterize the complex interactive systems of the heart, the EEG to characterize the billions of neurons in the brain, and so on. In each of these examples we have a measured signal ζ, which is a function of the phase space variables describing the system. If the vector $X(t)$ is the continuous representation of the system on the attractor at time t, then the experimental time series is given by $\zeta(X(t))$. If we discretize the observation time, then we can define the jth point ζ_j of the time series as $\zeta_j = \zeta(X(t_j))$, where $t_j = j\Delta t$, $j = 1, 2, ..., M$ and Δt is the resolution time of the measurement apparatus.

We now describe the attractor reconstruction technique in which the single time series ζ_j is used to reconstruct the attractor in phase space on which the evolution of the system unfolds. Of course, if ζ_j is truly an experimental time series, then whether or not there is an attractor to be reconstructed must be determined as part of the problem. The assumption implicit in ART is that such an attractor does in fact exist. The procedure was first applied by Packard et al.[21] to a time series generated by a set of differential equations proposed by Rössler:[22]

$$\frac{dX}{dt} = -(X + Y) \quad (2a)$$

$$\frac{dY}{dt} = X + aY \qquad (2b)$$

$$\frac{dZ}{dt} = bX + XZ - cZ \qquad (2c)$$

where a, b, and c are constants. For one set of parameter values, Farmer et al.[23] called the attractor solution the "funnel" for obvious reasons (see Figure 2). Equation (2) is one of the simplest sets of differential equation models possessing a chaotic attractor. Figure 3 depicts a projection of the attractor onto the (x,y)-plane for four different values of the control parameter c. Notice that as c is increased, the trajectory changes from a simple limit cycle with a single maximum (see Figure 3A), to one with two maxima (see Figure 3B) and so on until the asymptotic orbit becomes aperiodic (see Figure 3D).

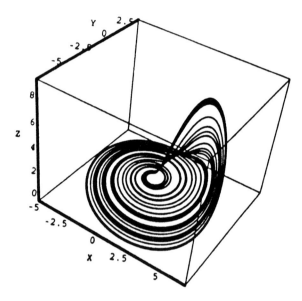

Figure 2 The "funnel" attractor solution to the Rössler Equations (2a, b, c) with parameter values $a = 0.36$, $b = 0.4$, and $c = 4.5$.

Packard et al.[21] playing the role of experimentalist, sampled the $X(t)$ coordinate of the Rössler attractor. They then noted a number of possible alternatives to the phase-space coordinates (x,y,z) that could give a faithful representation of the dynamics using the single $X(t)$ time series. One possible set was the time series itself and its first two time derivatives $X(t)$, $\dot{X}(t)$, and $\ddot{X}(t)$, but taking the time derivatives of experimental time series is notorious for generating noise. However, the set of variables $X(t)$, $X(t + \tau)$ and $X(t + 2\tau)$ where τ is a constant time interval does not introduce noise into the data and is a perfectly good representation of the original system. Note that implicit in this choice of variables is the idea that $X(t)$ is so strongly coupled to the other degrees of freedom that it contains dynamical information about these coordinates as well as about itself.

Figure 4A shows the projection of the Rössler chaotic attractor onto the (x,y)-plane taken from Figure 3D. In Figure 4B is shown the reconstruction of this projected part of the attractor using the sampled $X(t)$ time series in the $(X(t), X(t + \tau))$-plane. It is clear that the two attractors are not identical, but it is just as clear that the

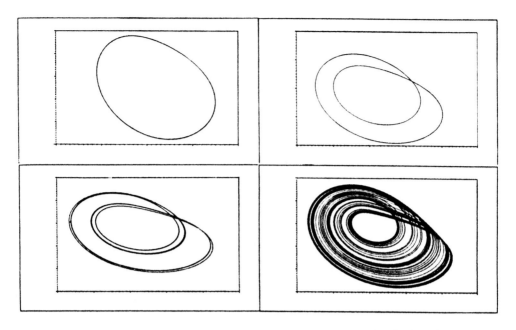

Figure 3 An x-y phase plot of the solution to the Rössler Equations (2a, b, c) with parameter values $a = 0.36$, $b = 0.40$, and at four different values of c: 2.5, 3.0, 3.3, 4.0.

reconstructed attractor retains the topological characteristics of the experimental attractor. One quantitative measure of the equivalence of the experimental and reconstructed attractors is the Lyapunov exponent associated with each one. This exponent can be determined by constructing a return map for each of the attractors and then applying the relation

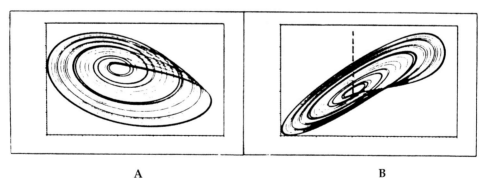

A B

Figure 4 A two-dimensional projection of the Rössler chaotic attractor (A) is compared with the reconstruction in the (x,y) plane of the attractor (B) from the time series $X(t)$, with $a = 0.36$, $b = 0.40$, and $c = 4.50$.

$$\sigma = \lim_{N \to \infty} \frac{1}{N} \sum_{j=1}^{N} \ln f'(\xi_j) \qquad (3)$$

where $f'(.)$ is the derivative of the return map with respect to its argument. A return map is obtained by constructing a Poincaré surface of section. In this example of an attractor projected onto a two-dimensional plane, the Poincaré surface of section is the intersection of the attractor with a line transverse to the attractor. We indicate this by the dashed line in Figure 4 and the measured data are the sequence of values $\{X_n\}$ denoting the crossing of the line by the attractor in the positive direction. These data are used to construct a next amplitude plot in which each amplitude X_{n+1} is plotted as a function of the preceding amplitude X_n, i.e., $X_{n+1} = f(X_n)$ where $f(.)$ is the mapping function. It is possible for such a plot to yield anything from a random spray of points to a well defined curve. If, in fact, we find a curve with a definite structure, then it may be possible to construct a return map for the attractor.

More generally, we assume that the system of interest, an EEG/ERP signal, can be described by N variables, where N is large but unknown, so that at any instant of time there is a point $X(t) = \{X_1(t), X_2(t), ..., X_N(t)\}$ in an N-dimensional phase space that completely characterizes the system. This point moves around in the phase space as the system evolves, in some cases approaching a fixed point or limit cycle asymptotically in time. In other cases, the motion appears to be purely random and one must distinguish between a system confined to a chaotic attractor and one driven by noise. To reiterate, in experiments one often records the output of a single detector, which selects a physical observable of the system for monitoring. In general the experimentalist (observer) does not know the size of the phase space since the important variables usually are not known and therefore he/she must extract as much information as possible from the single time series available, $\zeta(t)$ say. For sufficiently long times τ, one uses the embedding theorem of Takens[15] to construct the sequence of the displaced time series $\{\zeta(t), \zeta(t + \tau), ..., \zeta(t + [m - 1]\tau)\}$. This set of variables has been shown to possess the same amount of information as the N-dimensional phase space, provided that $m \geq 2d + 1$, where d is the dimension of the attractor for the original N-variable system. The condition on the embedding dimension m is often overly restrictive and the reconstructed attractor usually does not require m to be so large.

In treating data sets, the above reconstructed attractor can be written in the discrete form $\{\zeta(t_j), \zeta(t_j + \tau), ..., \zeta(t_j + [m - 1]\tau)\}$, where the discretization of the data has been made in terms of $t_j = j\Delta t$. The embedding theorem was proved by Takens[15] using topological arguments regarding the properties of the solutions to deterministic differential equations. More recently, Sauer et al.[24] were able to establish the same conditions on the embedding dimension, provided that the physical observable $\zeta(t)$ is generic in a theoretic sense and is restricted to the case where the attractor has a fractal dimension. As Sauer[25] points out, in the latter case, the embedding dimension must be strictly greater than twice the box-counting dimension of the attractor.

Takens presented the mathematical conditions under which one could expect to obtain an adequate reconstruction of the attractor for the underlying low-dimensional dynamical system. As with many mathematical proofs, however, the conditions are unrealistically severe, and one needs an infinitely long, noise-free time series which is sufficiently well sampled that the result is insensitive to the lag-time τ. Of course, with the finite length time series of real data records, the conditions of the mathematical theorem are never met and so one must test the ART under operational conditions.

Analysis of experimental time series has to a large degree been based almost exclusively on Taken's ART of the multidimensional signal and then on estimating its correlation dimension using the GPA. Since these two approaches are usually taken together we shall review GPA before we critique the technique.

10.2.2 Grassberger-Procaccia algorithm

Grassberger and Procaccia[16b,c] extended their original method, being inspired by the work of Packard et al.[21] and Takens,[15] to the embedding procedure of ART. Instead of using a scalar time series, they employ the m-dimensional vector

$$\xi(t_j) = \{\zeta(t_j), \zeta(t_j + \tau), \cdots, \zeta(t_j + [m-1]\tau)\} \qquad (4)$$

from m copies of the original observable. The m-dimensional correlation integral is

$$C_m(r) = \lim_{M \to \infty} \frac{2}{M(M-1)} \sum_{i>j=1}^{M} \Theta(r - |\xi(t_i) - \xi(t_j)|) \qquad (5)$$

where $\Theta(x)$ is the Heaviside unit step function which is equal to 0 for $x < 0$ and 1 for $x \geq 0$. For a chaotic (fractal) time series, the correlation function has the power-law form

$$C_m(r) \sim r^{\nu_m} \qquad (6)$$

where

$$\nu = \lim_{m \to \infty} \nu_m \qquad (7)$$

is the correlation dimension for the time series.

Grassberger and Procaccia argue that the correlation exponent (dimension) is a useful measure of the local properties of the attractor, whereas the fractal dimension is a purely geometrical measure and is rather insensitive to the local dynamical behavior of the trajectories on the attractor. The information dimension, σ, is somewhat sensitive to the local behavior of the orbits and is a lower bound on the Hausdorff dimension d. In fact, they observe that in general one has

$$\nu \leq \sigma \leq d. \qquad (8)$$

Thus, if the correlation integral obtained from the experimental data set has the power-law form of Equation (6), with $\nu < m$, one knows that the data set arises from deterministic chaos rather than completely random noise because uncorrelated noise yields

$$C_m(r) \sim r^m. \qquad (9)$$

Note that for a fixed point attractor $\nu_m = m = 0$; for periodic sequences, $\nu_m = m = 1$; for uncorrelated random noise, $\nu_m = m$, the embedding dimension, while for chaotic sequences $\nu < m$ is noninteger.

$C_m(r)$ is theoretically a monotonically increasing function from zero at $r = 0$ to the saturation value of unity whenever $r = \geq \max|\xi(t_i) - \xi(t_j)|$. Grassberger and Procaccia derive Equation (6) for m sufficiently large and r sufficiently small, i.e., the logarithm of the correlation integral as a function of the logarithm of correlation length has a linear region, called the *scaling region*, with slope v_m;

$$\ln C_m(r) = v_m \ln r + \ldots \tag{10}$$

Thus the extent of the scaling region can be found by taking the derivative of Equation (10) to obtain

$$\frac{d \ln C_m(r)}{d \ln r} = v_m. \tag{11}$$

One criteria for determining the derivative in the scaling region is that it not vary by more than some specified amount, say 10% (see, for example, Albano et al.[26]). This last step establishes the existence of v_m and perhaps v, but not necessarily its true value.

10.2.3 What is wrong with GPA?

As pointed out by Albano et al.[26] the number of data points required to determine the dimension of an attractor depends on the structure of the attractor, the distribution of points on the attractor, and the precision of the data. Let us consider the best possible situation, i.e., we generate our own data with known properties and see how well the GPA works.

The logarithm of the correlation function $C_m(r)$ vs. the logarithm of the scale r is depicted in Figure 5 for the Lorenz model for various embedding dimensions. We see that the slope of the curve is becoming constant with increasing embedding dimension. We find that by a least-square fitting of the high dimension curves (i.e., $m = 8, 9, 10$), we obtain $v = 2.059$, as shown in Figures 6 and 7. Swinney proposed a rule of thumb for the number of data points required to resolve the dimension of a given attractor. This is 10^v points of at least 0.5% accuracy for correlation dimensions up to 5. Unfortunately, no universally accepted criteria for the scaling region has been established, so Swinney's estimate might be somewhat optimistic. Our results for the Lorenz attractor are consistent with this rule.

Eckmann and Ruelle[27] argue that the number of pairs of points with mutual distance less than r [$N(r)$] for a fractal object is given by

$$N(r) \simeq \text{const.} \, r^v. \tag{12}$$

If R is the diameter of the reconstructed attractor, for M data points the number of pairs of points is the entire data set

$$N(R) \approx \frac{1}{2} M^2 \tag{13}$$

so that comparing this with Equation (12) the constant yields

$$N(r) \approx \frac{M^2}{2} \left(\frac{r}{R}\right)^v. \tag{14}$$

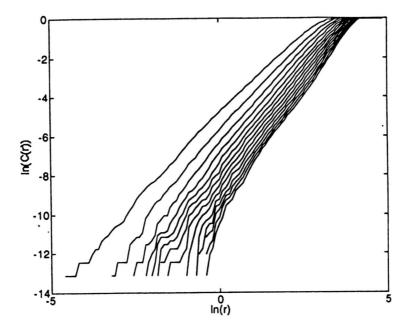

Figure 5 The time series $X(t)$ from the Lorenz model is segmented into 10^3 data points. The GP algorithm is used to evaluate the correlation function in embedding spaces of dimension m where $2 \leq m \leq 15$. The lag time in ART is selected to be $\tau = 2$.

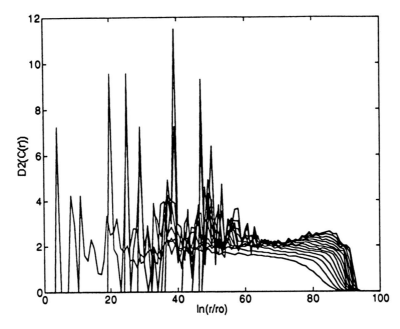

Figure 6 The correlation functions $C_m(r)$ from the Lorenz model in Figure 5 are differentiated with respect to $\ln r$. The plateau region indicates the existence of a correlation dimension.

Thus, to determine the correlation dimension from the logarithm of Equation (14) we need several values of $N(r)$ with $r < R$. Also we need $N(r) \gg 1$ to get good statistics. These restrictions impose the conditions

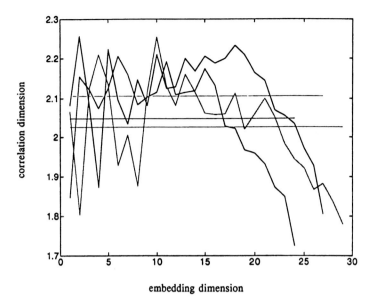

Figure 7 The plateau region for each embedding dimension is fit with a best mean-square horizontal line. Each line gives a correlation dimension. When the three values for $m = 8$ to 10 are averaged together we obtain $v = 2.0599$.

$$\frac{1}{2}M^2\left(\frac{r}{R}\right)^v \gg 1 \quad \text{and} \quad \frac{r}{R} = \rho \ll 1. \tag{15}$$

Taking logarithms of Equation (15), they find the requirement

$$2\log M > v \log(1/\rho) \tag{16}$$

from which it is clear that the GPA will not produce dimension larger than

$$v_{max} = \frac{2\log M}{\log(1/\rho)}. \tag{17}$$

For example, choosing $\rho = 0.1$, it is clear that for $M = 10^3$ data points that $v \le 6$, and if $M = 10^5$, then $v \le 10$. As Eckmann and Ruelle[27] comment, "Thus, if the GP algorithm produces a dimension of 6 for 10^3 points, the result is probably worthless." They go on to estimate that the number of data points necessary to determine the Lyapunov exponent is approximately equal to the square of that needed to determine the correlation dimension.

No universally acceptable criteria has been established for determining the slope of the above correlation function and, therefore, although it is relatively straightforward to obtain a value of v, it is extremely difficult to obtain an accurate value using the GPA. Although there are differences in v estimates obtained using different algorithms and different signal processing protocols, the ratio of vs seems to be robust.[26]

Another variable to which GPA is sensitive is the sampling interval, since the correlation dimension can be made artificially small by making the time between successive measurements arbitrarily small. So what is the optimal sampling time?

The short answer is that we do not know. The sample interval must not be too small, because then one essentially has the same point for successive data samples, and it must not be too large, because then the data are mutually independent. Estimates of this time have been defined in terms of the first zero of the autocorrelation function, the largest frequency of the power spectrum, and the first minimum of the mutual information, to name a few. But none has emerged as clearly superior to the others, and so to a large extent finding the sampling interval is more a matter of trial and error than theory in using the GPA.

Finally, there is the question of nonstationarity of the time series and therefore of the attractor dimension itself. Here we take cognizance of the possibility that the difficulty in determining the fractal dimension of an attractor may not necessarily be associated with experimental noise or the finiteness of the data set, but rather with the properties of the attractor changing over time. In fact, we anticipate that, for EEG data, if the time series is too long one must see a change in the attractor properties just because the level of cognitive activity changes over time. Thus, the change in ν over time may be an interesting way of characterizing EEG data, but this means that we must know how to handle nonstationary data. There is the traditional trade-off in signal processing between temporal resolution (epoch is too long and nonstationarity becomes important) and frequency resolution (epoch is too short).

10.2.4 Singular value decomposition

Singular value decomposition is a new name (see, for example, Broomhead and King[28]) and for a technique based on the Karhunen-Loéve theorem, see Loéve.[29] This theorem establishes that a random function can always be expanded as a decomposition on a set of orthonormal functions on the interval over which it is defined. The singular system approach developed by Broomhead and King is a sophisticated linear analysis of time series for the purpose of erecting a phase-space reconstruction. As they put it, we wish to construct an invariant distribution function or a measure of the set of data points in the phase space. The second moments of the measure centered on the mean are found by diagonalizing the covariance matrix

$$C = \frac{1}{M} \sum_{j=1}^{M} X(j) X^T(j). \tag{18}$$

This procedure is known as principal component analysis, or the derivation of a Karhunen-Loéve basis as described by Devijver and Kittler.[30] It is related to the singular value decomposition (Golub and van Loan[31]) of the $N \times m$ trajectory matrix, X, whose rows consist of the row vector $X^T(j)$. The covariance matrix is $X^T X$; therefore, the right singular vectors of X are the eigenvectors of the covariance matrix, whereas the corresponding singular values of X are the positive square roots of the eigenvalues of the covariance matrix.[28] The fundamental idea is that the mean-squared distance between points on the reconstructed attractor should be maximized.

The Broomhead and King[28] technique requires that the reconstruction consist of a linear projection from a set of n-dimensions, with $n > m$, the embedding dimension, onto the m-dimensional subspace. The basis is chosen to maximize the covariance matrix. The first basis vector $\mathbf{c}_0$ is chosen to maximize $<|\mathbf{w}\cdot\mathbf{c}_0|^2>$, and the second is chosen to maximize $<|\mathbf{w}\cdot\mathbf{c}_1|^2>$ subject to $\mathbf{c}_0\cdot\mathbf{c}_1 = 0$. The procedure is continued in a straightforward way and produces an orthonormal basis such that $<(\mathbf{c}_j\cdot\mathbf{w})(\mathbf{c}_k\cdot\mathbf{w})> = \delta_{jk}\lambda_j^2$, i.e., the random variables $X(t + jT) = \mathbf{e}_j\cdot\mathbf{w}$ are linearly independent. Fraser[32] emphasizes that this is not statistical or general independence but only linear independence. This analysis can be done on an experimental data set using a singular value decomposition routine, as available, for example, in Dongarra et al.[33] The numbers $\lambda_j = \langle|\mathbf{w}\cdot\mathbf{c}_j|^2\rangle^{1/2}$ are called the singular values (where $\lambda_0 \geq \lambda_1 \geq \lambda_2 \geq \ldots$).

Let us consider the time series defined by a difference vector, where the incoming signal Y has m degrees of freedom in it, though X is an n-vector because of the measurement process and is contaminated by some noise $\eta(j)$, i.e., $X = Y + \eta$. We assume the noise to be a zero-centered Gaussian process such that

$$\langle \eta_\alpha(j) \rangle = 0,$$
$$\langle \eta_\alpha(j)\eta_\beta(j') \rangle = \sigma_\eta^2 \delta_{\alpha\beta} \delta_{jj'}, \tag{19}$$

where again the brackets denote an average over an ensemble of realizations of the noise. In this case, the covariance matrix given by Equation (18) has the elements

$$C_{\alpha\beta} = \frac{1}{M}\sum_{j=1}^{M} \langle X_\alpha(j) X_\beta(j) \rangle \tag{20}$$

so that

$$C_{\alpha\beta} = \frac{1}{M}\sum_{j=1}^{M} Y_\alpha(j) Y_\beta(j) + \sigma_\eta^2 \delta_{\alpha\beta} \tag{21}$$

and the goal is to estimate the rank of the sample covariance matrix of the Ys. Note that this latter matrix should have rank m, the number of degrees of freedom of the incoming signal in the absence of noise.

From Equation (21) we can see that when the data are noisy then every singular value of the Y-covariance matrix is augmented by the strength of the noise σ_η^2. Thus, the noise dominates any component whose corresponding eigenvalue is comparable to σ_η^2, and Broomhead and King[28] argued that such components should be discarded. If the assumption about the additive noise is correct, then m eigenvalues should have values substantially above the noise floor σ_η^2, and $n - m$ eigenvalues will be comparable to the noise, i.e., lost in the noise. They explain that this procedure results in a reduction of the noise power in the reconstructed space by a factor m/n, where $(n - m)$ is the dimension of the discarded subspace and m is the dimension of the retained space. Thus, the significant eigenvalues $\lambda_0, \ldots, \lambda_{m-1}$, those above the noise floor, and the corresponding eigenvectors $\mathbf{c}_0, \ldots, \mathbf{c}_{m-1}$ represent the deterministic aspects of the time series, and the remaining eigenvalues and eigenvectors represent the noise.

Abarbanel[34] criticizes the use of this technique as not being sufficiently objective, as does Fraser.[32] They both suggest alternative procedures based on invariant measures of the statistics involving information theoretic approaches. One example is sufficient to illustrate that the clean separation of the deterministic subspace from that of the noise in terms of the magnitude of eigenvalues envisioned by Broomhead and King does not always occur in practice.

Wax and Kailath[35] superimpose two sinusoids ($m = 2$) with random phases in an $n = 7$ dimensional space with a signal-to-noise ratio of 10 db. Using 100 data points ($M = 100$), the seven eigenvalues of the covariance matrix are: 21.24, 2.17, 1.43, 1.10, 1.05, 0.94, and .073. It is clear that the first eigenvalue is substantially above the others, but the next few are not well separated. One would be hard pressed to deduce $m = 2$ from this spectrum of eigenvalues. The limitations of the singular system approach have also been examined by Brandstater et al.[36] and Meese et al.[37]

Of course, even after one has completed the singular value decomposition by rotating the coordinate system and deleting the highest components of the data vector, one still applies the GPA to the resulting reconstruction of the attractor in the embedding space.

10.2.5 Re-embedding technique

Fraedrich and Wang[38] propose a modification of the GPA that gives a reliable estimate of the correlation dimension with fewer data points than 10^v. The technique uses the singular value decomposition method in a particular way. First, the ART is applied to a measured time series, and an m-dimensional embedding is obtained. An equivalent space is obtained by a linear transformation using singular value decomposition to identify the dominant axis (linearly independent) in the embedding space. Second, the few principle eigenvectors obtained in the first step are used to form a subspace. This subspace is now treated as was the original data, which is to say it is embedded in a higher-dimensional space. This process is called re-embedding. The GPA is now applied to the re-embedded data.

To optimize the use of the available data, the lag-time τ for the first embedding is set equal to the sampling time Δt. The selection of the largest eigenvalues for the principle components, those that are above the noise floor, clearly reduces the noise level in the data. The idea is that the dimension of an attractor is an invariant measure, so it should not change under a linear transformation, i.e., under a homeomorphic mapping. The original choices of data vectors are not orthogonal, but it is possible to transform the embedding space into an equivalent space having orthogonal coordinates. These latter coordinates are those determined by singular value decomposition.

Fraedrich and Wang[38] demonstrate that the re-embedding idea worked quite well on both the output of the Lorenz model contaminated by noise and the output of the Mackey-Glass equation with noise added. Using 5×10^3 data points in the former case, they went from a situation using $m = 1 - 15$ in the ART, in which there was no plateau region to a clear plateau over a decade of r-values when the data was re-embedded. Similarly, in the latter case they were able to do the same for 10^4 data points. The latter example certainly does better than Swinney's rule of thumb, since the dimension of the Mackey-Glass attractor in the parameter region considered was $d = 5.0$. Fraedrich and Wang applied this technique to meteorological data sets that do not concern us here except to say that the results were consistent with those obtained using the mathematically generated data.

We apply this technique to EEG data in a subsequent section wherein we compare this approach with various other newly developed methods.

10.2.6 An alternative to Takens

An alternative to Takens' ART which has recently been emphasized in the EEG context by Dvorák[39] and Palus et al.[40] is the multichannel attractor reconstruction technique (MART) originally proposed by Eckmann and Ruelle.[41] Dvorák[39] shows that Takens' ART and GPA give biased and unsatisfactory estimates of the correlation dimension (exponent) of EEG registered in the standard state (relaxation, eyes closed). He argues that a bias in the GPA may be lifted by modifying the assumed form of the correlation function used in the GPA. The scaling law for the correlation integral is assumed to be

$$C_m(r) \sim [r(2-r)]^{v_m} \tag{22}$$

instead of r^{v_m}. This modification of the GPA seems to work particularly well at higher correlation dimensions since it reduces the bias in uncorrelated noise. This modification in the correlation function introduced a quadratic correlation to the fitting curve so that

$$D_m = d \ln C_m(r) / d \ln(r(2-r)) \tag{23}$$

is a much better estimate of v_m than is the one obtained from the ln$r$ derivative. This improvement in the fit is due to the extension in the plateau region due to stretching of the axis.

Dvorák[39] asserts that the main source of upward bias seems to be due to the application of Takens' ART for embedding signals with finite resolution. In actuality, the finite resolution has the effect of superimposing a white noise on the actual signal, a white noise that becomes more and more important as the embedding dimension is increased. Singular value decompensation was designed to handle this problem. A different way to overcome this problem is to use MART; that is, take each registration channel of the EEG as one component of the vector variable with dimension given by the number of registration leads. This was first done with EEG data by Destexhe et al.[42] Note that this approach obviates the discussion about lag time τ for embedding, since there is no embedding of a scalar-time series.

Figure 8 shows the dependence of the activated correlation dimensions on the number of channels considered in MART using the modified GPA. The saturated value of v_m is 5.68 ± 0.07 (mean and standard deviation of the estimates for $m = 12 - 16$). The analysis was repeated by Dvorák using different subsets of channels in the reconstruction, always with the same result. The insert in Figure 8 documents that in the same time the considered scaling region in the $\ln C_m(r)/\ln[r(2-r)]$ plot for high m-values never exceeds one order of magnitude. Note that Takens' ART deals with "local" information concerning the EEG time series taking place immediately beneath the registration electrode, but MART deals with "global" information taking place within a large portion of the whole brain cortex.

Palus et al.[40] extend MART to determine the spatio-temporal dynamics of human EEG. This is done by combining this approach with the singular value decomposition theory outlined in the previous section. In this way, one can obtain the spatial orthonormal modes covering the whole brain and determine how the signal spatial correlation evolves over time. This provides a measure of the linear complexity of the EEG time series.

It is possible that one can do better when this technique is further modified using the re-embedding technique to even further suppress the effect of noise, but this has not yet been done.

10.2.7 Overcoming the limitations: the distribution function

Judd[20a,b] proposed a new algorithm to estimate the fractal dimension of a data set using an invariant probability density. In this algorithm, the possibility of having a dimension that changes with scales is introduced. This would make it useful for data

chapter ten: Fractal Probability Density and EEG/ERP Time Series   287

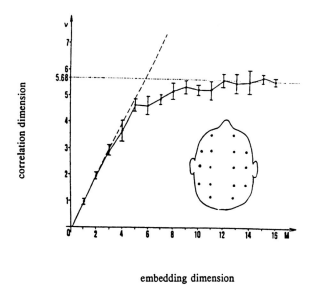

Figure 8 Saturation of the global correlation exponent of EEG with increasing number of scrutinized channels. (From Palus, M., Dvorák, I., and David, I., *Physica A*, 185, 433–438 (1992). With permission.)

that do not display a plateau or scaling region. Judd verifies the algorithm using data having Gaussian statistics and demonstrates that this technique has less demanding data requirements than the GPA. The data requirements to estimate even modest dimensions using conventional techniques on EEG/ERP time series are beyond what is available or what can be reasonably collected.

Note that the existence of a dimension implies that there is an invariant probability density that one can define in the phase space. If the points X and Y are selected independently from the attractor, then

$$P(r) = \text{Prob}(|X - Y| < r) \tag{24}$$

is the interpoint-distance distribution. Estimation of dimension using Equation (24) consists of two steps. First, an empirical distribution $\hat{P}(r)$ is obtained by means of a histogram. Second, an assumed form for Equation (24) is fit to the histogram.

Step 1: The enormous number of interpoint distances (M^2 for M data points) is efficiently handled by binning the interpoint distances to construct the histogram $\hat{P}(r)$. Choose a largest scale r_0 and then construct the bins to be uniform on a logarithm scale; that is, the sequence of scales that form the bins are given by $r_j = \lambda^j r_0$ where j is an positive integer and $0 < \lambda < 1$. The bins, then, are $[r_j, r_{j-1}]$, the cutoff is r_0, and the bin width is $w = \ln(1/\lambda)$. The optimal size of the bins is determined by choosing an initial width w_0 that is extremely small and then aggregating the bins until an optimum value of w is obtained.

Step 2: Given an r_0 and an optimal binning of the interpoint distances, the dimension d is estimated by fitting a particular functional form for the distribution to the histogram. Judd selected

$$P_j(d, \mathbf{a}) = \left(\frac{r_j}{r_0}\right)^d \left[a_0 + a_1 \frac{r_j}{r_0} + \cdots a_T \left(\frac{r_j}{r_0}\right)^T\right] \tag{25}$$

where T is the topological dimension which rarely exceeds two in practice. Here, Equation (25) is the weight of the interpoint distribution in the jth bin. We can then write the probability density p_j as

$$p_j(d, a) = P_j - P_{j+1} \tag{26}$$

where if b_j is the observed number of interpoint-distances in the jth bin, the maximum-likelihood estimates of the dimension d and the parameters a are given by maximizing the log-likelihood function

$$L(d, a) = \sum_{j=0}^{\infty} b_j \log p_j \tag{27}$$

subject to $p_j > 0$ and the normalization

$$\sum_{j=0}^{\infty} p_j = 1 \tag{28}$$

Smith[43] has shown that a Gaussian process has an asymptotic interpoint-distance distribution that is a special case of Equation (25). For a Gaussian measure of dimension D, Judd[20b] shows that the interpoint-distance distribution is

$$P(r) = Nr^D e^{-r^2/2} \left[1 + \frac{r^2}{D+2} + \frac{r^4}{(D+2)(D+4)} + \cdots \right] \tag{29}$$

with a normalization N, from which one obtains

$$\frac{d \ln P(r)}{d \ln r} = D - \frac{D}{D+2} r^2 + O(r^4). \tag{30}$$

This expression implies that a log-log plot of the correlation function of a Gaussian measure is necessarily curved, i.e., there is no linear scaling region because of the r^2 dependence of the derivative. Thus, it is clear that the GPA would not yield D for this process; however, if r_0 is the cutoff of the data, the dimension estimate is given by, to a first approximation,

$$\hat{d} = \frac{d \ln P(r_0)}{d \ln r_0} \cong D - \frac{D}{D+2} r_0^2$$

$$\hat{d} = D - \frac{D}{D+2} \exp[2 \ln r_0] \tag{31}$$

so that $\hat{d}$ converges to D exponentially in $\ln r_0$. With sufficiently large samples one should observe that dimension estimates are asymptotic to the true dimension as r_0 is decreased. Thus we see that the dimension changes as we decrease the cutoff.

The above algorithm has been tested by Judd using uniform, Poisson, Gaussian, fractal, and dynamical measures. In each of these cases, the effect of correlation present in samples is negligible on the described algorithm. He finds that the algorithm just fails at a 90% confidence level for a ten-dimensional Gaussian measure using 300 data points, but is successful for a dimension eight or less.

We have used this algorithm on the time series generated by the Lorenz attractor. In Figure 9 we depict the fit of Equation (29), denoted by the solid curve, to the histogram denoted by the astrices. Minimizing the log-likelihood function of Equation (27) yields a fractal dimension for the $X(t)$-time series of $d = 2.063$ with a χ^2 value of 0.33 for 40 degrees of freedom. Note the excellent agreement with the commonly accepted value of the dimension of the Lorenz attractor. It is also worth pointing out that this value has a much narrower confidence interval than the one depicted in Figure 6 obtained using the GPA.

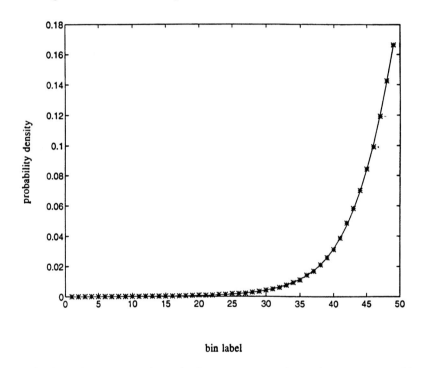

Figure 9 The fractal time series from the Lorenz attractor is used to construct a histogram for the probability of distances between pairs of data points. The asterisks denote the b_j and the solid curve depicts Equation (25) with a_0, a_1, a_2 having the maximum likelihood fit, Equation (27). The embedding dimension is $m = 10$, the fit values of the coefficients are $a_2 = -0.0522$, $a_1 = 0.222$, and $a_0 = 0.688$ for $M = 10^3$ data points and $\chi_{40}^2 = 0.326$.

10.2.8 Summary

Here we list the four main problems associated with the use of Takens' ART and the GPA as itemized by Judd:[20b]

1. In a sample trajectory consisting of M points, there are $M(M-1)/2$ interpoint distances, not all of which are independent.
2. The correlation function $C_m(r)$ is proportional to the number of interpoint distances less than r and it becomes more statistically correlated as r increases.

3. The so-called scaling region may not reflect information about the dimension of an attractor, since it may only reflect the large-scale properties of the attractor. This region completely ignores information about small distances which may be important.
4. In its standard implementation, there is no estimate of the error made in estimating the dimension.

The last three items on the list are obviated by considering the distribution of the interpoint distances rather than the correlation function. This leaves item 1 remaining and the challenge of obtaining truly independent samples. This may be accomplished at least in part by not using all the $M(M - 1)/2$ interpoint distances, but instead by randomly selecting a subset of them to construct the histogram. This random sampling hypothesis is tested in a subsequent section.

10.3 Processing brain wave data

The relationship between the neural physiology of the brain and the overall electrical signal measured at the brain's surface is not understood. We have, however, noted the similarity of the neuron to fractal networks elsewhere (see West[1] and West and Deering[44]). The electrical signals originate from the interconnections of the neurons through collections of dendritic tendrils interleaving the brain mass. These collections of dendrites generate signals that are correlated in space and time near the surface of the brain, and their propagation from one region of the brain's surface to another can actually be followed in real time. This signal is attenuated by the skull and scalp before it is measured by the EEG contacts.

One of the more dramatic results that has been obtained in recent years has to do with the relative degree of order in the electrical activity of the human cortex in an epileptic human patient and in normal persons engaged in various activities (see Figure 10). Babloyantz and Destexhe[46] used an EEG time series from a human patient undergoing a petit mal seizure to demonstrate the dramatic change in the neural chaotic attractor using the phase-space ART. Freeman[6] has induced an epileptic form of seizure in the prepyriform cortex of the cat, rat, and rabbit. The seizures closely resemble variants of psychomotor or petit mal epilepsy in humans. His dynamic model enables him to propose neural mechanisms for the seizures and to investigate the model structure of the chaotic attractor in transition from the normal to the seizure state. As we have pointed out, the attractor is a direct consequence of the deterministic nature of brain activity, and what distinguishes normal activity from that observed during epileptic seizures is a sudden drop in the dimensionality of the attractor. Babloyantz and Destexhe[46] determine the dimensionality of the brain's attractor to be 4.05 ± 0.5 in deep sleep and to have the much lower dimensionality of 2.05 ± 0.09 in the epileptic state. There is, however, some controversy over the values of these dimensionalities and the associated errors.

West[1] reviewed a number of experiments measuring the perceptual cognitive demands of a task. These experiments clearly indicate that the fractal dimension of the EEG increases with cognitive demand and suggest the existence of a cognitive attractor. These and subsequent studies establishing the fractal nature of EEG time series shall be shown to be of value when the time series include ERPs.

We concur with West[1] and Pritchard and Duke[47] in their belief that the saturation of the estimates of the fractal dimension, called *dimensional complexity* by some, with increasing value of embedding dimension is a strong indication that the time series is governed by a low-dimensional dynamical system having chaotic solutions. Of course, this is in the event that one can discriminate between a chaotic time series

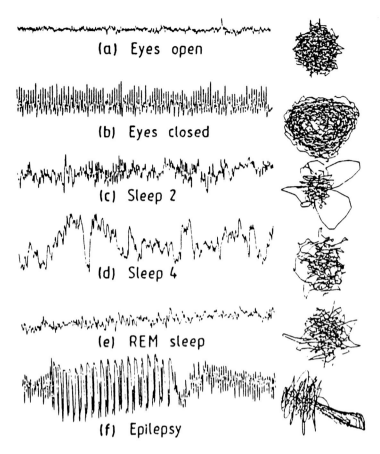

Figure 10 Typical episodes of the electrical activity of the human brain as recorded from the electroencephalogram (EEG) together with the corresponding phase portraits. These portraits are the two-dimensional projections of three-dimensional constructions. The EEG was recorded on an FM analog tape and processed offline (signal digitized in 12 bits, 250 Hz frequency, 4th order 120 Hz low pass filter). (From Babloyantz, A., and Destexhe, A., *Proc. Natl. Acad. Sci. U.S.A.*, 83, 3515–3517 (1986). With permission.)

and colored noise; the former being an intrinsic nonlinear property of the system, whereas the latter is due to interactions of the system with the environment. Pritchard and Duke, among others, find that the fractal dimension increases with processing load, as observed when the human EEG is modeled as a chaotic time series and the GPA is used to determine the dimension. We find the same result when more refined processing techniques are employed.

Jansen[48] points out that one should not put too much credence in the absolute value of the fractal dimension, but rather should focus on how this measure changes in response to different psychophysiological conditions. He mentions the response to drug therapy and the changes in the EEG due to cognitive activity; the latter is of particular interest here. It should be emphasized that Albano et al.[26] discovered that even though the estimates of the dimension are sensitive to different algorithms and different signal sampling protocols, the ratio of dimensions is not; therefore, how the fractal dimension changes as one learns may be robust against different ways of quantifying the learning process. In addition, some of the observed sensitivity is a consequence of the data processing technique, so we have implemented the more promising of these methods for quantifying the fractal behavior of the time series data.

10.3.1 GPA applied to EEG

Almost all we know regarding the qualitative mathematical properties of EEG time series has been obtained by using the GPA or variants of it. Let us summarize these results. First, an eyes-closed EEG appears to have a lower dimension than an eyes-open EEG (Dvorák et al.;[49] Nan and Jinghua;[8] Rapp et al.[50]) Second, giving a subject a mental task to perform increases dimensional complexity (Graf and Elbert;[51] Dvorák et al.;[49] and Rapp et al.[52]). Pritchard and Duke[47] confirmed these results and were the first to apply statistical measures to the computed dimensions to determine the statistical significance of the results. They concluded that their version of the GPA, "... clearly tapped some 'real' aspect of the data." Further, they randomly selected subsets of the EEG epochs and shuffled the data points. In every case, the randomized data sets yielded increased values of the dimensional complexity, as it should for a chaotic system. A system with colored noise would show no change in the GP-dimension between the randomized and unrandomized data. Gregson et al.[7] relate estimates of the dimensionality of the EEG to the complexity of task load in a visual scanning task. The fractal dimension using a modification of GPA was shown to increase as the task became more complicated. The effects are small but consistent with those of other investigators.

Finally, Arle and Simon[53] have shown that transient deterministic data in the EEG have a fractal dimension different from the background. In their analysis, two ERPs separated by 100 ms could not be distinguished; however, three such ERPs clearly change the local fractal dimension of the time series.

10.3.2 EEG time series

We now turn our attention to calculations of the fractal dimension using data collected at the University of North Texas using the lead placement shown in Figure 11. The sample was composed of 32 Anglo-American boys ages 8 to 10. Nineteen of the boys had been previously diagnosed as ADHD following DSM III-R criterion and met the following additional inclusion requirements: (1) identified as positive responders to methylphenidate; (2) currently taking a minimum of 10 mg of methylphenidate (Ritalin, not generic) but not time-released methylphenidate S-R; and (3) free from any other identified handicapping conditions. The mean age for the ADHD group ($n = 19$) was 9.65 years, while the mean age for the controls ($n = 13$) was 9.84. There was not a significant difference between the chronological ages of the two groups ($t = .615$, $p = .543$). Each subject was administered the Woodcock-Johnson Revised Tests of Cognitive Ability. There were no significance differences between the ADHD and controls for the seven cognitive cluster standard scores based on a Hotelling T-statistic ($F = 1.254$, $p = .311$). The range of standard scores for the ADHD group was 92 to 110, while the range for the control group was 100 to 121.

The subjects' brain electrical activity was recorded from 19 electrodes sewn into an Electrocap (Electrocap International, Inc.; Eaton, OH), then stored and analyzed using the Brain Atlas III (BioLogic Systems Corp.; Mundelein, IL). An extra channel was interpolated at Oz for a total of 20 electrode sites (see Figure 11). Linked ears served as the common reference for all electrode, and the ground was Fpz linked, with another electrode placed between Fpz and Fz in the Electrocap. Impedances were kept below 5 kΩ for each electrode, and the range of the highest and lowest impedance was kept to a maximum of 3 kΩ. Prior to each subject's data collection, the 19 amplifiers were recalibrated. The EEG was amplified 20,000 times with a bandpass of 1.0 to 15 Hz across all channels. A 60-Hz notch filter was also used. The EEG was digitized at 200 samples per second.

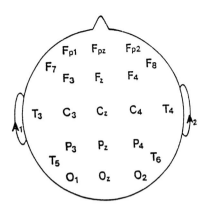

Figure 11 The location of the leads used in collecting EEG data in the experiments conducted at the University of North Texas.

A computerized version of the Letter Pair Test (Excellence in Learning, Inc.; 1991) was used in this study. The Letter Pair Test was based upon Naglieri and Das' Planning, Attention, Simultaneous, and Successive (PASS) model of intelligence.[54] This test was developed to measure selective attention and utilizes Posner's[55] idea for matching pairs of letters that are perceptually identical (AA, rr, TT, etc.) or semantically identical (Aa, hH, Bb, etc.). The test consists of two parts; in Part 1, the subject is required to select the pairs of letters that are physically the same (i.e., HH, ee, but not Tr or eR), and in Part 2, the subject is instructed to select only same-name letter pairs (i.e., Aa, kK, but not Be or St). Altogether, the subject is presented with four computer screens; each screen consists of 10 rows of 10 letters pairs. The first two screens presented are for physical-matched letter pairs; the last two screens are for name-matched letter pairs. In each part of Letter Pair Test, there is a total of 200 letter pairs; 50 (25%) of them are targets.

Following placement of the Electrocap on the subject's head and following the adjustment for impedances, the subject was seated in front of the Macintosh IIx computer and was given instructions for the Letter Pair Test. The subject was instructed to indicate the selection of the target by placing the pointer on the target (physical match or name match) and to click the mouse button. The subject was to begin the task at the upper left-hand corner of the first row and to continue from right to left through all the rows (see Figure 12). The subject first was to identify the physically matched pairs on the first two screens; he then was to continue with the identification of name-match letter pairs in the last two screens. EEG data was continuously recorded as the subject performed the Letter Pairs tasks. A custom interface (hardware and software) connected the Macintosh IIx computer via the modem port to the COM port of the Brain Atlas II. Each time the subject clicked the mouse of the computer, a pulse marker triggered the Brain Atlas to indicate the onset of the mouse click (selection of target) on the Oz channel. Performance data were collected by the Macintosh IIx computer; data collected included number of correct responses, number of misses, number of false alarms, and the completion time. All further computational analysis has been derived from raw EEG/ERP time series recorded from one control subject while he was performing the search of perceptually identical letters pairs.

In Figure 13 we plot the logarithm of the correlation function $C_m(r)$ vs. the logarithm of the interpoint distance r, as well as the derivative of $\ln C_m(r)$ vs. $\ln r$. Here we see the formation of a bump for higher embedding dimensions in nine of the recording channels, and even in those not shown, this bump is equally apparent.

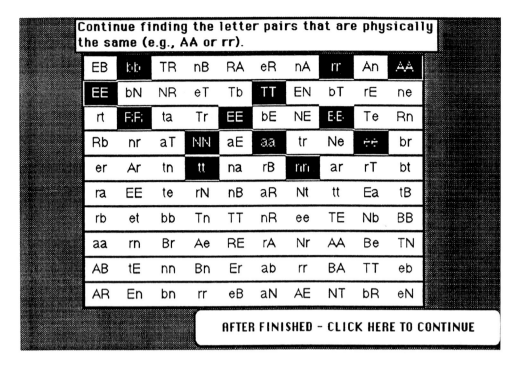

Figure 12 The computer screen for the Letter Pair Test used in the Miller et al. study.[2]

This bump is typical of all the 20 recording channels. The difficulties associated with finding an unambiguous linear scaling region is clear. The alternative of plotting the derivative of $\ln C_m(r)$ vs. $\ln r$ is shown in the bottom row of Figure 13. In this row, we see that no plateau region is in evidence and, therefore, no indication of a fractal dimension using traditional measures. From these typical results we conclude that the GPA is not of much help in determining the fractal dimension of the EEG time series data collected here.

10.3.3 Distribution approach to EEG

We now compare the results of the GPA used on the EEG time series data with that obtained by constructing the corresponding distribution function. We know from the analysis in Section 10.2.7 that the correlation function is actually an estimate of the distribution function for the dynamical process. In Figure 14A we depict the logarithm of the correlation function $C_m(r)$ vs. the logarithm of the interpoint distance r for embedding dimensions 5 to 20. For m small there appears to be a linear scaling region over a significant range of rs. However, as m increases, a bulge becomes evident and, just as for the results depicted in Figure 13, there is no apparent asymptotic linear scaling region. The EEG results in Figure 14A comes from the same individual as used in Figure 13, but we use a different minimal scale r_0 in the latter case. This choice makes the limitations of GPA abundantly clear.

In Figure 14B, the graph of the derivative of the logarithm of the correlation function with respect to the logarithm of the interpoint distance vs. the logarithm of the interpoint distance suggests the existence of a scaling region. However, the apparent plateau is not strictly horizontal, so one would have difficulty in obtaining a reliable correlation dimension from this figure.

In Figure 15, we depict the histogram from the EEG data in channel Cz with an embedding dimension $m = 10$ and a lag time $\tau = 4$. It is clear from this figure that the fractal distribution function is smooth. The histogram is fit by the Equation (25) in Figure 16 using the maximum log-likelihood function, Equation (27). The fractal dimension estimate from this fit is $d \cong 4.45$, with a χ^2 value of 0.734 for 69 degrees of freedom. This is a remarkably good fit to the histogram by the probability distribution and indicates that the EEG data can be well characterized by a fractal dimension. Thus, the distribution approach succeeds where GPA fails.

10.3.4 Distribution and MART applied to EEG

In the preceding section, we described the first ever application of the fractal distribution function approach to EEG time series data. The single channel results were very encouraging in regard to our being able to obtain a reliable estimate of the fractal dimension. For the present discussion, we now take a giant step and apply this method to the multichannel attractor reconstruction technique in which the embedding dimensions are replaced with EEG channels. In Figure 17, the MART is applied to EEG channels Fp1 through T6, and the logarithm of the correlation function $C_m(r)$ is plotted vs. the logarithm of the interpoint distance. Here again, there appears to be a linear scaling region when only a few of the EEG channels are used (equivalent to low embedding dimension); however, as more and more of the channels are included in the GPA processing, the familiar bulge becomes evident and the linear scaling region is apparently lost.

In Figure 18, the graph of the derivative of the logarithm of the correlation function with respect to the logarithm of the interpoint distance vs. the logarithm of the interpoint distance suggests the possible existence of a scaling region. The apparent plateau, however, is quite noisy and is not strictly horizontal. We would be hard pressed to obtain a reliable fractal dimension from this figure.

The probability density approach applied on five channels produced the histogram depicted in Figure 19, which is quite interesting since it reveals two maxima. This bimodal distribution suggests that there is more than one fractal dimension in the spatial EEG data, unlike the results obtained for single channel processing using the fractal probability density approach. We, therefore, must be circumspect in combining the EEG channels using MART. We separate the channels into those for the left hemisphere and those for the right hemisphere and apply MART to each hemisphere separately. In Figure 20, the multichannel embedding for the right lobe is applied to the EEG data. The logarithm of $C_m(r)$ vs. $\ln r$ in Figure 20A shows a nearly linear region for the eight dimensional embedding. The derivative of $\ln C_m(r)$ vs. $\ln(r/r_0)$ depicted in Figure 20B indicates a noisy plateau region. The histogram for these eight channels is shown in Figure 21, where we obtain a fractal dimension $d \cong 3.43$. Here again the distribution function approach succeeds where GPA fails.

In Figure 22, the multichannel embedding for the left lobe is applied to the EEG data. The logarithm of $C_m(r)$ vs. $\ln r$ in Figure 22A shows a nearly linear region for the eight dimensional embedding. The derivative of $\ln C_m(r)$ vs. $\ln(r/r_0)$ depicted in Figure 22B indicates a noisy plateau region. The histogram for these eight channels is shown in Figure 23, where we obtain a fractal dimension $d \cong 3.52$.

Here we have obtained results consistent with those of Nan and Jinghua:[8] the two-hemispheres of the brain in resting states are relatively symmetric in their dimensions. However, we have obtained this result using the fractal probability density, not the GPA as they did.

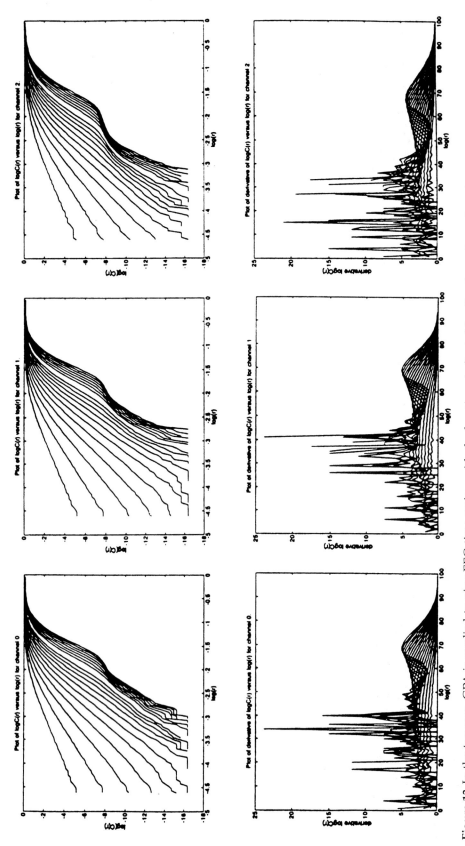

Figure 13 In the top row, GPA is applied to nine EEG time series, with the locations indicated by the channel numbers 0 to 5 and 17 to 19. The lack of a linear scaling region in all these data at high embedding dimension (m = 10 to 20) is clear.

chapter ten: Fractal Probability Density and EEG/ERP Time Series

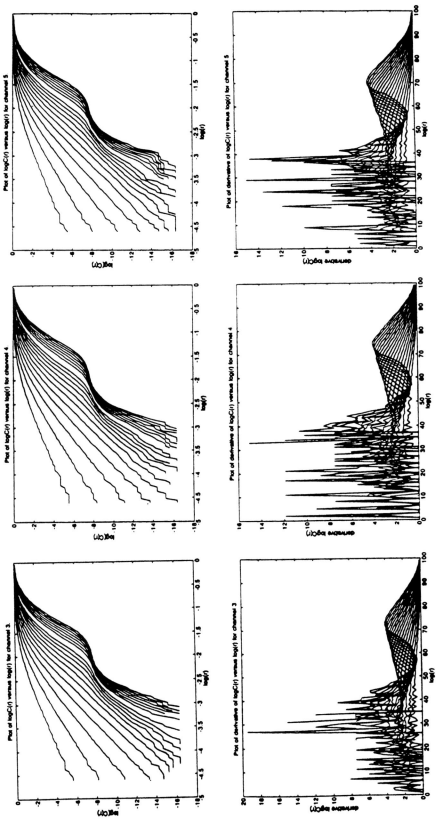

Figure 13 (continued) In the bottom row, the corresponding derivatives of $\ln C_m(r)$ vs. $\ln r$ are shown. The absence of a plateau in all channels is clear.

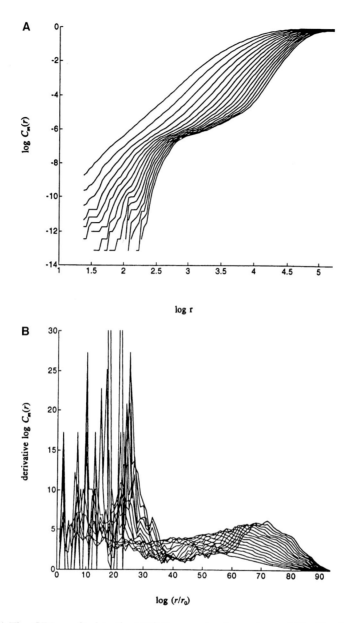

Figure 14 (A) The GPA applied to the EEG time series from channel Cz. The lack of a linear scaling region at higher embedding dimensions is the same as in Figure 13. (B) The derivative of log $C_m(r)$ vs. log r is depicted for the EEG time series in channel Cz. Note the approximate plateau region that is not quite horizontal and therefore does not give a reliable correlation dimension.

10.4 Measuring the event-related potential

The results of calculations of the degree of complexity of the EEG time series suggests that the erratic signals from the brain are correlated with the cognitive activity of the patient. The complex electrical signal and its change in shape and amplitude are related to such states as sleep, wakefulness, alertness, problem-solving, and hearing,

chapter ten: Fractal Probability Density and EEG/ERP Time Series 299

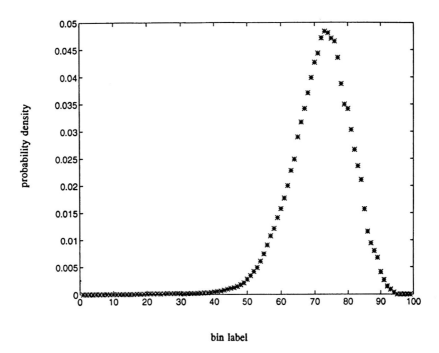

Figure 15 The normalized histogram for the EEG time series from channel Cz is depicted for an embedding dimension $m = 10$, $m = 10^3$ data points, and a log time $\tau = 4$.

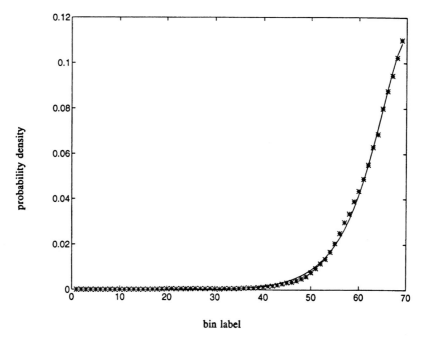

Figure 16 The histogram in Figure 15 is fit by the fractal distribution Equation (25) using the maximum log-likelihood function, Equation (27). The a-parameter values that minimize the difference $a_0 = 0.90$, $a_1 = -1.69 \times 10^{-6}$, and $a_2 = -1.51 \times 10^{-4}$ for a fractal dimension estimate $d = 4.544$. The χ^2 value is 0.734 for 69 degrees of freedom.

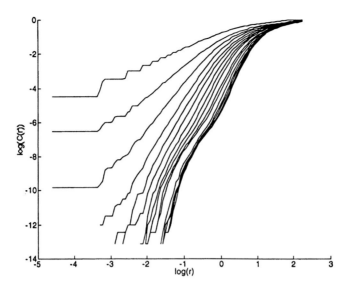

Figure 17 The GPA applied to EEG time series from channel 1 to channel 16. The lack of a linear scaling region when the many channels are included in MART resembles Figure 14A.

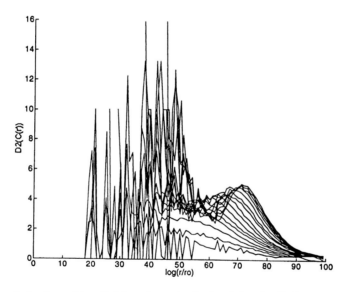

Figure 18 The derivative of log $C_m(r)$ vs. log r is depicted for the EEG time series shown in Figure 17. Note the narrow plateau region that is both noisy and not horizontal and therefore will not give a reliable correlation dimension.

as well as to several clinical conditions,[56] such as epilepsy,[57,58] and schizophrenia.[59] This in itself is not a new result; it has been known for some time that brain activity responds in a demonstrable way to external stimulation. The direct or peripheral deterministic simulation could be electrical, optical, acoustical, etc., depending on conditions of the experiment. This induced change in the brain's electrical activity is called an *evoked potential*. One can distinguish between an *evoked potential* and the spontaneous electrical activity of the brain in the following way:[60]

> "(a) It bears definite temporal relationship to the onset of the stimulus. In other words, it has a definite latent period

chapter ten: Fractal Probability Density and EEG/ERP Time Series

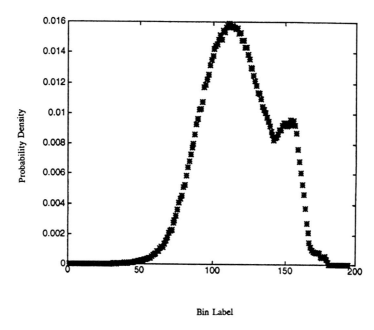

Figure 19 The normalized histogram for the MART EEG time series is depicted for eight channels. Note the smoothness of the histogram and the existence of two maxima.

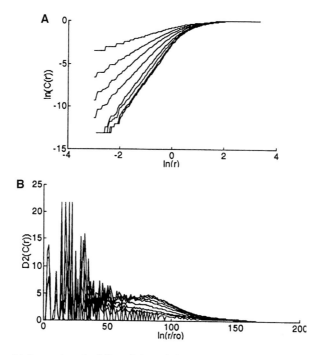

Figure 20 The multichannel embedding (MART) for the right lobe, channels 1, 5, 6, 10, 11, 15, 16, 18 are shown: (A) the GPA is applied to the eight channels; (B) the derivative of $\ln C_m(r)$ is plotted against $\ln(r/r_0)$.

determined by the conduction distance between the point of stimulation and the point of recording. (b) It has a definite pattern of responses characteristic of a specific system which is more or less predictable under similar conditions."

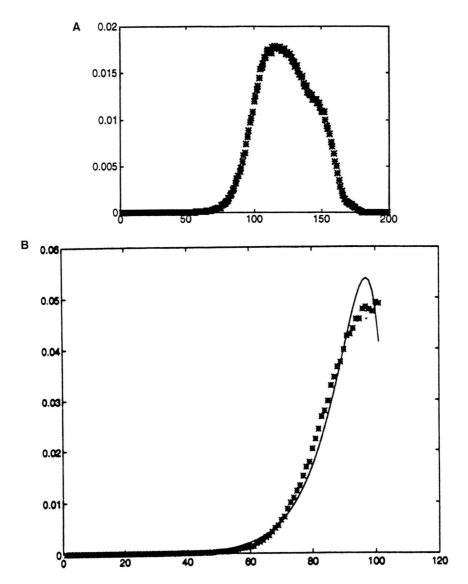

Figure 21 (A) The histogram for the eight channels in the right lobe is constructed. (B) The fit of the distribution to the histogram yields the coefficients $a_0 = -3.6 \times 10^{-3}$, $a_1 = 2.5 \times 10^{-3}$, and $a_2 = 10^{-2}$ with a fractal dimension of 3.43.

10.4.1 Average ERP

In another context, the evoked potential is also called the *event-related brain potential* (ERP). The ERP, like the EEG, is a series of voltage oscillations in the brain that are recorded on the scalp, but it differs in that it is a transient response to a discrete stimulus event. This characteristic of neural response to the environment has been proposed as an indicator of mental workload, in part because it is a noninvasive measure of cognitive information processing. Mental workload has been viewed as equivalent to the arousal level of the subject; it has been defined as a person's subjective experience of cognitive effort; it has been measured as the time taken to perform a task and conceptualized as the demand imposed upon the limited information-processing capabilities of the subject.[61] Wickens[62] describes workload in terms

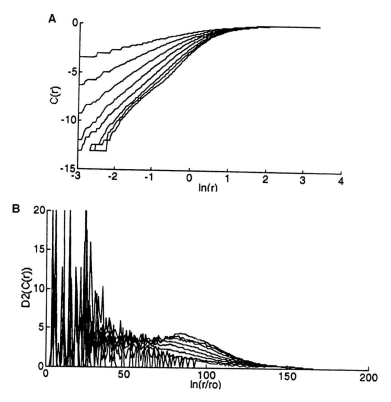

Figure 22 The multichannel embedding (MART) for the left lobe, channels 0, 2, 3, 7, 8, 12, 13, 17 are shown: (A) the GPA is applied to the eight channels; (B) the derivative of ln(r) is plotted against $\ln(r/r_0)$.

of the interaction between the difficulty of the task performed and the relative overlap of the common information-processing resources. The term "resource" is used here to describe instruments inherent in the organism that must be used in performing tasks, and, like other such concepts, is a hypothetical construct derived to account for variance in performance. In a sequence of experiments, which we discuss below, the ERP measures were found to reflect systematic differences in task workload and to vary closely with reaction time measures.[63] The classic of experiment is one in which a perceptual load is measured by monitoring a simulated air-traffic-control display for discrete events.[64-66] It was found that the information-processing resources allocated to the primary and secondary tasks are reciprocal; i.e., as the allocation of resources to the primary increases, those to the secondary decrease, thereby providing quantitative support for the casual observation that when the human mind is occupied with one task it often lacks the capacity to perform others.

As stated, the allocation of information processing resources was measured indirectly by assigning to an operator engaged in a primary task a secondary task that must be executed concurrently with it. The primary task in this collection of experiments was to track a target on a video screen using a joystick-controlled cursor, while maintaining a correct count of the number of occurrences of one tone from a Bernoulli (random) sequence of tones (secondary task). The amplitude of the positive component of the ERP elicited by these tones 300 ms after the initiation of a given tone (P300) was then examined to assess the sensitivity to the processing demands of the primary task. Wickens et al.[66] demonstrated that the amplitude of the P300 component was attenuated by the load imposed by the concurrent tracking task. The

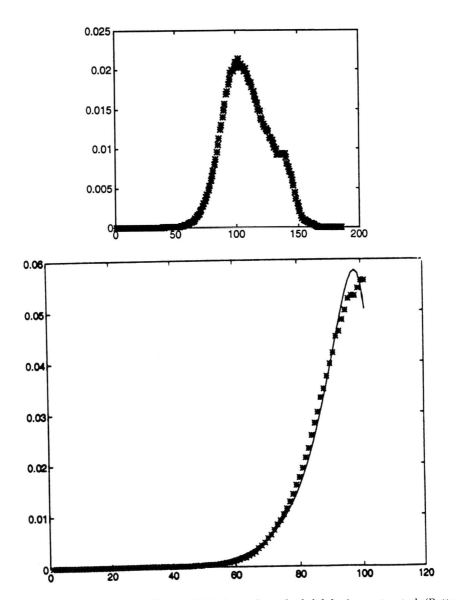

Figure 23 (Top) The histogram for the eight channels in the left lobe is constructed. (Bottom) The fit of the distribution to the histograms yields the coefficients $a_0 = -3.9 \times 10^{-3}$, $a_1 = 4.3 \times 10^{-3}$, and $a_2 = 8.9 \times 10^{-3}$ with a fractal dimension of 3.52.

workload was varied in these experiments in a number of ways: varying the number of "aircraft" tracked by the subject;[63] controlling the directional regularity of the target by randomly changing the size of the step in the movement of the aircraft as well as its direction;[66] varying the relationship between the movement of the joystick and that of the cursor;[66] randomly moving the cursor;[65] and others.[61]

It is not our intent to fully review the extensive results obtained by these scientists,[62,64,65,67] but rather to give a typical result and discuss its possible utility in the context of the reconstruction technique. Consider the tracking experiment conducted by Kramer[61] in which the subject must initially superpose the cursor so as to "capture" the target (see Figure 24). The control dynamics of the joystick consisted of two linear segments to yield the system output $X(t)$:

chapter ten: Fractal Probability Density and EEG/ERP Time Series

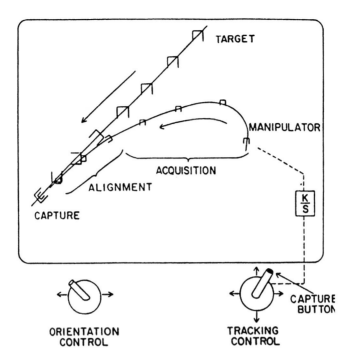

Figure 24 The temporal sequence of the target acquisition task (from upper right to lower left). (From Kramer, A. F., Wickens, C. D., and Donchin, E., *Hum. Factors*, 25, 597–621 (1983). With permission.)

$$X(t) = (1-\alpha)\int_0^t u(t')\, + \alpha \int_0^t dt' \int_0^{t'} u(t'')\, dt'' \tag{32}$$

where $u(t)$ is the instantaneous stick position and α is the level of difficulty. A first-order system is given by $\alpha = 0$, which has relatively easy tracking; the more difficult trials were conducted with $\alpha = 0.15$, i.e., a linear combination of first and second order dynamics. The average ERPs elicited by the counted tones are depicted in Figure 25. There is a clear separation in the ERP amplitude of the waveforms in the vicinity of 400 ms that can be attributed to the introduction of the tracking tasks and to the increase in difficulty of the dynamics. Donchin et al.[67] have suggested that the P300 wave form is a manifestation of neuronal activity that is involved whenever individuals update their internal model of the environment. Thus, the P300 is a promising candidate as an index of a selective aspect of cognitive workload that is related to the updating of an internal model of the task structure. If this is the case, then the allocation of processing resources to the tracking task would attenuate the ERP of the counting task, the attenuation increasing with certain increasing kinds of difficulty of the primary task but not with others, i.e., with those aspects that require an updating of the internal template of the subject.[61,63] In short,[61] "The P300 amplitude reflects only those aspects of performance that are related to updating the template in memory, and thus will not necessarily co-vary with other measures of secondary-task performance."

These results indicate that components of the ERP, specifically P300, can be used to quantify the resource allocation of complex cognitive tasks. More recently experiments have been done to explore the degree to which these results could be

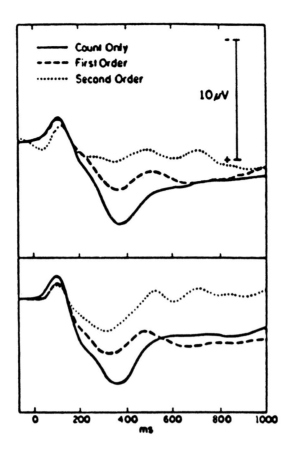

Figure 25 Top parietal grand average ERPs (11 subjects) elicited by counted tones in Experiment 1. The upper curve is after 90 practice episodes, the lower curve after 300 practice episodes. (From Kramer, A. F., Wickens, C. D., and Donchin, E., *Hum. Factors*, 25, 597–621 (1983). With permission.)

generalized to complex real-world tasks.[68] Student pilots flew a series of instrument flight rule (IFR) missions in a single-engine, fixed-base simulator and concurrently discriminated between two tones differing in frequency. The difficult flight was associated with increased deviations from the command altitude, heading, and glideslope, as well as with high subjective effort ratings. The P300 components of the ERP successfully discriminated among levels of task difficulty — as before, decreasing in amplitude with increased task demands.[68]

Figure 26 presents the grand average ERPs elicited by the target tones in the single- and dual-task conditions. It is clear that the same general features evident in Figure 25 arise here; that is, the P300 wave form attenuates with increasing task difficulty. In this study, the intermediate amplitude case was when the students were flying with no wind, turbulence, or subsystem failures, and the smallest was with high winds, turbulence, and a heading indicator failure. Kramer et al.[68] conclude that these results provide preliminary support for the assertion that components of the ERP can provide sensitive and reliable measures of the task demands imposed upon operators of complex, real-world systems. Here we reach a similar conclusion regarding the change in fractal dimension associated with the ERP.

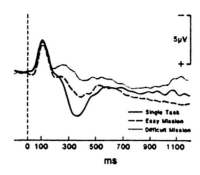

Figure 26 Parietal grand average ERPs overplotted for the tone discrimination task alone and for both of the flight mission. (From Kramer, A. F., Sirevaag, D. J., and Braune, R., *Hum. Factors*, 25, 145–160 (1987). With permission.)

10.4.2 Point estimate of ERP fractal dimension

Molnár and Skinner[11] developed a new method to calculate the dimensional complexity. Instead of the GPA, they use a "point estimate" of the fractal dimension in which only the local derivative of a point in the time series is used to calculate the local fractal dimension. These authors restrict the determination of the dimension to a short time series in the region of the point of interest, so that the technique is applicable to nonstationary data sets. The experimental data they analyze is for an "odd-ball" paradigm: there is a regular (periodic) background tone, then a random target tone for which the subject depresses a trigger. There were three configurations for the experiment: (1) control, where the subject was told to ignore the target tone, (2) target 10%, where the probability of a target tone occurring is 10%, and (3) target 30%, where the probability of a target tone occurring is 30%.

The point dimension is denoted by PD2 in Figure 27, and as we can see it changes in different regions of the ERP time series. Note that $3.0 \leq PD2 \leq 4.0$ when the subject ignores the target and drops below 3 when the target tone is attended to. The values of PD2 in the 200 to 400-ms interval after presentation of the target stimulus were 2.54 and 3.03 for the probabilities of 10 and 30%, respectively. These constitute a unit decrease in dimension from the baseline values. The significance of this dimension decrease is not yet understood. A dimensional increase evoked by either a "novel stimulus"[70] or "cognitive effort"[12] suggests that the system has increased in complexity so that a dimensional decrease may indicate a dynamical simplification. But why fewer degrees of freedom or less complexity would be necessary to process the same signal when it has meaning compared to when it is ignored, as in the present data, or why greater degrees of freedom or more complexity would be necessary to process the signal when it is unfamiliar and/or requires effort is simply not yet understood.

There does not as yet exist a proof that the PD2 quantity calculated from the data by Molnár and Skinner[11] approaches the usual fractal dimension in the long time limit. They will, of course, be the same if the irregularity in the time series is produced by the geometry of the underlying attractor in a low-dimensional dynamical system. However, it does appear that the PD2 algorithm enables one to study the event-related dimensional shifts in small EEG subepochs, i.e., points in the interval $\tau \times m$, where τ is the lag and m is the embedding dimension.

10.4.3 New technique for ERP fractal dimension

In the same spirit as Molnár and Skinner,[11] we wish to determine the local fractal dimension in the EEG/ERP time series; however, we will not evaluate the local

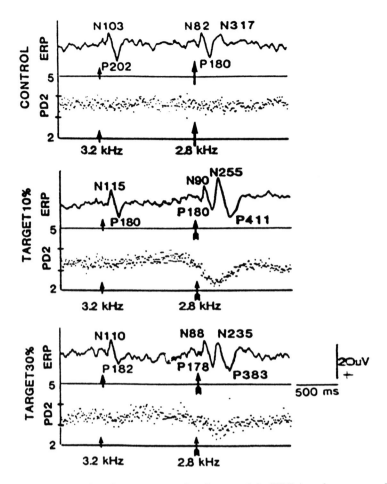

Figure 27 Grand means of auditory event-related potentials (ERPs) and corresponding point correlation dimension (PD2) values recorded from 11 subjects in an experiment in which the effect of changing the probability of the target presentation was studied. (Upper) Control experiment with 3.2 kHz background and 2.8 kHz control stimuli (control, $n = 340$). (Middle) Targets presented at 30% probability (target 10%, $n = 340$). (Lower) Targets presented at 30% probability (target 30%, $n = 210$). The PD2 values are displayed between 2 and 5 dimensions. (From Molnár, M., and Skinner, J. E., *Intl. J. Neurosci.*, 66, 263–276, (1992). With permission.)

derivative of the time series at a local time point as they did, but rather we will partition the time series into overlapping blocks and construct histograms for the distribution function using the data in each of the blocks. In this way we directly determine the local fractal dimension and the reliability of each estimate using the probability density function. In Figure 28, we show the fractal dimension for the left side of the brain as a function of time. Recall that the multichannel attractor reconstruction technique (MART) uses the separate channels as embedding dimensions. Here, we use blocks of 400 samples and the blocks overlap by 200 samples. Note that 200 samples corresponds to 1 s of data. The dimension was calculated twice: once using a selected number of randomly chosen interpoint distances and again using all pairs of interpoint distances. It is clear from comparing Figure 28A and 28B that, although the fractal sequences are not exactly the same, they both have the same range of values and degree of fluctuations. Thus, we are confident that the procedure we have adopted using a randomly selected subject of pairs is reliable.

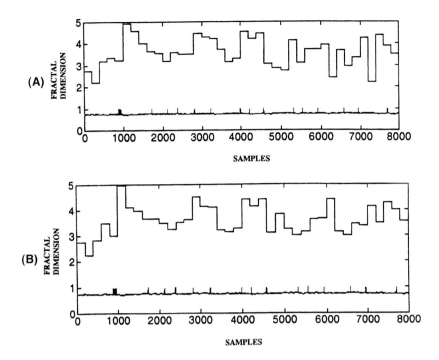

Figure 28 The fractal dimension for the left side of the brain using all eight channels (MART) and overlapping blocks of 400 samples. The blocks overlap by 200 samples (1 s) (A) 5×10^4 randomly chosen interpoint distances were used in determining the fractal dimension. (B) All the pairs of interpoint distances were used in determining the fractal dimension.

In Figure 29, the EEG/ERP time series from the occipital channel O1 is used, and the fractal dimension histogram using randomly selected interpoint distances is used to calculate the fractal dimension as a function of time. In the figure, the control channel (Oz) registers the response of the subject (mouse clicks) to visual stimulation. Between two pulse markers the subject is searching visually for the target stimuli: perceptually identical letter pairs. The change in the fractal dimension associated with the ERP is evident from the figure. The fractal dimension was calculated using 40,000 randomly chosen distances over blocks of 400 samples (2 s). The dimension plotted for the samples 0 to 200 was calculated using the samples in blocks 0 to 400, the dimension plotted for the samples 200 to 400 was calculated using the samples in blocks 200 to 600, and so forth. In this way the fractal dimension is seen to increase in a block prior to an event and drop in the interval containing the event. This is merely an artifact of the way we do the calculation. The point is that we can unambiguously determine the fractal dimension as a function of time and therefore, determine how it changes over time both in the presence and absence of learning.

10.5 Summary and plans for the future

In summary, we can say that the erratic nature of the EEG/ERP time series is due to chaos rather than noise, which means that the fluctuations are generated by some deterministic nonlinear dynamical process and contains useful information about the operation of the brain. We do not need to know what this underlying process is, however, since the fractal dimension has been shown to provide a quantitative measure of the degree of irregularity of the time series. We have shown an efficient

310 Fractal Geometry in Biological Systems: An Analytical Approach

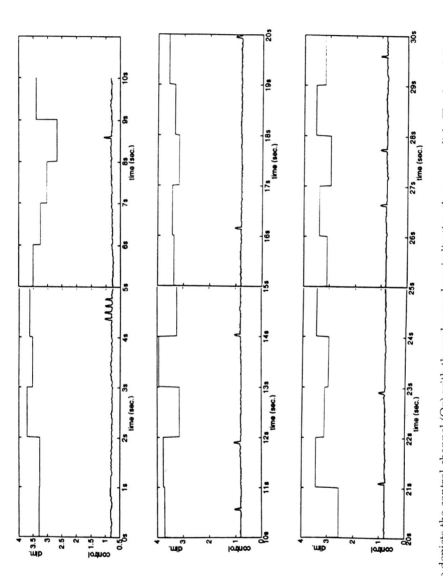

Figure 29 The bottom line depicts the control channel (Oz) with the pulse markers indicating the mouse click. The fractal dimension is graphed on the top line using the fractal probability density. The fractal dimension was calculated using 4×10^4 randomly chosen distances over blocks of 400 samples (2 s). The dimension plotted for the samples 0 to 200 was calculated using the samples in block 0 to 400, and so forth.

and reliable calculational algorithm for evaluating the fractal dimension of EEG/ERP time series based on the probability distribution function (see Section 10.2.8 for a technical summary). Here we merely stress that by using this technique we obtain the fractal dimension in near real time, enabling us to track the change in the cognitive attractor as an individual learns.

Acknowledgment

We thank the U.S. Navy for the Phase I SBIR (N93-187) that partially supported this research.

References

1. West, B. J., *Fractal Physiology and Chaos in Medicine*, World Scientific, New Jersey (1990).
2. Miller, D. C., Kavcic, V., and Leslie-Smith, J. E., Project titled *Methylphenidate Effects on Cognitive Event-Related Potentials* presented at (1) TENNET Conference in Montreal, May 1993; (2) American EEG Society Conference in New Orleans, October 1993; and (3) International Conference on ADHD, Amsterdam, June 1994.
3a. Mandelbrot, B. B., *Fractals, Form and Chance*, W.H. Freeman, San Francisco (1977).
3b. Mandelbrot, B. B., *The Fractal Geometry of Nature*, W.H. Freeman, San Francisco (1982).
4. West, B. J., Bhargava, V., and Goldberger, A. L., Beyond the principle of similitude: renormalization in the bronchial tree, *J. Appl. Physiol.*, 60, 189–197 (1986).
5. Goldberger, A. L., Bhargava, V., West, B. J., and Mandell, A. J., On a mechanism of cardiac electrical stability: the fractal hypothesis, *Biophys. J.*, 48, 525–528 (1985b).
6. Freeman, W. J., Simulation of chaotic EEG patterns with a dynamic model of the olfactory system, *Biol. Cybern.*, 56, 139–150 (1987).
7. Gregson, R. A. M., Campbell, E. A., and Gates, R. A., Cognitive load as a determinant of the dimensionality of the electroencephalogram: a replication study, *Biol. Psychol.*, 35, 165 178 (1992).
8. Nan, X. and Jinghua, X., The fractal dimension of EEG as a physical measure of conscious human brain activities, *Bull. Math. Biol.*, 50, 559–565 (1988).
9. Hoyert, M. S., Order and chaos in fixed-interval schedules of reinforcement, *J. Exp. Anal. Behav.*, 57, 339–363 (1992).
10. Siegler, R. S., Cognitive variability: a key to understanding cognitive development, *Curr. Dir. Psychol. Sci.*, 3, 1–5 (1994).
11. Molnár, M. and Skinner, J. E., Low-dimensional chaos in event related brain potentials, *Intl. J. Neurosci.*, 66, 263–276 (1992).
12. Rapp, P. E., Bashare, T. R., Martineire, J. R., Albano, A. M., Zimmerman, I. D., and Mees, A. I., Dynamics of brain electrical activity, *Brain Topogr.*, 2, 99–118 (1989).
13. Rapp, P. E., Bashore, T. R., Zimmerman, T. R., Martinerie, J. M., Albano, A. M., and Mees, A. I., Dynamical characterization of brain electrical characterization, in *The Ubiquity of Chaos*, Krasner, S., Ed., American Association for the Advancement of Science, Washington, D.C., pp. 10–22 (1990).
14. Ott, E., Grebogi, C., and Yorke, J. A., Controlling chaos, in *Chaos*, Campbell D. K., Ed., American Institute of Physics, New York, pp. 153–172 (1990).
15. Takens, F., in *Lecture Notes in Mathematics*, Rand, D. A. and Young, L. S., Eds., Springer-Verlag, Berlin (1981).
16a. Grassberger, P., and Procaccia, I., Measuring the strangeness of strange attractors, *Physica*, 9D, 189–208 (1983).
16b. Grassberger, P., and Procaccia, I., Characterization of strange attractors, *Phys. Rev. Lett.*, 50, 346 (1983).
16c. Grassberger, P., and Procaccia, I., Estimation of Kolmogorov entropy from a chaotic signal, *Phys. Rev.*, 28A, 2591 (1983).

17. Layne, S. P., Mayer-Kress, G., and Holzfuss, J. Problems associated with dimensional analysis of electroencephalogram data, in *Dimensions and Entropies in Chaotic Systems*, Mayer-Kress G., Ed., Springer-Verlag, Berlin, pp. 246–256 (1986).
18. Hanley, J., Electroencephalography in psychiatric disorders: parts I and II, in *Directions in Psychiatry*, vol 4, pp. 1–8 (1984).
19. Freeman, W. J., *Mass Action in the Nervous System*, Academic Press, New York, p. 489 (1975).
20a. Judd, K., An improved estimator of dimension and some comments on providing confidence intervals, *Physica D*, 56, 216–228 (1992).
20b. Judd, K., Estimating dimension from small samples, *Physica D*, 71, 421–429 (1994).
21. Packard, N. H., Crutchfield, J. P., Farmer, J. D., and Shaw, R. S., Geometry from a time series, *Phys. Rev. Lett.*, 45, 712–716 (1980).
22. Rössler, O. S., An equation for continuous chaos, *Phys. Lett.*, 57A, 397 (1976).
23. Farmer, D., Crutchfield, J., Froehling, H., Packard, N., and Shaw, R., Power spectra and mixing properties of strange attractors, *Ann. N.Y. Acad. Sci.*, 357, 453–472 (1980).
24. Sauer, T., Yorke, J. A., and Casdagli, M., Embeddology, *J. Stat. Physiol.*, 65, 579–616 (1991).
25. Sauer, T., A noise reduction method for signals from nonlinear systems, to appear in *Physica D*, 58, 193–201 (1992).
26. Albano, A. M., Mees, A. I., de Guzman, G. C., and Rapp, P. E., Data requirements for reliable estimation of correlation dimensions, in *Chaos in Biological Systems*, Degn, H., Holden A. V., and Olsen, L. F., Plenum Press, New York (1987).
27. Eckmann, J. P., and Ruelle, D. Fundamental limitations for estimating dimensions and Lyapunov exponents in dynamical systems, *Physica D*, 56, 185–187 (1992).
28. Broomhead, D. S., and King, G. P., Extracting qualitative dynamics from experimental data, *Physica D*, 20, 217 (1986).
29. Loéve, M., *Probability Theory*, D. Van Nostrand, New York (1955).
30. Devijver, P. and Kittler, J. *Pattern Recognition: A Statistical Approach*, Prentice-Hall, London (1982).
31. Golub, G. H., and van Loan, C. F., *Matrix Computation*, Johns Hopkins University Press, Baltimore (1985).
32. Fraser, A. M., Reconstructing attractors from scalar time series: a comparison of singular system and redundancy criteria, *Physica D*, 34, 391–404 (1989).
33. Dongarra, J. J., Moler, C. B., Bunch, J. R., and Stewart, G. W., *Linpak User's Guide*, SIAM, Philadelphia (1979).
34. Abarbanel, H. D. I., Prediction in chaotic nonlinear systems: time series analysis for nonperiodic evolution, Lectures at the NATO Advanced Research Workshop on Model Ecosystems and Their Changes (1989).
35. Wax, M., and Kailath, T., Detection of signals by information theoretic criteria, *IEEE Trans. Acoust., Speech, Signal Process*, ASSP-30, 387–392 (1985).
36. Brandstater, A., Swinney, H. L., Chapman, G. T., *Dimension and Entropies in Chaotic Systems*, Mayer-Kress, G., Ed., Springer-Verlag, Berlin (1986).
37. Meese, A. I., Rapp, P. E., Jennings, L. S., Singular-value decomposition and embedding dimension, *Phys. Rev. A*, 36, 340 (1987).
38. Fraedrich, K., and Wang, R., Estimating the correlation dimension of an attractor from noise and small data sets based on re-embedding, *Physica D*, 65, 373–398 (1993).
39. Dvorák, I., Takens vs. multichannel reconstruction in EEG correlation exponent estimates, *Phys. Lett. A*, 151, 225–233 (1990).
40. Palus, M., Dvorák, I., and David, I., Spatio-temporal dynamics of human EEG, *Physica A*, 185, 433–438 (1992).
41. Eckmann, J. P., and Ruelle, D., Ergodic theory of chaos and strange attractors, *Rev. Mod. Physiol.*, 57, 617–656 (1985).
42. Destexhe, A., Seprelchre, J. A., and Babloyantz, A., A comparative study of the experimental quantification of deterministic chaos, *Phys. Lett. A*, 2, 101–106 (1988).
43. Smith, R. L., Estimating dimension in noisy chaotic time series, *J. R. Stat. Soc. B*, 54, 229–251 (1992).

44. West, B. J., and Deering, W., Fractal physiology for physicists: Lévy statistics in biology, *Phys. Rep.*, 246, 1–100 (1994).
46. Babloyantz, A., and Destexhe, A., Low dimensional chaos in an instance of epilepsy, *Proc. Nat. Acad. Sci. U.S.A.*, 83, 3515–3517 (1986).
47. Pritchard, W. S., and Duke, D. W., Dimensional analysis of no-task human EEG using the Grassberger-Procaccia method, *Psychophysiology*, 29, 182–192 (1992).
48. Jansen, B. H., Quantitative analysis of electroencephalograms: is there chaos in the future?, *Intl. J. Biomed. Comput.*, 27, 95–123 (1991).
49. Dvorák, I., Siska, J., Wackerman, J., Hrudova, L. and Dostalek, C., Evidence for interpretation of the EEG as a deterministic chaotic process with a low dimension, *Act. Nerv. Super.*, 28, 228–231 (1986).
50. Rapp, P. E., Zimmerman, I. D., Albano, A. M., de Guzman, G. C., Greenbaum, N. N., Bashore, T. R., Experimental studies of chaotic neural behavior: cellular activity and electorencephalogram signals, in *Nonlinear Oscillations in Biology and Chemistry*, Othmer, H. G., Ed., Springer-Verlag, New York, pp. 175–205 (1987).
51. Graf, K. E., and Elbert, T., Dimensional analysis of the working EEG, in *Brain Dynamics. Progress and Perspectives*, Basar, E., and Bullock, T. H., Eds., Springer-Verlag, New York (1989).
52. Rapp, P. E., Zimmerman, I. D., Albano, A. M., de Guzman, G. C., and Greenbaum, N. N., Dynamics of spontaneous neural activity in the simian motor cortex: the dimension of chaotic neurons, *Phys. Lett.*, 110A, 335–338 (1985).
53. Arle, J. E., and Simon, R. H., An application of fractal dimension to the detection of transients in the electroencephalogram, *Electroencephalogr. Clin. Neurophys.*, 75, 296–305 (1990).
54. Das, J. P., Naglieri, J. A., and Kirby, J. R., *Assesment of Cognitive Processes: The PASS Theory of Intelligence*, Allyn and Bacon, Boston (1994).
55. Posner, M. I. and Keele, S. W., Decay of visual information from a single letter, *Science*, 158, 137–139 (1967).
56. Bullock, T. H., Orkand, R., and Grinnel, A., *Introduction to the Nervous Systems*, W. H. Freeman, San Francisco (1981).
57. Principe, J. C., and Smith, J. R., Microcomputer-based system for the detection and quantification of petit mal epilepsy, *Comput. Biol. Med.*, 12, 87–95 (1982).
58. Siegel, A., Grady, A. L., and Mirsky, A. F., Prediction of spike-wave bursts in disence epilepsy by EEG power-spectra signals, *Epilepsia*, 23, 47–60 (1982).
59. Itil, T. M. Qualitative and quantitative EEG findings in schizophrenia, *Schizophr. Bull.*, 3, 61–79 (1977).
60. Chang, H. T., The evoked potentials, in *Handbook of Physiology*, vol. I, Field, J., Ed., American Physical Society, Washington D.C. pp. 299–314 (1959).
61. Kramer, A. F., Wickens, C. D., Donchin, E., An analysis of the processing requirements of a complex perceptual-motor task, *Hum. Factors*, 25, 597–621 (1983).
62. Wickens, C. D., Human workload measurement, in *Mental Workload; Its Theory and Measurement*, Morey, N., Ed., Plenum Press, New York (1979).
63. Isreal, J. B., Wickens, C. D., Chesney, G. L., and Donchin, E., The event-related brain potential as an index of display-monitoring workload, *Hum. Factors*, 22, 211–224 (1980).
64. Donchin, E., Brain electrical correlates of pattern recognition, in *Signal Analysis and Pattern Recognition in Biomedical Engineering*, Inbar, G. F., Ed., John Wiley & Sons, New York (1975).
65. Isreal, J. B., Chesney, G. L., Wickens, C. D. and Donchin, E., P300 and tracking difficulty: evidence for multiple resources in dual-task performance, *Psychophysiology*, 17, 259–273 (1980).
66. Wickens, C. D., Isreal, J. B., and Donchin, E., The event related potential as an index of task workload, in *Proceeding of the Human Factors Society 21st Annual Meeting*, Neal, A. S., and Palasek, R. F., Eds., San Francisco, October (1977).
67. Donchin, E., Ritter, W., and McCallum, W. C., Cognitive psychophysiology: the endogenous components of the ERP, in *Event-Related Brain Potentials in Man*, Callaway, E., Tueting, P., and Koslow, S. H., Eds., Academic Press, New York (1978).

68. Kramer, A. F., Sirevaag, D. J., and Braune, R., A psychophysiological assessment of operator workload during simulated flight mission, *Hum. Factors*, 25, 145–160 (1987).
70. Skinner, J. E., Martin, J. L., Landisman, C. E., Mommer, M. M., Fulton, A., Mitra, M. Burton, W.D., and Saltzberg, B., Chaotic attractors in a model of neocortex: dimensionalities of olfactory bulb surface potentials are spatially uniform and event related, in *Chaos in Brain Function*, Baser, E., Ed., Springer-Verlag, Berlin, pp. 119–134 (1990).

Section VI

Advanced Topics in Fractal Geometry

chapter eleven

Fractal Geometry

Tamás Vicsek

> *Scientists will (I am sure) be surprised and delighted to find that not a few shapes they had to call* grainy, hydralike, in between, pimply, pocky, ramified, seaweedy, strange, tangled, tortuous, wiggly, wispy, wrinkled, *and the like can hence forth be approached in rigorous and vigorous quantitative fashion.*
>
> B.B. Mandelbrot

11.1 Introduction

In this chapter an overview of the basics of fractal geometry will be given for those who are interested in applying fractals to biological systems.

During the last decade it has been widely recognized by researchers working in diverse areas of science that many of the structures commonly observed possess a rather special kind of geometrical complexity. This awareness is largely due to the activity of Benoit Mandelbrot[1] who called attention to the particular geometrical properties of such objects as the shores of continents, the branches of trees, or the surface of clouds. He coined the name *fractal* for these complex shapes to express that they can be characterized by a noninteger (fractal) dimensionality. With the development of research in this direction, the list of examples of fractals has become very long and includes structures from microscopic aggregates to the clusters of galaxies. Objects of biological origin are many times fractal-like.

Before starting a more detailed description of fractal geometry, let us first consider a simple example. Figure 1 shows a cluster of particles which can be used for demonstrating the main features of fractals. This object was proposed to describe diffusion-limited growth[2] and has a loopless branching structure reminiscent of many shapes of biological origin. Imagine concentric circles of radii R centered at the middle of the cluster. For such an object it can be shown that the number of particles in a circle of radius R scales as

$$N(R) \sim R^D, \tag{1}$$

where $D < d$ is a noninteger number called the *fractal dimension*. Naturally, for a real object the above scaling holds only for length scales between a lower and an upper cutoff. Obviously, for a regular object embedded into a d-dimensional Euclidean

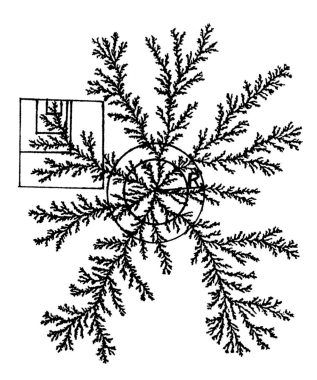

Figure 1 A typical stochastic fractal generated in a computer using the diffusion-limited aggregation model.[2]

space, Equation (1) would have the form $N(R) \sim R^d$ expressing the fact that the volume of a d-dimensional object grows with its linear size R as R^d. Clusters having a nontrivial D are typically self similar. This property means that a larger part of the cluster after being reduced "looks the same" as a smaller part of the cluster before reduction. This remarkable feature of fractals can be examined visually on Figure 1, where parts of different sizes (included in rectangular boxes) can be compared from this point of view.

In the following we shall describe the major type of self-similar and self-affine fractals (with examples) and give a sort introduction to multifractals. The reader can find further details on these topics in a number of recent books.[1,3-8] In particular, a comprehensive overview of fractal growth is reviewed in Reference 7.

11.2 Definitions

11.2.1 Fractals as mathematical and biological objects

In addition to the self-similarity mentioned above, a characteristic property of fractals is related to their volume with respect to their linear size. To demonstrate this we first need to introduce a few notions. We call the Euclidean dimension d of the space in which the fractal can be embedded the *embedding dimension*. Furthermore, d has to be the smallest such dimension. Obviously, the volume of a fractal (or any object), $V(l)$, can be measured by covering it with d-dimensional balls of radius l. Then the expression

$$V(l) = N(l) l^d \qquad (2)$$

gives an estimate of the volume, where $N(l)$ is the number of balls needed to cover the object completely and l is much smaller than the linear size L of the whole structure. The structure is regarded as covered if the region occupied by the balls includes it entirely. The phrase "number of balls needed to cover" corresponds to the requirement that $N(l)$ should be the smallest number of balls with which the covering can be achieved. For ordinary objects, $V(l)$ quickly attains a constant value, while for fractals typically $V(l) \to 0$ as $l \to 0$. On the other hand, the surface of fractals may be anomalously large with respect to L.

There is an alternative way to determine $N(l)$ which is equivalent to the definition given above. Consider a d-dimensional hypercubic lattice of lattice spacing l which occupies the same region of space where the object is located. Then the number of boxes (mesh units) of volume l^d which overlap with the structure can be used as a definition for $N(l)$ as well. This approach is called *box counting*.

Returning to the cluster shown in Figure 1, we can say that it can be embedded into a plane ($d = 2$). Measuring the total length of its branches (corresponding to the surface in a two-dimensional space) we would find that it tends to grow almost indefinitely with the decreasing length l of the measuring sticks. At the same time, the measured "area" of the cluster (volume in $d = 2$) goes to zero if we determine it by using discs of decreasing radius. The reason for this is rooted in the extremely complicated, self-similar character of the cluster. Therefore, such a collection of branches will be definitely much "longer" than a line, but will have an infinitely small area; it is neither a one- nor a two-dimensional object.

Thus, the volume of a finite geometrical structure measured according to Equation (2) may go to zero with the decreasing size of the covering balls while, simultaneously, its measured surface may diverge following a power law instead of the better behaving exponential convergence. In general, we call a physical object fractal if, when measuring its volume, surface, or length with d, $d - 1$, etc. dimensional hyperballs, it is not possible to obtain a well converging finite measure for these quantities as l changes over several orders of magnitude.

It is possible to construct mathematical objects which satisfy the criterion of self-similarity exactly, and their measured volume depends on l even if l or (l/L) becomes smaller than any finite value. Figure 2 gives examples of how one can construct such fractals using an iteration procedure. Usually one starts with a simple initial configuration of units (Figure 2A) or with a geometrical object (Figure 2B). Then, in the growing case this simple seed configuration (Figure 2A, $k = 2$) is added repeatedly to itself in such a way that the seed configuration is regarded as a unit, and in the new structure these units are arranged with respect to each other according to the same symmetry as the original units in the seed configuration. In the next stage, the previous configuration is always looked at as the seed. The construction of Figure 2B is based on division of the original object, and it can be well understood how the subsequent replacement of the squares with five smaller squares leads to a self-similar, scale invariant structure.

One can generate many possible patterns by this technique; the fractal shown in Figure 2 was chosen just because it has an open branching structure analogous to many observed growing fractals.[9] Only the first couple of steps (up to $k = 3$) of the construction are shown. Mathematical fractals are produced after an infinite number of such iteration. In this $k \to \infty$ limit, the fractal displayed in Figure 2A becomes infinitely large, while the details of Figure 2B become so fine that the picture seems to "evaporate" and cannot be seen any more. Our example shows a connected construction, but disconnected objects distributed in a nontrivial way in space can also form a fractal.

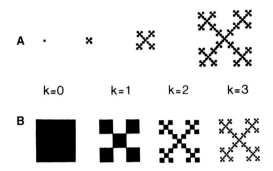

Figure 2 The figures in (A) demonstrate how one can generate a growing fractal using an iteration procedure. In (B), an anlogous structure is constructed by subsequent divisions of the original square. Both procedures lead to fractals for $k \to \infty$ with the same dimension $D \simeq 1.465$. (From Vicsek, T., *Fractal Growth Phenomena*, World Scientific, Singapore, 1992. With permission.)

In any real system there is always a lower cutoff of the length scale; in our case, this is represented by the size of the particles. In addition, a real object has a finite linear size which inevitably introduces an upper cutoff of the scale on which fractal scaling can be observed. This leads us to the conclusion that, in contrast to the mathematical fractals, for fractals observed in natural phenomena (including biology) the anomalous scaling of the volume can be observed only between two well defined length scales. A possible definition for a biological fractal can be based on the requirement that a power law scaling of $N(l)$ has to hold over at least two orders of magnitude.

11.2.2 Definitions

Because of the two main types of fractals demonstrated in Figure 2, one typically uses two related approaches to define and determine the fractal dimension D. For fractals having fixed L and details on very small length scale, D is defined through the scaling of $N(l)$ as a function of decreasing l, where $N(l)$ is the number of d-dimensional balls of diameter l needed to cover the structure.

In the case of growing fractals, where there exists a smallest typical size a, one cuts out d-dimensional regions of linear size L from the object, and the volume, $V(L)$, of the fractal within these regions is considered as a function of the linear size L of the object. When determining $V(L)$, the structure is covered by balls or boxes of unit volume ($l = a = 1$ is usually assumed); therefore, $V(L) = N(L)$, where $N(L)$ is the number of such balls.

The fact that an object is a mathematical fractal, then, means that $N(l)$ diverges as $l \to 0$ or $L \to \infty$, respectively, according to a noninteger exponent.

Correspondingly, for fractals having a finite size and infinitely small ramifications we have

$$N(l) \sim l^{-D} \tag{3}$$

with

$$D = \lim_{l \to 0} \frac{\ln N(l)}{\ln(1/l)} \tag{4}$$

while

$$N(L) \sim L^D \tag{5}$$

and

$$D = \lim_{L \to \infty} \frac{\ln N(L)}{\ln(L)} \tag{6}$$

for the growing case, where $l = 1$. Here, as well as in the following expressions, the symbol $\sim$ means that the proportionality factor, not written out in Equation (3), is independent of l.

Obviously, the above definitions for nonfractal objects give a trivial value for D coinciding with the embedding Euclidean dimension d. For example, the area (corresponding to the volume $V(L)$ in $d = 2$) of a circle grows as its squared radius grows, which according to Equation (6) results in $D = 2$.

Now we are in the position to calculate the dimension of the objects shown in Figure 2. It is evident from the figure that for the growing case

$$N(L) = 5^k \quad \text{with} \quad L = 3^k \tag{7}$$

where k is the number of iterations completed. From here, using Equation (6) we get the value $D = \ln 5 / \ln 3 = 1.465...$, which is a number between $d = 1$ and $d = 2$ just as we expected.

11.2.3 Useful rules

In this section we mention a few rules which can be useful in predicting various properties related to the fractal structure of an object. Of course, because of the great variety of self-similar geometries, the number of possible exceptions is not small and the rules listed below should be regarded, at least in part, as starting points for more accurate conclusions.

1. Many times it is the projection of a fractal that is of interest or can be experimentally studied (e.g., a picture of a fractal embedded into $d = 3$). In general, projecting a $D < d_s$ dimensional fractal onto a d_s dimensional surface results in a structure with the same fractal dimension $D_p = D$. For $D \geq d_s$, the projection fills the surface, $D_p = d_s$.
2. It follows from (1) that for $D < d_s$ the density correlations $c(r)$ (see next section) within the projected image decay as a power law with an exponent $d_s - D$ instead of $d - D$, which is the exponent characterizing the algebraic decay of $c(r)$ in d.
3. Cutting out a d_s dimensional slice (cross-section) of a D-dimensional fractal embedded into a d-dimensional space usually leads to a $D + d_s - d$ dimensional object. This seems to be true for self-affine fractals (next section) as well, with D being their local dimension.
4. Consider two sets A and B having fractal dimensions D_A and D_B, respectively. Multiplying them together results in a fractal with $D = D_A + D_B$. As a simple example, imagine a fractal made of parallel sticks arranged in such a way that

its cross-section is the fractal shown in Figure 2B. The dimension of this object is $D = 1 + \ln 5/\ln 3$.
5. The union of two fractal sets A and B with $D_A > D_B$ has the dimension $D = D_A$.
6. The fractal dimension of the intersection of two fractals with D_A and D_B is given by $D_{A \cap B} = D_A + D_B - d$. To see this, consider a box of linear size L within the overlapping region of two growing stochastic fractals. The density of A and B particles is respectively proportional to L^{D_A}/L^d and L^{D_B}/L^d. The number of overlapping sites $N \sim L^{D_{A \cap B}}$ is proportional to these densities and to the volume of the box which leads to the above given relation. The rule concerning intersections of fractals with smooth hypersurfaces (rule 3) is a special case of the present one.
7. The distribution of empty regions (holes) in a fractal of dimension D scales as a function of their linear size with an exponent $-D - 1$.

11.3 Types of fractals

One of the most fascinating aspects of fractals is the extremely rich variety of possible realizations of such geometrical objects. This fact raises the question of classification, and in the book of Mandelbrot[1] and in following publications many kinds of fractal structures have been described.

11.3.1 Deterministic and random fractals

Since fluctuations are always present in natural processes, they never lead to structures with perfect symmetry. Instead, fractals in nature are more or less random with no high level of symmetry. Yet, it is of interest to investigate simple, idealized fractal constructions, because the main features of fractal geometry can be effectively demonstrated using them as examples.

Figure 2A shows a typical fractal generated by a deterministic rule. When constructing such growing mathematical fractals, one starts with an object (particle) of linear size a. In the first step ($k = 1$), $n - 1$ copies of this seed object are added to the original one so that the linear size of the resulting configuration becomes ra, where $r > 1$. In this way, the number of particles and the linear size of the structure become n^2 and $r^2 a$, respectively. In the $k > 1$ step, each particle is replaced by the $k = 1$ configuration. Thus, the kth configuration is made of n units being identical to the $k - 1$th cluster. The $k \to \infty$ limit results in a deterministic mathematical fractal. At the end of this section, a few examples are given for the types of fractals described above and in the following.

The fractal dimension for such objects can be obtained readily taking into account that for $L = r^k a$ the volume (the number of particles) of the structure is $N(L) = n^k$

$$D = \frac{\ln n}{\ln r} \qquad (8)$$

which is an exact expression for D. The construction of deterministic fractals generated by subsequent divisions of a starting object proceeds in an analogous manner.

Self-similarity can be directly checked for a deterministic fractal constructed by iteration, but in the case of random structures one needs other methods to detect the fractal character of a given object. In fact, random fractals are self similar only in a statistical sense (not exactly), and to describe them it is more appropriate to use the term *scale invariance* than *self-similarity*. Naturally, for demonstrating the presence of fractal scaling, one can use the definition based on covering the given structure with balls of varying radii; however, this would be a rather troublesome procedure. It is more effective to calculate the so-called density-density or pair correlation function

$$c(\vec{r}) = \frac{1}{V} \sum_{\vec{r}'} \rho(\vec{r} + \vec{r}') \rho(\vec{r}') \qquad (9)$$

which is the expectation value of the event that two points separated by $\vec{r}$ belong to the structure. For growing fractals the volume of the object is $V = N$, where N is the number of particles in the cluster, and Equation (9) gives the probability of finding a particle at the position $\vec{r} + \vec{r}'$, if there is one at $\vec{r}'$. In Equation (9), ρ is the local density, i.e., $\rho(\vec{r}) = 1$ if the point $\vec{r}$ belongs to the object; otherwise, it is equal to zero. Ordinary fractals are typically isotropic (the correlations are not dependent on the direction), which means that the density correlations depend only on the distance r so that $c(\vec{r}) = c(r)$.

Now we can use the pair correlation function introduced above as a criterion for fractal geometry. An object is nontrivially scale invariant if its correlation function determined according to Equation (9) is unchanged up to a constant under rescaling of lengths by an arbitrary factor b:

$$c(br) \sim b^{-\alpha} c(r) \qquad (10)$$

with α a noninteger number larger than zero and less than d. It can be shown that the only function which satisfies Equation (10) is the power-law dependence of $c(r)$ on r

$$c(r) \sim r^{-\alpha} \qquad (11)$$

corresponding to an algebraic decay of the local density within a random fractal, since the pair-correlation function is proportional to the density distribution around a given point. Let us calculate the number of particles $N(L)$ within a sphere of radius L from their density distribution

$$N(L) \sim \int_0^L c(r) d^d r \sim L^{d-\alpha}, \qquad (12)$$

where the summation in Equation (9) has been replaced by integration. Comparing Equation (12) with Equation (5) we arrive at the relation

$$D = d - \alpha \qquad (13)$$

which is a result widely used for the determination of D from the density correlations within a random fractal.

11.3.2 Self-similar fractals

There are three major types of fractals when their scaling behavior is considered. Self-similar fractals are invariant under isotropic rescaling of the coordinates, while for self-affine fractals, scale invariance holds for affine (anisotropic) transformation.

324 Fractal Geometry in Biological Systems: An Analytical Approach

Up to this point, the former case has been discussed primarily, and in this section examples for such self-similar fractals will be given.

Example 1. One of the simplest and best known fractals is the so called triadic Cantor set, which is a finite size fractal consisting of disconnected parts embedded into one-dimensional space ($d = 1$). Its construction based on the subsequent division of intervals generated on the unit interval [0,1] is demonstrated in Figure 3. First, [0,1] is replaced by two intervals of length 1/3. Next, this rule is applied to the two newly created intervals, and the procedure is repeated *ad infinitum*. As a result, we obtain a deterministic fractal, and to calculate its fractal dimension we can use Equation (8). Obviously, for the present example, $n = 2$ and the reduction factor is $1/r = 1/3$. Therefore, Equation (8) gives, for the dimension of the triadic Cantor set, $D = \ln 2/\ln 3 = 0.6309...$, which is an irrational number less than 1. In general, Cantor sets with various n and r can be constructed. For example, keeping $n = 2$ and changing r between 2 and ∞, any fractal dimension $0 \leq D \leq 1$ can be produced. On the other hand, various Cantor sets with the same fractal dimension can be constructed as well. The two sets $n = 2, r = 4$ and $n = 3, r = 9$ have the same fractal dimension $D = 1/2$, but different overall appearance.

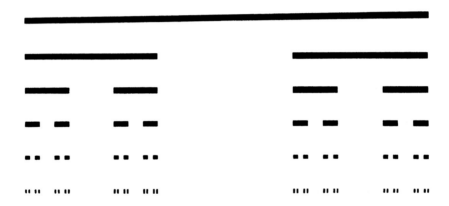

Figure 3 This figure shows perhaps the best known fractal embedded into one dimension. The triadic Cantor set is generated on the unit interval by replacing each of the intervals obtained at a given stage with two shorter ones. (From Vicsek, T., *Fractal Growth Phenomena*, World Scientific, Singapore, 1992. With permission.)

Example 2. One of the standard ways to construct a fractal surface is to replace the starting object with a single connected object of larger surface (made of reduced parts of the original one) and to repeat this procedure using the reduced parts as originals. In two dimensions, this method leads to a line (coastline) of infinite length with a fractal dimension larger than 1. Let us consider again the unit interval and replace it with a curve consisting of five intervals of unit length as shown in Figure 4. The fractal dimension is obtained from Equation (8) and is equal to $D = \ln 5/\ln 3 \simeq 1.465$, which exactly coincides with D of the fractal shown in Figure 2. In fact, the structure generated by this method is also analogous to that of Figure 2.

Using a related procedure (Figure 5), it is possible to define a single curve which can cover the unit square, i.e., it has a dimension equal to 2. For this so-called Peano curve $D = \ln 9/\ln 3 = 2$, and in the limit of $k \to \infty$ it establishes a continuous correspondence between the straight line and the plane. The Peano curve is a very peculiar construction (it has infinitely fine details being arbitrarily close to each other but does not have any intersections); however, it is not a fractal according to the definition given earlier. This is indicated by the absence of empty regions.

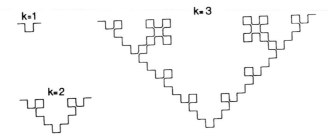

Figure 4 Fractal curves may have a dimension larger than 1. The total length of this growing line grows faster than the linear extension of the structure at each stage. This curve has the same fractal dimension as the objects shown in Figure 2. (From Vicsek, T., *Fractal Growth Phenomena*, World Scientific, Singapore, 1992. With permission.)

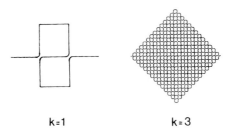

Figure 5 When one applies Equation (8) to the above displayed Peano curve, the result for the fractal dimension is $D = d = 2$. Moreover, the convergence of the measured "area" of the structure converges exponentially fast to that of a standard two-dimensional object. This means that this construction does not lead to a fractal according to the definition given in Section 11.2. (From Vicsek, T., *Fractal Growth Phenomena*, World Scientific, Singapore, 1992. With permission.)

Example 3. The construction presented in Figure 6 leads to a fractal that is both growing and nonuniform. To grow this fractal, one adds to the four main tips of the already existing configuration a part of it, in the following manner. The part to be added is the configuration itself minus one of the the main branches growing out from the center vertically. Moreover, this part has to be rotated and reduced in an appropriate way (see Figure 6) before attachment (reduction by a factor b is needed when the horizontal branches are updated).

Example 4. The random motion of a particle represents a particularly simple example of stochastic processes leading to growing fractal structures. A widely studied case is one in which the particle undergoes a random walk (Brownian or diffusional motion), making steps of lengths distributed according to a Gaussian in randomly selected directions. Such processes can be described in terms of the mean squared distance $R^2 = \langle R^2(t) \rangle$ made by the particles during a given time interval t. For random walks $R^2 \sim t$ independently of d, which means that the Brownian trajectory is a random fractal in spaces with $d > 2$. Indeed, measuring the volume of the trajectory by the total number of places visited by the particle making t steps, ($N(R) \sim t$), the above expression is equivalent to

$$N(R) \sim R^2 \tag{15}$$

and comparing Equation (15) with Equation (5), we conclude that for random walks $D = 2 < d$ if $d > 2$. In this case, rather unusually, the fractal dimension is an integer

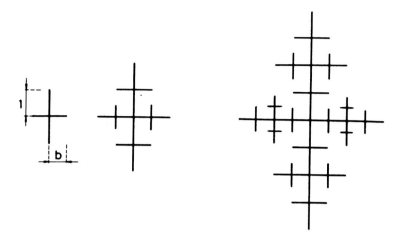

Figure 6 Nonuniform fractals are generated by iteration procedures using different rescaling factors for the same seed configuration. This nonuniform fractal grows by adding to the four principal tips of the $(n - 1)$th configuration the structure itself without the lower main stem. This addition has to be done by applying appropriate rotation and shrinking to keep the ratio of the corresponding branches equal to $b < 1$. (From Vicsek, T., *Fractal Growth Phenomena*, World Scientific, Singapore, 1992. With permission.)

number; however, the fact that it is definitely smaller than the embedding dimension indicates that the object must be nontrivially scale invariant.

11.3.3 Self-affine fractals

In many physically relevant cases, the structure of the objects is such that it is invariant under dilation transformation only if the lengths are rescaled by direction-dependent factors. These anisotropic fractals are called *self-affine*[1,10,11] (also see further recent publications on the subject[6,12,13]).

Single-valued, nowhere-differentiable functions represent a simple and typical form in which self-affine fractals appear. If such a function $F(x)$ has the property

$$F(x) \simeq b^{-H} F(bx) \tag{16}$$

it is self-affine, where $H > 0$ is some exponent. Equation (16) expresses the fact that the function is invariant under the following rescaling: shrinking along the x-axis by a factor $1/b$, followed by rescaling of values of the function (measured in a direction perpendicular to the direction in which the argumentum is changed) by a different factor equal to b^{-H}. In other words, by shrinking the function using the appropriate direction-dependent factors, the function is rescaled onto itself. For some deterministic self-affine functions this can be done exactly, while for random functions the above considerations are valid in a stochastic sense (expressed by using the sign $\simeq$).

A definition of self-affinity equivalent to Equation (16) is given by the expression for the height correlation function $c(\Delta x)$

$$c(\Delta x) = \left\langle \left[F(x + \Delta x) - F(x) \right]^2 \right\rangle \sim \Delta x^{2H} \tag{17}$$

which can be easily used for the determination of the exponent H. In addition to functions satisfying Equations (16) and (17), there are also self-affine fractals different from single-valued functions (see Example 6 below).

Let us first construct a deterministic self-affine model, in order to have an object which can be treated exactly.[11] An actual construction of such a bounded, self-affine function on the unit interval is demonstrated in Figure 7. The object is generated by a recursive procedure by replacing the intervals of the previous configuration with the generator having the form of an asymmetric letter z made of four intervals. However, the replacement this time should be done in a manner different from the earlier practice. Here, every interval is regarded as a diagonal of a rectangle becoming increasingly elongated during the iteration. The basis of the rectangle is divided into four equal parts and the z-shaped generator replaces the diagonal in such a way that its turnovers are always at analogous positions (at the first quarter and the middle of the basis). The function becomes self-affine in the $k \to \infty$ limit.

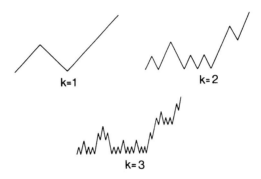

Figure 7 Self-affine functions can be generated by iteration procedures. The single-valued character of the function is preserved by an appropriate distortion of the z-shaped generator ($k = 1$) of the structure. (From Vicsek, T., *Fractal Growth Phenomena*, World Scientific, Singapore, 1992. With permission.)

As we shall see, self-affine fractals do not have a unique fractal dimension. Instead, their global behavior is characterized by an integer dimension D_G smaller than the embedding dimension, while the local properties can be described using a local fractal dimension D_L. To show this, we shall concentrate on functions of a single scalar variable. Such a function is, for example, the plot of the distances measured from the origin as a function of time t, $X(t)$, of a Brownian particle diffusing in one dimension. It is obvious that a so-called fractional Brownian plot for which $\langle X_H^2(t) \rangle \sim t^{2H}$ satisfies Equation (17).

The local dimension can be obtained from the following considerations. During a time interval δt, $|\max[X_H(t)] - \min[X_H(t)]|$ is of the order of $(\delta t)^H$. Covering the part of $X_H(t)$ on the interval δt by squares of side δt requires on the order of $(\delta t)^H/\delta t = \delta t^{H-1}$ squares. Therefore, covering $X_H(t)$ on the interval [0,1] requires

$$N(\delta t) \sim \frac{\delta t^{H-1}}{\delta t} = \delta t^{-(2-H)} \tag{18}$$

which according to Equation (3) leads to $D_L = 2 - H$. On the other hand, since $0 < H < 1$, $X_H(t)/t \to 0$ as $t \to \infty$. Correspondingly, when covering the function with boxes having a linear size larger than some t_c it will behave as a one-dimensional object ($D_G = 1$).

This means that by rescaling a bounded self-affine function, or, in other words, changing the units in which the distances are measured, an extra dimension called *global dimension* could be found. This raises the question of choosing the appropriate units to measure F and t. To see both the local and global behaviors, one should make a choice for the unit of both F and t. Then a quantity t_c can be defined through

$$\langle |F(t+t_c) - F(t)| \rangle \sim |t_c|. \tag{19}$$

This t_c can be called the *crossover scale* for the given process. The most essential fact about t_c is that it depends on the units which happen to be selected for F and t; therefore, the position of the crossover in general is not intrinsic. D_L is manifested on length scales smaller than t_c, while D_G corresponds to the behavior for $t_c \gg t_c$.

An important consequence of the above statement is that for self-affine structures with a lower or upper length scale, changing the units may lead to losing the possibility of detecting the local fractal or the global trivial scaling. Indeed, if the units we chose are such that t_c becomes the same order as the lower cutoff length, a local fractal dimension cannot be observed. This is the case, for example, for the record of a one-dimensional random walk on a lattice if the same unit is used for the increments of F and the time t, since then the lower cutoff and the crossover scale coincide. Similar arguments are valid for the detectability of a trivial global dimension.

Before going to further examples, we will provide a further basic feature of fractional Brownian motion. Calculating the Fourier spectrum of a fractional Brownian function one finds that the coefficients of the series, $A(f)$, are independent Gaussian random variables, and their absolute value scales with the frequency f according to a power law

$$|A(f)| \sim f^{-H-\frac{1}{2}}. \tag{20}$$

Example 6. So far we have discussed single-valued self-affine functions, because they seem to be more relevant from the point of view of applications than other possible self-affine structures. The example given here represents another type of self-affine objects, constructed in a spirit related to that used to generate some of the self-similar fractals. It is constructed by dividing the unit square into anisotropic subunits which serve as seeds for further divisions. Figure 8 demonstrates the actual procedure analogous to that used for generating the fractal shown in Figure 2B. For this example, the scales are chosen in such a way that the crossover scale is the same as the side of the unit square, and because of this the trivial global scaling is not manifested. On the other hand, the structure has a local fractal dimension, following from rule 4 of Section 11.2. Let us assume that in the kth step the sides of the rectangles are $l_1^{(k)} = l_1^k$ and $l_2^{(k)} = l_2^k$, where $l_1 < l_2$. Since the fractal shown in Figure 8 can be generated by multiplying two Cantor sets of dimensions $D_1 = \ln 2/\ln(1/l_1)$ and $D_2 = \ln 2/\ln(1/l_2)$, respectively, we obtain for the local fractal dimension of the resulting structure $D_L = \ln 2/\ln(1/l_1) + \ln 2/\ln(1/l_2)$.

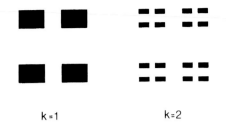

Figure 8 Generating a disconnected self-affine fractal embedded into two dimensions using elongated rectangles instead of squares during its construction. (From Vicsek, T., *Fractal Growth Phenomena*, World Scientific, Singapore, 1992. With permission.)

11.3.4 Fat fractals

The most important feature of structures discussed in the previous sections was that they had a fractal dimension strictly smaller than the embedding dimension d. There are, however, structures for which $D = d$, but which still exhibit a fractal behavior[1] in the following sense. When one calculates the volume $V(l)$ of such fat fractals using balls of decreasing size l, it converges to a finite value algebraically with a noninteger exponent. This is in contrast to ordinary or thin fractals, where $V(l) \to 0$ if $l \to 0$. Fat fractals are not self similar!

In general, the resolution-dependent volume of an object can be written in the form

$$V(l) = V(0) + f(l) \tag{21}$$

where $V(0) = V_0$ is the volume in the limit $l \to 0$. For thin (ordinary) fractals $V(0) = 0$, and $f(l) \sim l^{d-D}$ with $D < d$. For fat fractals, $V(0) > 0$, but $f(l)$ — as in the case of the thin ones — follows a power law with an exponent that can be regarded as a quantity characterizing the scaling properties of the structure. This fact can be expressed in the form

$$V(l) \simeq V(0) + Al^\beta \tag{22}$$

where A is a constant and β is an exponent quantifying fractal properties.

The example to be discussed below is constructed by generalizing the procedure leading to Cantor sets; however, there are many natural systems in which fat fractals are expected to occur. It has been shown that fat fractals can be associated with chaotic parameter values beyond the period-doubling transition to chaos, chaotic orbits of Hamiltonian systems, or ballistic aggregation clusters. In addition, such biological objects as bronchia in the lung or coral colonies are most likely to have the structure of fat fractals.

Example 7. As an illustration of fat fractals, consider the following modified Cantor set. In the original version (Example 1) first the central third of an interval is deleted, then the central third of the remaining intervals, and so on *ad infinitum*. To "fatten" this thin fractal, instead delete the central $\frac{1}{3}$, then $\frac{1}{9}$, then $\frac{1}{27}$, etc., of each remaining interval. The resulting set is topologically equivalent to the classical Cantor set, but the holes decrease in size sufficiently fast so that the limiting set has nonzero Lebesque measure and a dimension equal to 1.

This example can be extended to arbitrary embedding dimension. Suppose that we cut out a piece of volume v_1 from the unit hypercube of dimension d in such a way that the resulting structure is made of the 2^d hypercubes remaining at the corners of the starting object. Next, from each of these cubes a similar piece of relative volume v_2 is cut out, and this process is repeated infinitely many times. The total volume remaining after the kth iteration is

$$V_k = (1-v_1)(1-v_2)\ldots(1-v_k) = \prod_0^k (1-v_k). \tag{23}$$

V_k decreases as $k \to \infty$ to a limiting value V. For v_k fixed, one has $V = 0$; however, if $\sum_0^\infty v_k < \infty$, the limiting volume is positive, $\prod_0^\infty (1-v_k) > 0$.

11.4 Multifractals

In the previous sections, complex geometrical structures were discussed which could be interpreted in terms of a single fractal dimension. The present section is mainly concerned with the development of a formalism for the description of the situation when a singular distribution is defined on a fractal. As we shall see, the structure of fractals plays an essential role in the various processes they are involved in, and, as a result, one needs infinitely many dimension-type exponents to characterize these distributions, as was first recognized by Mandelbrot.[14] This idea was later developed further by several authors.[15-17]

It is typical for a large class of phenomena in nature that the behavior of a system is determined by the spatial distribution of a scalar quantity, e.g., concentration, electric potential, probability, etc. For simpler geometries, this distribution function and its derivatives are relatively smooth, and they usually contain only a few (or none) singularities, where the word "singular" corresponds to a local power-law behavior of the function. (In other words, we call a function singular in the region surrounding point $\bar{x}$ if its local integral diverges or vanishes with a noninteger exponent when the region of integration goes to zero.) In the case of fractals, the situation is quite different: a process in nature involving a fractal may lead to a spatial distribution of the relevant quantities which possesses infinitely many singularities.

As an example, consider an isolated, charged object. If this object has sharp tips, the electric field around these tips becomes very large in accord with the behavior of the solution of the Laplace equation for the potential. In the case of charging the branching fractals produced in the $k \to \infty$ limit of constructions shown in Figures 1 or 2, one has an infinite number of tips and corresponding singularities of the electric field. Moreover, tips being at different positions in general have different local environments (configuration of the object in the region surrounding the given tip) which affect the strength of singularity associated with that position.

11.4.1 Definitions

The above discussed time independent distributions defined on a fractal substrate are called *fractal measures*. In general, a fractal measure possesses an infinite number of singularities of infinitely many types. The term *multifractality* expresses the fact that points corresponding to a given type of singularity typically form a fractal subset whose dimension depends on the type of singularity. It is perhaps difficult to imagine how such an extremely complex distribution can appear as a result of simple processes. Let us, therefore, use a deterministic construction to demonstrate the mechanisms by which fractal measures can be generated.

We shall assume that the measure denoted by $\mu(\vec{x})$ is normalized so that its total amount on the fractal is equal to 1. Then, the fractal measure can be regarded as a probability distribution. Consider now a deterministic recursive (multiplicative) process generating a nonuniform fractal with varying weights or probabilities attributed to each rescaled part in a manner demonstrated in Figure 9. In the first step, the starting object (square) is replaced by its three smaller copies with corresponding reduction factors $1/r_1$, $1/r_2$, and $1/r_3$. In addition, the three newly created objects are given a weight factor (probability) denoted by P_1, P_2, and P_3. In the next step ($k = 2$), the same procedure is repeated for each of the squares, treating them as starting objects. To obtain a fractal measure, this process has to be continued till $k \to \infty$. It can be seen from Figure 9 that as the recursion advances a very complex situation emerges both concerning the size and the weight distribution of the objects.

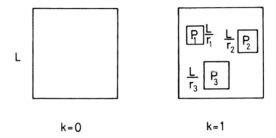

Figure 9 This figure shows a general case of constructing a fractal measure defined on a nonuniform fractal support embedded into two dimensions. The multifractal is obtained after infinitely many recursions. The sizes are rescaled by the factors r_i, while the corresponding weights or probabilities are given by P_i. (From Vicsek, T., *Fractal Growth Phenomena*, World Scientific, Singapore, 1992. With permission.)

To illustrate multifractality more directly, we consider a simple multifractal distribution which possesses the relevant features of many general stochastic fractal measures. The measure will be defined on the unit interval instead of a fractal support, but this fact does not change its characteristic scaling properties.

Consider the unit interval divided into three equal parts of length 1/3, with the corresponding weights or probabilities P_1, P_1, and P_2, where $P_2 = 1 - 2P_1$. We shall assume that the probability of the middle interval is larger than that of the two others having equal weights, i.e., $P_2 > P_1$. In the next step ($k = 2$), each of the three intervals is further divided and the probability is redistributed within the nine new intervals according to the proportions used in the first step. This construction, which corresponds to the special case $n = 3$ and $1/r_j \equiv 1/3$, is illustrated in Figure 10. In a few more steps, the distribution becomes so inhomogeneous that its structure becomes visible only on a logarithmic scale. The density distribution in the limit $k \to \infty$ turns into a single-valued, everywhere discontinuous function.

In order to describe a fractal measure in a quantitative way, we imagine that a d-dimensional hypercubic lattice with a lattice constant l is put on the fractal, and denote by p_i the probability associated with the ith box of volume l^d, where

$$p_i = \int_{i\text{th box}} d\mu(\vec{x}) \tag{24}$$

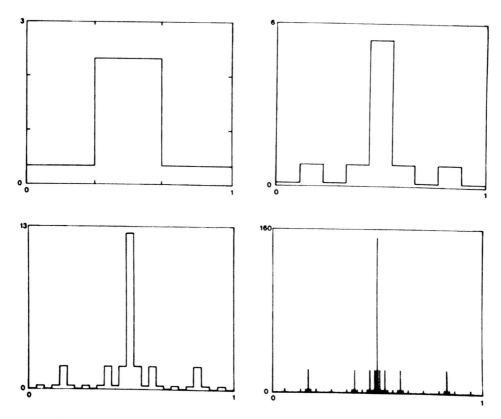

Figure 10 The first steps of constructing a relatively simple fractal measure on the unit interval. (From Vicsek, T., *Fractal Growth Phenomena*, World Scientific, Singapore, 1992. With permission.)

with $\sum_i p_i = 1$. First, we are interested in the behavior of p_i as a function of the box size. This dimensionless unit will be denoted by $\varepsilon = l/L$, and scaling of various quantities in the limit $\varepsilon \to 0$ (corresponding to either $l \to 0$ or, in the growing case, $L \to \infty$) will be studied. Trivially, for a homogeneous structure with a uniform distribution (density) $p_i(\varepsilon) \sim \varepsilon^d$. In the case of a uniform fractal ($r_i \equiv r$) of dimension D with a uniform distribution on it $p_i(\varepsilon) \sim \alpha$, since, for a distribution with a uniform density, p_i corresponds to the volume of the support in the ith box. (The term *support* is used for the fractal on which the measure is defined.)

In short, for a general multifractal the spatial behavior of the measure p is such that with $\varepsilon \to 0$ it converges to a nowhere continuous "function" (so that it is not even practical to call it a function). The "function" p/ε is almost everywhere zero or nearly zero or tends to infinity. There are two closely related ways of describing such a complex distribution. The main point is that we would like to associate a simple continuous function with the much more complex measure, just as we could define a single noninteger number (D) when characterizing the geometry of ordinary fractals.

In the first approach[18] we cover the measure with a grid of box size ε^d and determine the number of boxes $N_p(\varepsilon)dp$ having a measure between p and $p + dp$. Then, we make a plot of the quantity $\ln N_p(\varepsilon)dp / \ln(1/\varepsilon)$ vs. $\ln p / \ln \varepsilon$ for fixed values of ε as a function of p. If these histograms in the $\varepsilon \to 0$ limit converge to a unique, nontrivial function, the measure is a multifractal.

In the second approach (to be used in the following) we write the box probabilities for each ε in the following form

$$p(\varepsilon) \sim \varepsilon^\alpha, \quad (25)$$

where $\varepsilon \ll 1$ and α can take on a range of values depending on the given region of the measure. The noninteger exponent α corresponds to the strength of the local singularity of the measure[17] and is also called the *crowding index* or Hölder exponent. Although α depends on the actual position on the fractal, there usually are many boxes with the same index α. In general, the number of such boxes scales with ε as

$$N_\alpha(\varepsilon) \sim \varepsilon^{-f(\alpha)} \quad (26)$$

where, in view of Equation (3), $f(\alpha)$ is the fractal dimension of the subset of boxes characterized by the exponent α. The exponent α can take on values from the interval $[\alpha_\infty, \alpha_{-\infty}]$, and $f(\alpha)$ is usually a single-humped function with a maximum

$$\max_\alpha f(\alpha) = D. \quad (27)$$

The $f(\alpha)$ spectrum of an ordinary uniform fractal is a single point on the $f - \alpha$ plane.

Obviously, the two approaches are intimately related; their equivalence can be seen from the analogy between $\ln N_p(\varepsilon)dp/\ln(1/\varepsilon)$ and $f(\alpha)$. Furthermore, when making the histograms, we use $\ln p / \ln \varepsilon$, which is essentially the same as α of the second approach.

11.4.2 Multifractal formalism

In order to determine $f(\alpha)$ for a given distribution, it is useful to introduce a few quantities which are more directly related to the observable properties of the measure. Then, relations (multifractal formalism) among these quantities and $f(\alpha)$ make it possible to obtain a complete description of a fractal measure.

An important quantity which can be determined from the weights p_i is the sum over all boxes of the qth power of box probabilities

$$\chi_q(\varepsilon) \equiv \sum_i p_i^q \quad (28)$$

for $-\infty < q < \infty$. For $q = 0$, Equation (28) gives $N(\varepsilon)$, the number of boxes of size ε needed to cover the fractal support (the region where $p_i \neq 0$). Therefore,

$$\chi_0(\varepsilon) = N(\varepsilon) \sim \varepsilon^{-D} \quad (30)$$

where D is the dimension of the support. Since the distribution is normalized, $\chi_1(\varepsilon) = 1$.

Because of the complexity of multifractals, the scaling of $\chi_q(\varepsilon)$ for $\varepsilon \to 0$ generally depends on q in a nontrivial way

$$\chi_q(\varepsilon) \sim \varepsilon^{(q-1)D_q} \quad (31)$$

where D_q is the so-called order q generalized dimension. It can be shown that the D_q values monotonically decrease with growing q. From the comparison of Equation (30) with Equation (31), it follows that $D_0 = D$. In the simple case of uniform fractals with a uniform distribution, all D_q are equal to D.

The distribution of p_i is extremely inhomogeneous concerning both their values and the number of boxes with the same p_i. As a result, when $\varepsilon \to 0$ the dominant contribution to the sum of Equation (28) comes from a subset of all possible boxes. This subset forms a fractal with a fractal dimension f_q depending on the actual value of q, thus,

$$N_q(\varepsilon) \sim \varepsilon^{-f_q} \tag{32}$$

where $N_q(\varepsilon)$ is the number of boxes giving the essential contribution in Equation (28). In addition, all of these boxes have the same $p_i = p_q$. We shall denote the singularity strength for boxes with probability p_q by α_q, i.e.,

$$p_q \sim \varepsilon^{\alpha_q} \tag{33}$$

for $\varepsilon \to 0$. The f_q and α_q spectra defined by Equations (32) and (33) provide an alternative description of a fractal measure with regard to Equations (25) and (26).

What is the origin of the fact that a given q selects a fractal subset of dimension f_q with a corresponding crowding index α_q? We have to consider terms in the sum for $\chi_q(\varepsilon)$ of the form $N_q p_q^q \sim \varepsilon^{-f_q} \varepsilon^{q\alpha_q}$. Obviously, in the limit $\varepsilon \to 0$ for each fixed q this quantity will have a very sharp peak at a well defined pair of α_q, f_q values, minimizing the expression $q\alpha_q - f_q$.

There is a relation between D_q and the spectra f_q and α_q which can be easily obtained taking into account that

$$\chi_q(\varepsilon) \simeq N_q(\varepsilon) p_q^q, \tag{34}$$

since the essential contribution to $\chi_q(\varepsilon)$ comes from boxes with p_q. Inserting Equations (31), (32), and (33) into Equation (34), we get

$$(q-1)D_q = q\alpha_q - f_q. \tag{35}$$

The above expression is consistent with the earlier observation that $D = D_0 = f_0$. Moreover, since f is finite we have $D_{\pm\infty} = \alpha_{\pm\infty}$.

Now it is possible to relate the generalized dimensions to the $f(\alpha)$ spectrum. First we express $\chi_q(\varepsilon)$ for $\varepsilon \to 0$ through $f(\alpha)$ using Equations (25), (26), and (28):

$$\chi_q(\varepsilon) \sim \int_{\alpha_\infty}^{\alpha_{-\infty}} \varepsilon^{q\alpha' - f(\alpha')} \, d\alpha', \tag{36}$$

where α is a quasicontinuous variable. For $\varepsilon \ll 1$, the integral will be dominated by the value of α which minimizes the exponent. This leads to the condition

$$\frac{df(\alpha)}{d\alpha}\bigg|_{\alpha_q} = q, \tag{37}$$

where α_q is the value of α for which $q\alpha - f(\alpha)$ is minimal. We see that Equation (36) is consistent with Equation (35) if, in addition to Equation (37),

$$f_q = f(\alpha_q). \tag{38}$$

Thus, knowing $f(\alpha)$ we can find α_q from Equation (37) and then f_q from Equation (38). Alternatively, if D_q is given, one can calculate α_q from the relation

$$\alpha_q = \frac{d}{dq}\left[(q-1)D_q\right] \tag{39}$$

which can be obtained from Equation (35) using Equations (37) and (38). If we know α_q and D_q, f_q can be determined from Equation (35) and, finally, $f(\alpha)$ from Equation (38). Therefore, the spectra D_q and $f(\alpha)$ represent equivalent descriptions of multifractals, since they are Legendre transforms of each other.

For the evaluation of experimental data, the following procedure is usually applied. Using an appropriate normalization of the observable quantities, the set of p_i values is determined. Then the generalized dimensions are obtained from

$$D_q = \lim_{\varepsilon \to 0}\left[\frac{1}{q-1}\frac{\ln \sum_i p_i^q}{\ln \varepsilon}\right] \tag{40}$$

with the help of the corresponding log-log plot. Finally, the resulting plot of D_q vs. q is numerically derived to obtain $f(\alpha)$.

For the order $q = 1$ generalized dimension, for which we have from Equation (35) and its derivative taken at $q = 1$ the following relation:

$$D_1 = \alpha_1 = f_1. \tag{41}$$

Furthermore, by taking the $q \to 1$ limit (using l'Hospital's rule) in Equation (40), one gets

$$-\sum_i p_i \ln p_i \sim D_1 \ln(1/\varepsilon). \tag{42}$$

The left-hand side of Equation (42) is a familiar expression from information theory and corresponds to the amount of information associated with the distribution of p_i values. Therefore, according to Equation (42), D_1 describes the scaling of information as $\varepsilon \to 0$. This is why D_1 is called the information dimension. A distribution with $D_1 < D$ is necessarily a fractal measure.

11.4.3 Recursive multifractals

The construction of recursive multifractals is illustrated in Figure 10. Because of the nature of the construction, the fractal structure (support) on which the measure is defined can be divided into n parts, each being a rescaled version of the whole support by a factor $1/r_j$. The total measure associated with the jth such part is P_j. Therefore,

$$\chi_{q,j}(\varepsilon) = \sum_i p_{j,i}^q = P_j^q \chi_q(\varepsilon r_j), \qquad (43)$$

where $\chi_{q,j}(\varepsilon)$ is the quantity defined by Equation (28) evaluated for the jth part and $p_{j,i}$ is the probability of the ith box in the jth part. For the whole system

$$\chi_q(\varepsilon) = \sum_{j=1}^n \chi_{q,j}(\varepsilon). \qquad (44)$$

Using Equations (31) and (43) in Equation (44), we get

$$\sum_{j=1}^n P_j^q \, r_j^{(q-1)D_q} = 1, \qquad (45)$$

which is an implicit equation for the generalized dimensions D_q. Depending on the particular choice for q or the P_j and r_j values in Equation (45), various special cases can be recovered. As expected, for $q = 0$, Equation (45) provides the fractal dimension of the support. On the other hand, for general q but identical rescaling factors $r_j \equiv r$, Equation (45) can be solved for D_q

$$D_q = \frac{1}{q-1} \frac{\ln\left(\sum_{j=1}^n P_j^q\right)}{\ln(1/r)}. \qquad (46)$$

Example 8. Application of Equation (46) to the multifractal shown in Figure 10 yields an explicit expression for D_q through the probabilities P_1 and P_2:

$$D_q = \frac{1}{(q-1)\ln(1/3)} \ln(2P_1^q + P_2^q). \qquad (47)$$

From the above equation α_q can be obtained using Equation (39):

$$\alpha_q = \frac{1}{\ln(1/3)} \frac{2P_1^q \ln P_1 + P_2^q \ln P_2}{2P_1^q + P_2^q}. \qquad (48)$$

Example 9. We consider a hierarchical network of resistors shown in Figure 11. In this example, the flow of current can be associated with fluid flow in a network of blood vessels. At the kth step of the iteration procedure, each bond of length $L = 1$ is replaced with a configuration having a linear size equal to $r = 3$ and made of $n = 6$ bonds of unit length. This growing fractal in the $k \to \infty$ limit has a fractal dimension $D = \ln 6/\ln 3$. To complete the model, we assume that the bonds have the same resistivity, and the voltage between the two terminal bonds is equal to $V_0 = 1$. Then, in the unit cell there will be two bonds with voltage drop $V_1 = 1/3$ and four others with $V_2 = 1/6$. The sum of all voltage drops is not equal to 1; therefore, the distribution is not normalized (it can be normalized by simply choosing $V_0 = 3/4$). The multifractal properties can be deduced from[19]

$$\sum_i v_i^q \sim L^{-(q-1)D_q} \tag{49}$$

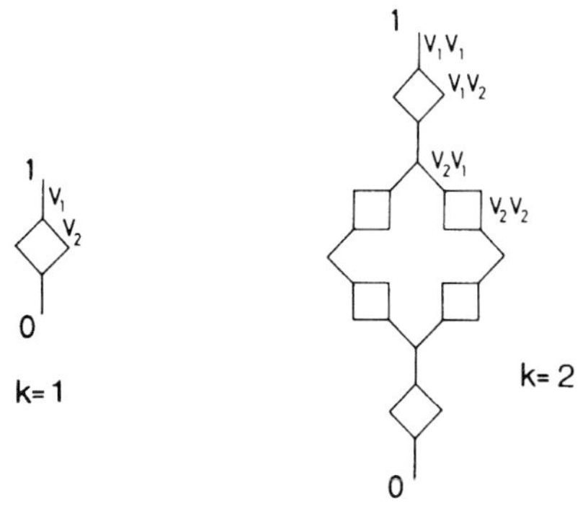

Figure 11 This is a construction of a growing fractal network of resistors. The distribution of voltage drops exhibits multifractal scaling. (From Vicsek, T., *Fractal Growth Phenomena*, World Scientific, Singapore, 1992. With permission.)

where v_i is the voltage drop across the ith bond, and $L = (l/\varepsilon) = 3^k$ is the linear size of the network. Because of the hierarchical structure of the model, the sum of the v_i^q after completing the kth recursion is

$$\sum_m N_m v_m^q = \left(2V_1^q + 4V_2^q\right)^k \tag{50}$$

In the above expression

$$N_m = 2^{k+m} \binom{k}{m} \tag{51}$$

is the number of bonds with the voltage drop

$$v_m = V_1^{k-m} V_2^m \tag{52}$$

with $0 \leq m \leq k$. Substituting Equation (50) into Equation (49) and using the actual voltage values, we get

$$D_q = \frac{1}{q-1} \frac{\ln\left[2\left(\frac{1}{3}\right)^q + 4\left(\frac{1}{6}\right)^q\right]}{\ln(1/3)} \tag{53}$$

showing that the moments of the distribution of voltage drops can be described by an infinite number of generalized dimensions.

11.4.4 Mass multifractality

The results discussed above were obtained for the general case when an inhomogeneous measure was defined on the support. The special case of a distribution with constant density on a nonuniform fractal, however, deserves particular attention. In this case, it is the geometry of the fractal only which is described by the formalism.

The fact that nonuniform fractals with a uniform distribution also can be described in terms of generalized dimensions has already been indicated by Equation (45). This equation valid for a recursive fractal, leads to a D_q spectrum even if the weight factors P_j are defined in such a way that the measure of a newly created part becomes the same as its volume, i.e., the probability distribution is uniform on the support. In the case of the multifractal shown in Figure 9, this can be achieved by choosing $P_j = (1/r_j)/\Sigma_j (1/r_j)$ which together with Equation (45) results in the nontrivial dependence of D_q on q.

Let us now consider the problem for the case of growing fractals. We assume that the structures grow on a lattice and are built up by identical particles. The actual linear size and mass of the cluster will be denoted by L and M, respectively. The structure is to be covered by boxes of size l such that

$$a \ll l \ll L, \tag{54}$$

where a is the lattice constant. (In some cases, the condition $a/l \to 0$ has to be satisfied.) One can then determine the mass M_i of the ith nonempty box. The mass index or singularity exponent α of this box is defined by

$$M_i \sim M(l/L)^\alpha \tag{55}$$

for $l/L \ll 1$. Boxes with the same mass index α form a fractal subset of dimension $f(\alpha)$. Their number $N(\alpha)$ is, therefore, related to $f(\alpha)$ via

$$N(\alpha) \sim (l/L)^{-f(\alpha)}. \tag{56}$$

If there exists a set of different mass indices, the growing structure will be called a *geometrical* or *mass multifractal*,[20,21] since the measure generating the spectrum is a uniform mass distribution (Lebesque measure). Thus, $f(\alpha)$ characterizes the geometry of the system directly.

To determine the generalized dimensions, one can use the scaling relation of Equation (31), which in the present notation has the form

$$\chi_q(M,l,L) \equiv \sum_i M_i^q \sim M^q (l/L)^{(q-1)D_q}. \tag{57}$$

As for general multifractals, knowing D_q we can determine $f(\alpha)$ using Equations (35) and (37).

At this point a few important comments should be made: (1) nonuniform recursive fractals are multifractals in a geometrical sense; (2) since mass multifractality as defined above is a consequence of local density fluctuations, it should be most pronounced in inhomogeneous growth processes; and (3) a necessary condition for observing this phenomenon is the existence of three well separated length scales which may require the linear size L to be close to the largest cluster size ever produced in numerical simulations.

11.5 Methods for determining fractal dimensions

When one tries to determine the fractal dimension of biological structures in practice, it usually turns out that the direct application of definitions for D given in the previous sections is ineffective or cannot be accomplished. Instead, one is led to measure or calculate quantities which can be shown to be related to the fractal dimension of the objects. Three main approaches are used for the determination of these quantities: experimental, computer, and theoretical.

11.5.1 Experimental methods for measuring fractal dimensions

A number of experimental techniques have been used to measure the fractal dimension of scale invariant structures grown in various experiments. The most widely applied methods can be divided into the following categories: (1) digital image processing of two-dimensional pictures, (2) scattering experiments, and (3) direct measurement of dimension-dependent physical properties.

Digitizing the image of a fractal object is a standard way of obtaining quantitative data about geometrical shapes. The information is picked up by a scanner or an ordinary video camera and transmitted into the memory of a computer (typically a PC). The data are stored in the form of a two-dimensional array of pixels whose nonzero (equal to zero) elements correspond to regions occupied (not occupied) by the image. Once they are in the computer, the data can be evaluated using the methods described in the next section, where calculation of D for computer-generated clusters is discussed.

The only principal question related to processing of pictures arises if two-dimensional images of objects embedded into three dimensions are considered. It has already been mentioned that the fractal dimension of the projection of an object onto a $(d-m)$-dimensional plane is the same as its original fractal dimension if $D < d - m$.

Scattering experiments represent a powerful method to measure the fractal dimension of structures. Depending on the characteristic length scales associated with the object to be studied, light, X-ray, or neutron scattering can be used to reveal fractal properties. There are a number of possibilities to carry out a scattering experiment. One can investigate the structure factor of a single fractal object, scattering by many clusters growing in time, or the scattered beam from a fractal surface, etc. In scattering experiments, a beam of intensity I_0 is directed on the sample, and the scattered intensity is measured as the function of the angle θ between the incident and the scattered beam. Let us denote the difference between the wave vectors corresponding to these beams denoted by $\vec{q} = \vec{k}_1 - \vec{k}_2$. In the case of small θ (small angle scattering), the main contribution to the scattered intensity comes from quasi-elastic processes with $|\vec{k}_1| = |\vec{k}| = k = 2\pi/\lambda$, where λ is the wavelength of the incident beam. Therefore,

$$q = |\vec{q}| = 2k\sin(\theta/2). \tag{58}$$

It is useful to identify a single scatterer with a corresponding form factor $P(q)$ and to separate the scattered intensity into two factors

$$I(q) = \rho_0 P(q)[1 + S(q)], \quad (59)$$

where ρ_0 is the average density in the sample, $S(q)$ is the structure factor, and $P(q)$ corresponds to the elementary units of the structure having a radius r_0. It can be shown that for $qr_0 \ll 1$ the form factor is approximately constant (Guinier regime), while for $qr_0 \gg 1$, $P(q) \sim q^{-4}$, which is called the *Porod law*.

According to the theory of scattering (see, for example, Reference 22), the structure factor $S(q)$ is the Fourier transform of the density-density correlation function $c(r)$ defined by Equation (14). From this and after some algebra, it is possible to determine that, in the relevant region, the structure function of fractals of dimension D scales as

$$I(q) \simeq S(q) \sim q^{-D} \quad \text{for} \quad 1/R \ll q \ll 1/r_0, \quad (60)$$

Measurements of *physical properties* of fractal objects also can be used for the experimental determination of D. A number of methods have been suggested, most of them based on electrical properties including measurements of current, electromagnetic power dissipation, and frequency dependence of the complex impedance of fractal interfaces. These methods typically provide an indirect estimate of D and have been used less extensively than the above discussed approaches.

11.5.2 Evaluation of numerical data

Throughout this section we assume that the information about the stochastic structures is stored in the form of d-dimensional arrays that correspond to the values of a function given at the nodes (or sites) of some underlying lattice. In the case of studying geometrical scaling only, the value of the function attributed to a point with given coordinates (the point being defined through the indexes of the array) is either 1 (the point belongs to the fractal) or 0 (the site is empty). When multifractal properties are investigated, the site function takes on arbitrary values. In general, such discrete sets of numbers are obtained by two main methods: (1) by digitizing pictures taken from objects produced in experiments, or (2) by numerical procedures used for the simulation of various biological structures. For convenience, in the following we shall frequently use the terminology *particle* for a lattice site which belongs to the fractal (or is filled) and *cluster* for the objects made of connected particles.

Below we discuss how to measure D for a single object. To make the estimates more accurate one usually calculates the fractal dimension for many clusters and averages over the results.

Perhaps the most practical method is to determine the number of particles $N(R) = R^D$ within a region of linear size R and to obtain the fractal dimension D from the slopes of the plots $\ln N(R)$ vs. $\ln R$. If the centers of the regions of radius R are the particles of the cluster, than $N(R)$ is equivalent to the integral of the density correlation function. In practice, one chooses a subset of randomly selected particles of the fractal (as many as needed for a reasonable statistics) and determines $\langle N(R) \rangle$ for a sequence of growing R (or counts the number of particles in boxes of linear size L). In order to avoid undesirable effects caused by anomalous contributions appearing at the edge of the cluster, one should not choose particles as centers close to the boundary region. The situation is shown in Figure 12. Typically there is a deviation from scaling for small and large scales.

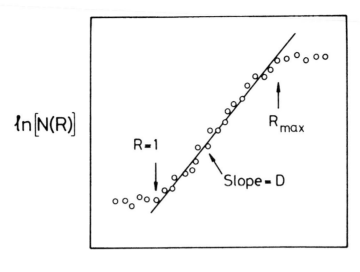

Figure 12 Schematic log-log plot of the numerically determined number of particles $N(R)$ belonging to a fractal and being within a sphere of radius R. If R is smaller than the particle size or larger than the linear size of the structure, a trivial behavior is observed. The fractal dimension is obtained by fitting a straight line to the data in the scaling region. (From Vicsek, T., *Fractal Growth Phenomena*, World Scientific, Singapore, 1992. With permission.)

The roughness exponent H corresponding to self-affine fractals is usually determined from Equation (17). An alternative method is to investigate the scaling of the standard deviation $\sigma(l) = [\langle F^2(x) \rangle_x - \langle F(x) \rangle_x^2]^{1/2}$ of the self-affine fuction F

$$\langle \sigma(l) \rangle \sim l^H, \tag{61}$$

where the left-hand side is the average of the standard deviation of the function F calculated for regions of linear size l. The roughness exponent H can be calculated by determining $\sigma(l)$ for parts of the interfaces for various l. An averaging should be made over the segments of the same length and the results plotted on a double logarithmic plot as a function of l. Further techniques are discussed in Reference 23.

Sometimes there is a large, slowly decaying correction to the simple power law behavior of $N(R)$ or $c(r)$. This correction may have various origins and forms. For example, if the growth takes place along a surface, the presence of the surface usually has an effect on the overall behavior of the quantities used for determining D. To extract the information concerning D, one assumes a special functional dependence of the correction. Then, instead of fitting a straight line to the $N(R)$ data, one fits a curve of the following form

$$N(R) \simeq AR^D [1 + f(R)], \tag{62}$$

where A is a constant and $F(R)$ is a function which is typically chosen to be decaying as an exponential or a power law (or a combination of these).

Selecting the most appropriate variables when plotting the results is another effective way to obtain more accurate data. Many times there already exists a theoretical result or a good guess of other source for the value of the fractal dimension. In such cases it is the deviation from this value which can be a quantity of interest. A common procedure is to plot for example $\ln[N(R)/R_g^D]$ vs. $\ln R$, where D_g is a guess for the fractal dimension. If the true D is approximately equal to D_g, the straight part of the plot is close to a horizontal line, and any deviation can be magnified.

In case of a fractal measure defined on a digitized structure, there is a weight or probability p attributed to each particle. The generalized dimensions D_q for such objects can be obtained using a procedure analogous to the box-counting method described in the section on multifractals; however, in most of the cases it is more practical to use a method similar to that discussed above for determining D from the scaling of $\langle N(R)\rangle$. The procedure follows.

One has to choose the centers of the regions of radius R within which χ_q is determined with a probability p corresponding to the measure or weight of this center site. This is done to obtain considerably improved estimates for negative q. To realize such choices, one attributes to each site (of weight p) an interval of length p on the unit interval and generates a random number between 0 and 1; then choose the site to which the interval onto which the random number "fell". Next, one calculates χ_q by noticing that such a choice of sites and the following averaging over the sites corresponds to taking an average weighted by p. Thus, to get the estimate of χ_q one should calculate $\chi_q \simeq \langle \sum_i p_i^{q-1} \rangle$, where i is the index of points within the region of radius R surrounding the jth site chosen during the averaging. Then the generalized dimensions can be determined from the slopes of plots of $\ln \chi_q(R)$ vs. $\ln R$, since, according to Equation (31),

$$\ln \chi_q(R) \sim (q-1)D_q \ln R. \tag{61}$$

The $f(\alpha)$ spectrum is obtained from D_q by Legendre transformation; see Equations (35), (38), and (39).

References

1. Mandelbrot, B. B., *The Fractal Geometry of Nature*, W. H. Freeman, San Francisco, 1982.
2. Witten, T. A. and Sander, L. M., *Phys. Rev.*, B27, 5686, 1983.
3. Stanley, H. E. and Ostrowsky, N., Eds., *On Growth and Form*, Kluwer, Dordrecht, 1986.
4. Feder, J., *Fractals*, Plenum Press, New York, 1988.
5. Avnir, D., Ed., *Fractal Approach to Heterogeneous Chemistry*, John Wiley & Sons, New York, 1989.
6. Family, F. and Vicsek, T., Eds., *Dynamics of Fractal Surfaces*, World Scientific, Singapore, 1991.
7. Vicsek, T., *Fractal Growth Phenomena*, World Scientific, Singapore, 1992.
8. Bunde, A. and Havlin, S., Eds., *Fractals in Sciences*, Springer-Verlag, Berlin, 1994.
9. Vicsek, T., *J. Phys.*, A16, L647, 1983.
10. Mandelbrot, B. B., *Phys. Scr.*, 32, 257, 1985.
11. Mandelbrot, B. B., in *Fractals in Physics*, Pietronero, L. and Tosatti, E., Eds., Elsevier, Amsterdam, 1986, p. 3.
12. Jullien, R., Kertész, J., Meakin P., and Wolf, D., Eds., *Surface Disordering*, Nova Science, New York, 1992.
13. Kertész, J. and Vicsek, T., in *Fractals in Sciences*, Bunde, A. and Havlin, S., Eds., Springer-Verlag, Berlin, 1994.
14. Mandelbrot, B. B., *J. Fluid. Mech.*, 62, 331, 1974.
15. Hentschel, H. G. E. and Procaccia, I., *Physica*, 8D, 435, 1983.

16. Frisch, U. and Parisi, G., in *Turbulence and Predictability in Geophysical Fluid Dynamics and Climate Dynamics* (International School of Physics "Enrico Fermi", course LXXXVIII), Ghil, M., Benzi, R., and Parisi, G., Eds., North-Holland, Amsterdam, 1984, p. 84.
17. Halsey, T. C., Jensen, M. H., Kadanoff, L. P., Procacia, I., and Shraiman, B. I., *Phys. Rev.*, A33, 1141, 1986.
18. Mandelbrot, B. B., in *Random Fluctuations and Pattern Growth*, Stanley, H. E. and Ostrowsky, N., Eds., Kluwer, Dordrecht, 1988.
19. de Arcangelis, L., Redner, S., and Coniglio, A., *Phys. Rev.*, B34, 4656, 1986.
20. Tél, T. and Vicsek, T., *J. Phys.*, A20, L835, 1987.
21. Vicsek, T., Family, F., and Meakin, P., *Europhys. Lett.*, 12, 217, 1990.
22. Squires, G. L., *Introduction to the Theory of Thermal Neutron Scattering*, Cambridge University Press, London, 1978.
23. Cox, L. B. and Wang, J. S. Y., *Fractals*, 1, 87, 1993.

Index

Index

A

Acetylcholine, 49
Acid protease, 74
Actin, 251, 253
Activation energy barriers, 7, 37
   distribution, 34–35, 36, 37, 43–46, 47, 53
      combinations of, 48–49
      dependence on voltage, ions, and ligands, 49–50
Active sites, 8, 97–98, 99, 104
Adenine, 16, 17
Adenylate kinase, 74, 80
ADHD, 268, 292
ADP, 89
Adrenal gland
   chimeras of the, 195, 196, 203
   patches within, 201
Agar media, 129
Age-specific failure rate, 41, 50
Agglutinin, 80
Aggregation clusters, 113, 209, 230, 231
Air-traffic-control simulation, 303–307
Alanine, 131
Alcohol dehydrogenase, 34, 86, 87
Algae, 225
Allosteric effects, 104–111, 119
Alzheimer's disease, 10
Ameboid ulcers, 219
Amino acids, stimulation of bacterial colony morphogenesis, 129
Analog-to-digital conversion rate, 39, 40
Anderson localization, 70
Anisotropic scaling, 146, 147, 169, 233. *See also* Self-affinity
Anlagen, 197, 198
Antipersistent data, 235
Area-perimeter relation, 230–231
ART. *See* Attractor reconstruction technique
Arterial tree, 250, 251
Artificial intelligence, 9, 205
Asparagine, 130
Aspartic acid, 130
Asphyxia, 273
Asymptotic fractal, 211, 214, 241
Asymptotic scaling exponent, 261, 262
Atomic charge density, 116
Atomic orbital, hybridized state of, 112–114
Attractor, 268–269, 273
   biochemical, 88–89, 90
   chaotic, 271, 276, 278
   cognitive, 290
   and correlation exponent, 279
   data points necessary, 274, 280
   funnel, 276
Attractor reconstruction technique (ART), 271–272
   critique, 272–290
   multichannel (MART), 285–286
   of Takens, 275–278, 286
      problems associated with, 289–290
Automatic segmentation, 217

B

Bacillus subtilis, 128, 149, 155, 161
   colony morphology, 137
      effect of serrawettin, 159, 164
      effect of terrestrial factor, 160
      phase diagram, 157
      similarity to DLA model, 156, 168
   microscopic generating process, 149–150
   population dynamics, 162, 165, 166, 167, 168
Backbone of proteins, 105, 113
Bacterial cells
   diffusion coefficient, 167
   Malthusian growth, 165
   molecular diversity, 132–133
   morphogenesis, 127–169
      Liesegang-ring-like, 159
   mutants, 132
      and colony morphology, 158–159
Bacterial colonies
   culture of. *See* Culture conditions
   diversity of morphology, 8, 132–136
      branch generation, 152–155
      experimental approach, 155–161
      random pattern growth, 148–155
   formation, 128–132
   fractal growth, 168–169
   fusion of two, 166, 167
   interface, 167, 168, 169
   mutant cells in, 132, 158–159
   population dynamics, 161–169
   pure culture preparation, 127

vertical sections, 133
Bacterial differentiation, 150–151, 154
Bacterial serine protease A, roughness of surface, 94
Bacterial strains, 128
Bacteriochlorophy 1-A-protein, 74
Ballistic aggregation clusters, 113, 209, 329
Ball-rolling algorithm, 66
Baroreflex, 254–255
Barr, S., 2
Base pairs. *See* DNA sequences
Bence-Jones protein, 69
Benzene group, 114
Berger, 272
Bergmann glial cells, 182
Bernoulli sequence, 303
Bicarbonate ion, 31
Bifurcation points of biochemical reactions, 88
Binarization, 174
Bins, 240
Biochemical attractors, and metabolite regulation, 88–89, 90
Biophysical Society, 39
Blastomeres, 190
Blobs (correlation volume), 64
Blood flow, 184, 250. *See also* Cardiovascular system
Blood pressure, 254–255
Boltzmann constant, 71, 81
Boltzmann energy factor, 34
Bond angle, 112, 113
Bonding parameter, 116, 117, 118
Bond lengths, 115
Border. *See* Perimeter
Bovine proteins trypsin inhibitor (BPTI), 69
Box-counting dimension, 61, 92, 278
Box-counting method, 143
  with chimeras, 198
  for colony pattern, 144
  interbeat interval time series, 258–259
  overview, 174, 175, 176, 230, 319
  for retinal occlusion, 220
  for tumor shape, 215
Boxes scales, 59–60
BPTI (bovine proteins trypsin inhibitor), 69
Brain Atlas III, 292, 293
Brain structure, 270
Brain wave activity. *See* Electroencephalogram; Electroencephalogram/event-related potentials
Branched polymers, 65–66
Branching
  bacterial colonies
    microscopic generating process, 152–155
    repulsion effect, 157–158
    screening effect, 157
  cardiac muscle, 250, 251, 252, 253
  dendritic. *See* Dendritic branching
  dense-branching morphology, 162
  polymer molecules, 65–66
  vasculature, 218
Bromopropene, 114

Bronchioles, 113, 329
Brownian motion, 83, 144, 250, 325–326, 328
  one-dimensional curve of, 145–146, 147, 148, 149, 209, 327
  and surface diffusion, 99
Brownian noise, 259, 261
Brown noise, 259, 263, 264. *See also* Noise
Bruxism, 224
Buffon's needle, 228

C

Calcium ion channels, 49–50
Calixarene, 114
Cancer, 215–217
  cell morphology, 210–211, 212, 213
  classification of malignant cells, 211–212, 217
  diagnosis of, 205, 212
  tumors shape, 212, 214, 215–217
Cantor dust, 7
Cantor set, 7, 11, 324, 328, 329
Carbonic anhydrase B, 74, 80
Carbon monoxide, 34
Carboxypeptidase A, 74, 78, 80
  active site probability distribution, 101
  roughness of surface, 94
  surface mass exponent, 98
Cardiac death, 263
Cardiac muscle, branching of, 10, 250, 251, 252, 253
Cardiovascular system, 10
  fractals and the, 249–264
  self-similarity of, 249
Caton, 272
Cell area, vs. time, 181
Cell death, 195, 203
Cell membrane, voltage across, 31
Cell morphology. *See* Bacterial cells, morphogenesis; Bacterial colonies, diversity of morphology; Cells, malignant
Cell movement, 195
  swarming, 129, 149–151, 158
Cell proliferation, 9, 195
  markers, 205
  orientation during, 198, 199, 201
Cells, 127–185
  malignant, 210, 213, 237. *See also* Tumors, shape
    classification of, 211–212, 214, 217
    morphology of, 210–211, 212, 213
    pattern formation mechanics, 193
    premalignant, 210, 215
Cellular automata (CA), 219, 221, 222, 223
Cellular replication models, 233
Censored intercept method, 228
Central nervous system (CNS), 9, 182, 274
Central vein occlusion, 218, 219, 220
Chaos
  definition, 51–52, 88, 264
  in proteins, 58
  theoretical analysis for biochemical reactions, 88–90, 119
Chaotic attractor, 271, 276, 278
Chaotic time series, 290

Index

Charge-coupled device (CCD), 149, 152
Chemical factors, in bacterial culture, 160–161
Chemical kinetics, 102–104
Chemoluminescence, 34
Chemotaxis, 193
Cheyne-Stokes respiration, 257, 260
Chimeras
    box-counting method, 198
    conservation, 194
    definition, 193–194
    mosaic pattern in tissues, 189–203
Chiral, 114
Chloride ion channel, 31, 48
Chordae tendineae, 251
Churchill's Illustrated Medical Dictionary, 205
Chymotrypsin, 74, 78
    active site probability distribution, 101
    multifractal spectrum, 97
    roughness of surface, 94
    surface mass exponent, 98
Circle
    departure from circularity, 210
    as structural microenvironment of reaction, 114
Citrate synthase, 87
Clostridium tetani, colony morphology, 134, 136, 138
Cluster, 340
Cluster aggregates, 113, 209, 230, 231, 329
CNS (central nervous system), 9, 182, 274
Cobalt, 115, 117, 118
α-Cobtratoxin, roughness of surface, 94
Coding regions of DNA sequence, 7, 19–21, 23–26
    finder algorithm, 19, 20
    linguistic analysis of, 23–26
    patch size, 19, 20
    percent of total genome length, 16
    "words" in, 24
Coding sequence finder (CSF), 19, 20
Cognitive activity, 269, 270, 291, 298–299
    allocation of tasks during, 303–307
    mental workload, 302–303
Cognitive attractor, 290
Cognitive variability, 271
Colored noise, 291, 292. *See also* Noise
Complexity, 173, 269. *See also* Scaling
    dimensional, 290
    of EEG time series, 298
Concanavalin A, 74, 80
Concentration of polymer solution, 63–65
Confocal microscopy of mosaic pattern, 203
Conformational entropy
    calculation of, 80–82
    of protein chain, 78–84
    relationship between, and fractal dimension, 82–84
Conformational state change, 7–8. *See also* Activation energy barriers; Allosteric effects
    discrete, in Markov model, 36, 37
    energy source for, 31
    ligand binding, 7
    models of, 104–105
    multistep, 108–110
    one-step, 106–108
    stable, local minima as, 33
    unknowns of the, 32
Congestive heart failure, 249, 263, 264
Connectivity, of polymeric chains, 62
Conservation, in chimeras, 194
Construction principle of fractals, 112
Contact surface method, 98
Control parameters, 89
Cooperativity, 104, 111
Coordinate skipping method, 228
Coral colonies, 113, 329
Coriolis force, 159–160
Cornea, 7, 9, 31, 218
    herpes simplex virus ulcers, 218–219, 221, 222–223, 230
Corneal endothelial cells, 31, 39–40, 41, 42
Corneal endothelium
    anatomy, 31
    ion channels in the, 7, 31–54
Coronary artery tree, 250, 251
Correlation dimension, 84–88, 119, 279, 281, 282, 298, 308
Correlation function, 280, 281, 293, 296, 298
    power-law form of, 231–232, 279
    scaling law for the, 286
Correlation volumes (blobs), 64
Corrugation, surface, 62, 90, 93
Cortical pyramidal neurons, 182
Corticosteroids, 219, 221
Covariance matrix, 283, 284, 285
Critical point, 16, 249
    fluctuations in heartbeat, 27
    and nucleotide content, 17
Crossover scale, 261, 262, 263, 328
Crowding index, 59, 333
Crown ether, 114
Crystal growth, dendritic, 144, 145
CSF (coding sequence finder), 19, 20
Cultivation temperature, 129
Culture conditions, 128–132
    genetic approach, 158–159
    identification of chemical factors, 160–161
    nutrients, 155–157
    physico-chemical factors, 155
    for pure culture, 127
    spatial arrangement, 157–158
    substrate construction, 158
    terrestrial factor, 159–160, 162
Cyclodepsipeptides, 160
Cyclodextrin, 113, 114
Cytochrome, hybrid orbitals of, 113
Cytochrome B_5, 74, 80
Cytochrome C, 74, 80
Cytochrome C_3, roughness of surface, 94
Cytosine, 16, 17

D

Data evaluation, 340–342
DBM (dense-branching morphology), 162

Debye-Waller factor, 35, 67
Degree of freedom, 180
Degree of learning, 268
Dendritic branching
  bacterial colony morphology, 149, 150
  crystal growth, 144, 145
  herpes simplex virus, 219
  microcropic generating process, 150–151
  neurons, 179, 185
  total length vs. time, 181
Dendritic ulcers, 219
Dense-branching morphology (DBM), 162
Density correlations, 323, 340
Deoxyribose sugar group, 16
Determinist dynamical system, 85, 86–87, 264, 309
Deterministic chaos, 88, 264, 292
Deterministic fractals, 322–323, 330
Detrended fluctuation analysis (DFA), 15, 18–19, 26
  of interbeat interval time series, 254, 258–259, 262, 263
Diabetic proliferative retinopathy, 218
Diagnosis, 205, 212, 215
Differentiation
  bacterial, 150–151, 154
  definition, 180
  mammalian embryonic development, 189–191
  markers, 205
  neurons and glial cells, 179–181
Diffusion
  of bacterial cells, 167
  in neovascularization, 218
  in pattern formation, 191
Diffusion-limited aggregation (DLA), 8–9, 209, 225, 317–318
  long-range correlation in, 20–21
  as neuronal growth mechanism, 178
  in radius of gyration method, 143, 146, 157
  in retinal vasculature, 217–218
  simulation vs. bacterial colony morphology, 156–157, 158, 168
  surface diffusion on, 99–100
Digital imaging, 39, 40, 239, 252, 339
Digital processing, 267
Dihydrofolate, 69
Dihydrofolate reductase, 74
Dilation method, 174, 175, 176, 198, 228–230
Dimension. *See* Fractal dimension; Topological dimension
Dimensional complexity, 290
Dimorphic transition, 150, 154
Dirac delta function, 66
Discretization of data, 278
Displacement, DNA walk, 17, 18
Distribution of kinetic rate constants, 44–46, 47, 48, 49
Disulfide bonds, 58
DLA. *See* Diffusion-limited aggregation
DNA coiling, 225
DNA sequences, 16–17
  coding and noncoding, 7, 15–27
  linguistic analysis, 23–26
  mosaic structure, 23
  patchiness, 20
  repetitive, 23
  as source of bacterial colony heterogeneity, 132–133
  uncorrelated, 21, 23
DNA walk, 17, 18
DSM III-R criterion, 292
Dynamical interpretation, ion channel kinetics, 43, 44, 50–51
Dynamical measures, 289
Dysplasia, 215, 216, 217

E

ECTI (epithelial-connective tissue interface), 215, 216, 217
Ectoderm cells, 190, 191
Eden-like pattern, 156–157
EEG. *See* Electroencephalogram; Signal (EEG)
Effective kinetic rate constant, 36, 40–43, 50, 53
Effective medium algorithm (EMA), 62, 71
Egg, during fertilization, 189
Eigenvector, 184, 283, 285
Einstein's theory of Brownian motion. *See* Brownian motion
Elastase, 74, 80, 87
Elastomer, 69
Electrocap, 292, 293
Electrocardiogram, 250, 252, 260
Electroencephalogram (EEG)
  distribution approach, 294–298
    and MART, 295–298
  estimation of Kolmogorov entropy, 87–88, 119
  GPA applied to, 292, 293–294, 296–298
  overview, 272–273
  processing brain-wave data, 290–298, 299, 300, 301, 302
  recordings
    epileptic seizure, 291
    sleep, 291
  time series, 292–294
    complexity of, 298
    histogram, 295, 299, 301, 302
Electroencephalogram/event-related potentials (EEG/ERP), 267–311
  and learning, 270
  measures of, 10, 268–269
  and performance, 270–271
  summary and future research, 309, 311
Electroencephalographic signal. *See* Signal (EEG)
Electronic information parameters, 118
Electron spin relaxation rate, 76
Electron spin resonance (ESR), fracton dimensions of proteins, 71
Electrostatic forces, in channel proteins, 52
EMA (effective medium algorithm), 62, 71
Embedding dimension, 85, 86, 318–319, 321
Embryonic development, 9, 189–191
  models of organogenesis, 197–200
Enantiomer, 129
Endoderm cells, 190, 191

Index

Energy barriers. *See* Activation energy barriers
Energy minima, 33
Entanglement
  in branched polymers, 65
  in polymeric chains, 62–63
Entropy
  in multifractal analysis, 217
  in short-range correlations in language, 24
Enzymatic kinetics, 8
  fractal approach for, 102–111
  and protein conformation, 57–119
  rate of reaction, 99–100
Enzyme catalysis, 119
  stereospecificity of, 114
  structural parameters, 114–118
Enzyme model design, 111–118, 119
Enzymes
  allosteric effects of, 104–111
  f-α spectrum, 96, 97, 119
  during fertilization, 189
  surfaces, fractal reactions of, 99–102
Epiblast, 190
Epidemics, dynamics of, 267
Epilepsy, 299
Epileptic seizures, 263, 267
  fractal dimension of EEG/ERP during, 269
  petit mal, 290
Epithelial-connective tissue interface (ECTI), 215, 216, 217
Epithelial dysplasia, 215, 216, 217
Equally spaced test-lines method, 228
Error-correction mechanism, in mosaic pattern formation, 195
Escherichia coli, 74
  colony morphology, similarity to DLA model, 168
  swarming, 129, 151
Esophageal carcinoma, 212
ESR (electron spin resonance), 71
Estradiol-porphyrin, 115
Estrone-porphyrin, 115
Ethinyl estradiol-porphyrin, 115
Euclidean geometry, 4, 11, 173, 206, 207, 240
Eugene Onegin, 36
Evaluation of numerical data, 340–342
Event-related potentials (ERP)
  average, 302–307
  EEG/ERP, 267–311
  new technique for, 307–309
  point estimate of fractal dimension, 307
Evoked potential, 300–301
Evolution
  adaptive fractal behavior of heart rate dynamics, 263
  of bacteria, 128, 129
  genome complexity and, 21, 26
Exocrine glands, 225
Exons, 17. *See also* Coding regions of DNA sequence
Expansive outline, 212
Expert systems, histopathological diagnosis, 9

Exponential decay. *See* Power-law decay

F

Falconer, K., 127
f-α spectrum, 96, 97, 119
Fast Fourier transform (FFT), 15, 233–234, 256
  of coding and noncoding sequences, 20, 26
Fat fractals, 8, 58–59, 329–330
  examples in nature of, 113
  and hybrid orbitals, 113
  lack of self-similarity, 329
  proteins as, 91–93, 95, 98
FB. *See* Fractional Brownian surfaces
Ferredoxin, 76, 80
Ferric iron, low-spin, 76
Fertilization, 189–190
Fetal distress syndrome, 263
FFT. *See* Fast Fourier transform
Fineness of scale, 79
Finite Lebesque measure, 58
Fisher equation, 165, 166, 167
Fixed point, 278
Flagellated bacteria, swarming, 149–151
Flavodoxin, 69, 74, 80
Flory theory, 63, 67
Fluorescein angiograms, 218, 219
Fluorescence energy transfer method, 105
Form factor, 210, 211
Fourier transform. *See* Fast Fourier transform
Fractal analysis
  of ion channel kinetics, 38–42
    physical interpretation of, 42–51
    summary of, 53–54
    vs. Markov approach, 52–53
  of proteins, 57–58
    chain conformation, 71–84
    chain structure, 62–68
    review of research, 105
  of tumors, sample location, 215
Fractal dimension. *See also* Fractal fragmentation; Power-law
  of active site distribution on enzyme surface, 101
  confidence interval for, 272
  correlation between, and chain length, 75–80
  determination of, 226–242
    in chimeras, 198–199
    length method, 174–176
    mass method, 176–177
  of EEG, 272, 292, 295, 299
    during epileptic seizure, 269, 290–291
  empirical nature, 174
  of ERP
    new technique for, 307–309
    point esimate of, 307
  generalized, 60, 334
  of glial cells, 178, 179–181
  as indicator of learning, 268. *See also* Electroencephalogram/event-related potentials
  and learning, 270
  local connected, 217, 218, 220, 236–237

mass. *See* Mass fractal dimension
  of metals, 115
  and morphology
    of branching bacterial colony, 156
    factors affecting, 177
    of neurons and glial cells, 177–181, 182, 184
    studies, 178–183, 184–185
  of neurons, 179–181
  overview, 6, 41, 57, 140, 141–143, 173–177, 207, 317–322
  and performance, 270–271
  of a planar curve, 76
  of proteins, 74–75, 78, 79–80, 119
    enzyme islands, 101
    surfaces, 90–99
  range of, 174, 240
  reduction of, 238, 239, 321
  relationship between, and conformational entropy, 82–84
  of retinal vasculature, 218
  of steroid-metalloporphyrin, 115–116
  of tumor shape, 215, 216, 217
  types, 61
  vs. L-systems growth, 181
  vs. time, 181
Fractal dimension of the walk, 61
Fractal fragmentation, 199, 200, 201, 202, 209, 232–233
Fractal geometry
  advanced topics in, 317–342
  application in pathology, 205–242
  atomic vibration, 69
  and cancer, 215–217
    cell morphology, 210–211, 212, 213
    tumors shape, 212, 214–215
  as "excess" from topological geometry, 4
  of the heart, 250–251, 263. *See also* Interbeat interval time series
  limits to biological application, 57, 239–242
  overview, 3–11, 57, 136, 139–148, 173–177, 269
Fractal growth, 180–181
Fractal harmonics, 236
Fractal landscape, 17, 252. *See also* Roughness
Fractal measure, 330, 331–332
Fractal object, 269
Fractal probability density
  as confidence interval for fractal dimension, 272
  and EEG/ERP, 267–311
  invariant, 286–287
Fractals. *See also* Self-affinity; Self-similarity
  asymptotic, 211, 214, 241
  in cardiovascular physiology, 250
  deterministic and random, 322–323
  and enzyme model design, 112
  fat. *See* Fat fractals
  growing, 319–321, 338
  of hydrid orbitals, 112–114
  ion channels as, 39
  polymeric, 62, 67
  principles, 112
  projection onto a plane, 239, 321
  properties of, 38–39
  random, 209
  self-exact, 206, 207
  texture. *See* Lacunarity; Succolarity
Fractal volume, 4
Fractional Brownian motion, 145–146, 234, 235, 239
Fractional Brownian noises, 234
Fractional Brownian surfaces (FB), 91, 93, 145–146, 147
Fracton dimension (spectral), 58, 61–62
  of proteins, 68–71, 109
Fracture of materials, 233
Freeman model, 269
Frequency resolution, of EEG signal processing, 279, 286
Fructose-6-phosphate, 89
Funnel attractor, 276

G

Gait, human, 27
Gas exchange, 250
Gastrulation, 192
Gaussian statistics, 38, 101
  interbeat increment, 256
  and invariant probability density, 287, 288, 289
  polymer chain, 66
  white noise, 235, 259, 263, 264, 284
Gelation, 225
Gelation/percolation threshold, 65
GenBank database, 15
  redundancy analysis of, 24–25
  systematic analysis of, 19–21, 26
Generalized dimensions, 60, 334
Generator, 5, 112, 139, 195
  determination of morphology, 112, 143
  structure skeleton as, 113
Genetic approach to bacterial culture morphology, 158–159
Genome. *See* DNA sequences
Genotype nomenclature, of bacterial strains, 128
Gibrat law, 26
Gleick, James, 15, 205
Glial cells, 9
  Bergmann, 182
  fractal dimensions of, 178, 179–181
  fractal studies of morphology, 182
  function, 185
  growth and differentiation, 179–181
  surface-to-volume ratio, 182
Global dimension, 328
Global relaxation models, 66–67
Globular proteins, 33–35, 52, 84–88, 119
Glucose, in bacterial media, 129
d-Glucose-6-phosphate isomerase, 75
Glyceraldehyde-3-phosphate dehydrogenase, 69, 71, 74, 80
  allosteric constants, 111
Glycine, 58, 68
Glycolysis, 88, 89, 90
Golgi epithelial cells, 182
GPA. *See* Grassberger-Procaccia algorithm
GRAIL neural net algorithm, 19

Grassberger-Procaccia algorithm (GPA), 84–86, 87, 271–272, 278, 279–283
  applied to EEG, 292, 293–294, 296–298
Grey scale, 174, 206, 236, 239, 240
Grid method. *See* Box-counting method
Guanine, 16, 17
Guinier regime, 340

H

Hamiltonian system, 275, 329
Haplotaxis, 193
Harmonic analysis, 273
Harrison, L.G., 189
Hausdorff, Felix, 5
Hausdorff-Besicovitch dimension, 194
Hausdorff dimension, 91, 279
  of Cantor dust, 7
  equation, 6
  of Koch snowflake, 6
  of polymeric chains, 67–68
  of Sierpinski gasket, 6
Heart. *See also* Cardiovascular system; Heartbeat
  cardiac muscle, 10
  fractals and the, 249–264
Heartbeat, 254
  beat-to-beat regulation, 249, 260
  dynamics of, 251–259
  healthy vs. diseased, 10, 27, 255, 256, 259, 260
    time series, 259–263
  heart rate dynamics, 263–264
Heart failure, congestive, 249, 263, 264
Heart valve, 10
Heaviside function, 85
α-Helices, 58, 73
Hemagglutinin, 87
Heme, 76
Hemerythrin, 74, 80
Hemoglobin, 69
  allosteric constants, 111
  carbon monoxide binding curve, 105
  conformational entropy, 80
  fractal dimension of, 74, 78, 80
  hybrid orbitals, 113
  oxygen binding curve, 104, 105
  roughness of surface, 94
  surface mass exponent, 98
Hemoproteins, 78
Hen egg-white. *See* Lysozyme
Herpes simplex virus (HSV), 9
  corneal ulcers, 218–219, 221, 222–223
  surface and mass fractal dimensions of, 230
Herpetic gingivostomatitis, 218
Hexokinase A, 75, 80
Hierarchical structure, 336–337
Hill coefficient, 8
  and allosteric effects of proteins and enzymes, 104, 105, 110
  and fracton dimension, 71, 76
  nonconstancy of, 111
  relationship with fractal dimension, 108

Hill equation, oxygen-binding curves of hemoglobin, 104
Hill-Wyman method, 241
His-Purkinje geometry, 250, 253
Histopathology
  current problems in, 205–206
  definition, 205
  uniformity in diagnosis, 9
Hölder exponent, 59, 333
Homeostasis of physiological processes, 263
Human beta globin chromosome (HUMHBB), 18, 21
Human Genome Project, 23
HUMRETBLAS, 24
Hurst analysis, 258
Hurst exponent, 145
Hurst's rescaled range analysis, 235
Hybrid orbitals, 112–114
Hydrogen bonds
  and fractal quality of proteins, 82
  and fracton dimension, 71
Hypercubic lattice, 59, 65, 319
Hyperscaling law, 65
Hypoblast, 190

I

Image digitalization. *See* Digital imaging
Immunoglobulin, 74, 80, 94
Inclusion principle of fractals, 112
Increment of interbeat interval. *See* Interbeat interval time series, increment of
Incubation time, of bacterial colony, 130, 132
Induced fit, 35
Infiltrative outline, 212
Information dimension, 279, 335
Information parameters, 117–118
  local vs. global, 286, 327, 328
Information processing, 302–307
Initiator, 5, 139
Inner cell mass, 190
Instrument flight rule, 306
Interbeat interval time series
  detrended fluctuation analysis, 254, 258–259, 262, 263
  in healthy vs. diseased heart, 259–263, 260
  increment of, 254–257
  long-range power-law correlations, 252
  power spectrum for the, 257
  scaling, 10, 252
Interpoint distance distribution, 287–288, 293, 296
Interpolated yardstick method, 228
Intersection, of two fractals, 322
Introns, 17, 18
Invariant probability density, 286–287
Invasion percolation, 225
Invertebrates, 26
Ion channel kinetics, 31–54. *See also* Conformational state change; Markov model
  fractal analysis of, 38–42
    physical interpretation of, 42–51

and memory, 41
open vs. closed probablities, 39–42, 45–51, 53
Ion channels, 31, 32. *See also* Conformational state change
   current in, 32, 39
   energy structure of, 33
   fractal form of, 39
   mRNA of, 32
   unknowns of, 32–33
Iron, 115, 117, 118
Iron protein, oxidized high potential, 74
Iron, proteins containing, 76
Isolates, bacterial strains, 128
Iteration procedure, 224, 319–320. *See also* Initiator

K

Karhunen-Loéve theorem, 283
Kinetic rate constant, 36, 40–43, 50, 53
   distribution, 44–46, 47, 48, 49
Kinetic theory of pattern formation, 192–193
KNF (Koshland-Nemethy-Filmer model), 104, 105
Koch, Robert, 127
Koch curve, 112, 139–140, 141, 195–197, 198, 207, 208, 226
Koch snowflake, 5–6, 114, 180
Koch triadic island, 176, 177, 180
Kolmogorov entropy, 85, 86, 87–88, 119
Korčak's empirical law, 232–233
Koshland-Nemethy-Filmer model, 104, 105

L

L-arabinose-binding protein, 74
Lacunarity, 183–184, 218, 238
Langmuir-Hinshelwood reaction, 100
Languages, linguistic analysis, 23–26
Laplace diffusion equation, 217, 330
Laplace transform, 34, 44
Laplacian field, 156, 157
Laryngeal carcinoma, 212
Lattice
   hypercubic, 59, 65, 319
   semilattice, 97
   square, 102
Learning
   degree of, 268
   fractal dimensions of, 270, 291
Legendre transforms, 335
Length methods, 3–4, 174–176, 226–230
Letter Pair Test, 293, 294
Lévy stable distribution, 255–256
Lévy walk, 15, 18
   classic, 21, 22
   generalized, 18, 21–22
   and mosaic nature of DNA structure, 23
Liesegang-ring-like morphogenesis, 159
Ligand binding, 7, 31
   effect on scaling, 49, 50
   and protein structure, 35
   and protein surface roughness, 62, 97–98
   reaction rate, 34
Ligand diffusion, and protein surface roughness, 62
Ligands
   and activation energy barrier distribution, 49–50
   macrocyclic, 114
Limit cycle, 278
Limit ring, 89
Limit ring surface, 89
Lindenmayer, Aristid, 8
Lindenmayer systems. *See* L-systems growth
Linear polymers, 63–64, 65, 67
Lineweaver-Burk plot, 241
Linguistic analysis, 23–26
Liver
   chimeras of the, 194, 195, 197, 203
   patches within, 201
   regeneration, 200
Local connected fractal dimension, 217, 218, 220, 236–237
Local slope. *See* Log-log plots, slope of
Log-logistic approach, 242
Log-log plots
   data points
      equally spaced, 240
      gaps in. *See* Lacunarity
   distributions of kinetic rates, 46, 47, 48
   effective kinetic rate constant of ion channel, 41, 42
   slope of, 4, 72, 175, 177, 240, 341
Long-range power-law correlations, 15–16, 23, 250
   of coding and noncoding sequences, 18–21, 26, 249
   of interbeat interval time series, 249, 252, 256, 259, 260–261, 264
Lorenz attractor, 280, 281, 285, 289
L-systems growth, 8–9, 178, 180, 181, 225
Lung
   bronchioles, 113, 329
   flow of air in, 184
   gas exchange, 225
   pulmonary function, 250
Lyapunov exponent, 277, 282
Lysozyme, 69, 74, 78, 80
   active site probability distribution, 101
   multifractal spectrum, 97
   roughness of surface, 94
   surface mass exponent, 98

M

Macrocyclic ligand, 114
Macromolecules, 57
Magnetic resonance imaging (MRI), of mosaic pattern, 203
Malthusian growth, 165
Mammalian development, 189–191
Mandelbrot, B.B., 10, 57, 317

Manganese, 115, 117, 118
Markov chain, 17, 249
Markov model, 7, 36–38
 power-law distributions, 46, 49
 vs. fractal ion channel kinetics, 52–53
MART. *See* Multichannel attractor reconstruction technique
Mass behavior, 153
Mass distributions, 183–184
Mass exponents
 by aggregation model, 98
 for protein surfaces, 95–99
 by simulation, 97
Mass fractal dimension, 67–68, 199, 200, 201, 202, 230
Mass method, of fractal dimension determination, 176–177
Mass multifractality, 338–339
Mass-radius method, 231, 232, 238
Match principle of fractals, 112
Mathematical forms, 10
Matushita model, 97
Maximum entropy method, 236
Maximum likelihood indicator, 236
Maximum log-likelihood function, 295
Measurement, scale of. *See* Scaling
Mechanochemical model of pattern formation, 193
 effect of errors on pattern, 203
 vs. reaction-diffusion model, 195
Media. *See* Culture conditions
Memory, 267
Menger sponge, 7, 11
Mental workload, 302–303
Mesodermal cells, 191
Metabolite regulation
 and biochemical attractors, 88–89
 switching, 90
Metalloporphyrin, 113, 114
 steroid-, 115–116
Metals, fractal dimensions of, 115
Methylphenidate, 268, 292
Mexican Hat kernel, 174
Micelles, 114
Michaelis-Menten equation, 104
Micrococcus luteus, colony morphology, 134
Minkowski-Bouligand dimension, 91–92
Minkowski sausage, 91, 92, 228–230
Molecular dynamics, 33–34
Molecular topological indices, 116–117
Molecules, 15–119
Moments, 94
 of mass distribution, 183–184
 q, 237, 333–334
 of stable distributions, 38
 statistical, 238
Monod-Wyman-Changeux model (MWC), 104–105
Monte Carlo method
 with branched polymers, 65
 calculation of conformational entropy, 81, 82, 83
 mass exponents, 96
 and protein chain conformations, 76, 79, 119

Rosenbluth Monte Carlo Method, 23
s-p enrichment, 95
 in surface phenomena, 99, 100
Morphogen
 gradient, 191–192
 in reaction-diffusion theory, 192–193
Morphogenesis, by bacterial cells, 127–169. *See also* Bacterial colonies
Morphogenic unit, 149
Morphology, mathematical, 205. *See also* Bacterial colonies, diversity of morphology; Cells, malignant
Morula, 190
Mosaic pattern, 193–197
 in chimeras, 200–203
 computer simulation of, 197, 199
 conservation and regulation, 194, 201, 203
 error-correction mechanism, 195
 future studies, 203
 patch oscillation, 198, 200, 201
 in tissues from chimeras, 189–203
Multichannel attractor reconstruction technique (MART), 272, 285–286
 applied to EEG, 295–298, 300, 301, 303, 308, 309
Multichannel embedding, 295, 303
Multifractal analysis, 8, 9, 119
Multifractal formalism, 333–335
Multifractals, 59–61, 236–237, 322, 330–339
 mass multifractality, 338–339
 of protein surfaces, 93–95, 97, 119
 recursive, 335–338
 thermodynamic properties of proteins, 84, 119
 of tumor shape, 215, 216, 217
Multifractal spectrum, 97
Multifractility, definition, 59
Multiplication principle of fractals, 112
Multistep conformational change, 108–110
Mutual information, 283
MWC (Monod-Wyman-Changeux model), 104–105
Myocardial infarction, 250
Myocardium, 250, 253
Myofilaments, 251, 253
Myoglobin
 and absorption of blue light, 34
 conformational entropy, 80
 fractal dimension of, 69, 74, 78, 80
 hybrid orbitals, 113
 mass dimension of, 68
 modal densities, 70
 molecular dynamics, 33–34
 roughness of surface, 94
Myohemerythrin, 74
Myosin, 251, 253

N

NADH, 89
Naphthalene group, 114
Natural language, 24
Neocortical astroglial cells, 182

Neoplastic growth, 9. See also Cells, malignant; Tumors shape
Nerve cells. See Central nervous system
Neural network, 52. See also Neurons
Neural physiology, 290
Neural response, 302
Neurite extensions, 8
Neuroblastoma, 47, 49
Neurons. See also Electroencephalogram
    cortical pyramidal, 182
    fractal dimensions of, 179–181
    function, 9
    growth and differentiation, 8, 179–181
    interspike interval distributions in, 267
    morphology, 176, 177, 178, 184
    physiology, 184–185, 290
Nitrogen, 117
NMR. See Nuclear magnetic resonance
Noise, 268
    brown, 259, 263, 264
    Brownian, 259, 261
    colored, 291, 292
    in EEGs, 273
    $1/f$, 224, 259, 260, 273
    fractional Brownian, 234
    in the Grassberger-Procaccia algorithm, 279
    in heartbeat interval fluctuations, 258
    in re-embedding technique, 285
    separation from signal, 272
    in singular value decomposition, 284–285
    white, in Gaussian form, 235, 259, 263, 264
Nomenclature, of bacterial strains, 128
Noncoding regions of DNA sequence, 7, 15, 21–27
    linguistic analysis of, 23–26
    outlook for the future, 26–27
    words in, 24
Nonstationary processes, of human heartbeat, 254, 257, 258
Nuclear magnetic resonance (NMR)
    measurement of motions, 35
    protein ion channel labelling, 53
    protons vs. deuterons, 34
Nuclear pleomorphism, 210
Nucleotide distance, 18
Nucleotides, 16. See also DNA sequences
Nutrient diffusion field, 156
Nutrients, in bacterial cultures, 155–157

O

$1/f$ noise, 224, 259, 260, 273
One-step conformational change, 106–108
Oral carcinoma, 212, 214, 215–217
Oral hygiene, 224
Orbitals, hybrid, 112–114
Organogenesis, 191, 197–200
Organs, 249–311
Orthogonality, 112, 113
Orthonormal functions, 283
Osmotic pressure, 63–65
Ovomucoid, 69

Oxidized high potential iron protein, 74
Oxygen, 34

P

Pair correlation function, 231–232
Papain, 74, 80
Parasympathetic stimulation, 253, 260
Parenchyma proliferation, 9
Particle, 340
Patch clamp, 31–32, 37, 39, 50, 53
Patch oscillation, in mosaic pattern, 198, 200, 201
Pathology, applications of fractal geometry in, 205–242
Pattern formation
    definition, 193
    mechanisms of, 191–193
Peano curve, 113, 324, 325
Peitgen, H.-O., 31
PEP, 90
Peptide bonds, 58
Peptone, 129, 131, 168
Perbas, 210
Percolation
    /gelation threshold, 65
    of herpes simplex virus, 225
    invasion, 225
    model, 65
    self-avoiding invasion, 218
    theories of, 9
Percolation clusters, 65, 103
Performance, fractal dimensions and, 270–271
Perimeter
    area-perimeter relation, 230–231
    length
        increase by generator, 5–6
        nuclear, 210, 211
    roughness of, 177
    tumor outlines, 212, 214–215
Perimeter trace method, 175, 176
Periodontal disease, 9–10, 221–222, 224, 225
Peroxidase reaction, 88
Persistent data, 235, 236
Petit mal epilepsy, 290
Phase diagram, pattern change in bacterial colonies, 156, 157
Phase space, 85
Phenotype nomenclature, of bacterial strains, 128
Phonon, 70, 76
Phosphofructokinase, 89
Phosphoglycerate kinase, 75, 80, 87
Physical properties, of fractal objects, 340
Physico-chemical factors, of bacterial medium, 155–157, 158, 159–161
Plastocyanin, 69
Plateau, 295, 297
β-Pleated sheets, 58, 73
    surface corrugation, 62, 90, 93
Plunk constant, 71
Poincaré surface of section, 277–278
Poisson distribution, 289
Polyelectrolytes, 114

Index

Polygon, as initiator, 5
Polygonal harmonics, 236
Polyisoprene, 65, 67
Polymeric fractal, 62
Polymer molecules, 62
  branched, 65–66
  enzyme catalysis of cyclodextrin, 114
  fractal aspects of, 63–65
  linear, 63–64, 65, 67
  proteins as "collapsed", 66
  solution concentration, 63–65
  star, 64–65
Polynucleotide chain, 16
Polypeptide, structure, 58
Polystyrene, 63–64
Population dynamics, of bacterial colonies, 161–169
Porod law, 340
Positional informational model, 191, 192
Potassium ion channel, 31, 39, 40, 47, 49
Potential energy functions, 33
Power-law, 17–18, 40–41. *See also* Fractal dimension; Scaling function
  and correlation function, 231–232, 279
  definition, 57–58, 206–207
  distributions, 46–47
  and heartbeat fluctuation, 27
Power-law correlations. *See* Long-range power-law correlations
Power-law decay, 16, 24, 321, 341
Power-law scaling, 46–51
Power spectrum, 233, 234, 240, 257, 283
Prealbumin, 74, 80, 94
Prefractals, 180
Premalignant stage of cancer, 210, 215
Primitive streak, 191
Principle component analysis, 184, 283–284
Probability density. *See* Fractal probability density
Procaccia algorithm. *See* Grassberger-Procaccia algorithm
Product release, and protein surface roughness, 62
Projection of fractal onto a plane, 239, 321
Protein conformation
  and enzymatic kinetics, 57–119
  fractal analysis, 71–84
Protein Data Bank, 101
Proteins. *See also* Ion channel kinetics; Polymer molecules
  allosteric effects of, 104–111
  four classes of, 73, 75
  globular
    correlative dimension for, 84–88, 119
    energy structure of, 33–35
    production of, 52
  ion channel. *See* Ion channels
  motions between, 33–34
  properties, 35–36
    and physical interpretation of the Markov model, 37–38
  structure, 58
    and conformational state, 104
    fractal analysis of chains, 62–68, 71–84, 119

  universal aspects of, 66–67
  surfaces, 8, 90–99
  thermodynamic properties, 84, 119
  tritium-loaded, 34
  vibration of, 68–69, 70, 119
Proteus mirabilis, 130–131
  colony morphology, 150, 152
  differentiation, 154
  flagella, 154
  growing branch tip, 155
  microcropic generating process, 149–151, 153
  morphogenesis, Liesegang-ring-like, 159
  swarming, 129
Psychomotor, 290
Pulmonary function, 250. *See also* Lung
Pure culture, preparation of, 127
Purine base, 16, 17
Pyrimidine base, 16, 17
Pyruvate kinase, 75, 89, 111

Q

q moments, 237, 333–334
QRS complex, 250
Quasi-periodic, 269
Quaternary structure of protein, 104

R

Radioactive decay, 34
Radius of gyration, 143–145, 231
  in active site distribution, 101
  of proteins, 66, 119
Raman electron spin relaxation rate, 76
Raman scattering, 105
Random fractals, 322–323
Randomness
  in EEG/ERP signal, 278
  mimicking of. *See* Chaos
  in natural systems, 225, 264
  predictability in, 11
Random pattern growth, 148–155
Random self-avoiding walks (RSAW), calculation of conformational entropy, 80, 81, 82
Random walk model, 17, 18, 21, 144, 250
  biased, 23
  in chemical kinetics, 103
  fractal correlations in physiological systems, 263
  of human writings, 23
  interbeat interval time series, 259, 261
  multistep conformational change, 108
  and spectral dimension, 61
  and surface diffusion, 99, 100
Reaction-diffusion theory, 192–193, 195, 202
  effect of errors on pattern, 203
Reaction kinetics, 99–100, 103, 110–111
Reaction rate
  chemical kinetics, 103
  and energy barrier, 34–35
  enzymatic kinetics, 99–100
  one-step conformational change, 106, 107–108
Recursion, 11, 224, 225, 331

Recursive multifractals, 335–338
Reduction of dimension, 238, 239, 321
Redundancy analysis, 24, 25
Re-embedding technique, 285
Reference edge, 191
Regulation, in chimeras, 194
Relative dispersion analysis, 236
Renyi entropy, 60, 86
Repetitive DNA sequences, 23
Repulsion effect, 157–158
Residue number, 86, 87, 96, 97
Resolution, of frequency, 279, 286
Resolution-dependent measurements, 210, 213, 228
Retina
   fractal dimension studies, 182–183
   photoreceptors in, 31
   vasculature, 217–218, 219, 220
Retinal vein occlusion, 219
Retinol binding protein, roughness of surface, 94
Retinopathy, 218
Retroviruses, 23
Rhodanese, 74
Ribonuclease A, 69, 74
Ribosomal protein, roughness of surface, 94
Richardson effect, 206, 210, 226
Richardson method, 174–175
Richter, P.H., 31
Ritalin (methylphenidate), 268, 292
Robustness
   of EEG signal, 273
   as evolutionary advantage of fractal systems, 225
   of fractal model of mosaic pattern, 195, 203
Root-mean-square fluctuation, 17, 234–235, 258
Rosenbluth Monte-Carlo method, 23
Rosenbluth-Rosenbluth weighting factor, 81
Rössler chaotic attractor, 276, 277
Roughness, 341
   of heartbeat time-scape, 252, 254, 259
   of perimeter, 177
   of protein surface, 62, 90, 93, 94, 97–98, 119
Roughness-length method, 236
Round colony, 133–134
Rubredoxin, 74
Rucker, R., 267

S

Salmonella
   anatum, 138
   colony morphology, 133, 135, 151
      similarity to DLA model, 168
   microscopic generating process, 152–153
   repulsion effect, 158
   swarming, 129, 151
   typhimurium, 135, 151, 152, 168
Salts, effect on bacterial colony morphogenesis, 129
Sandbox method, 232
SAS (solvent accessible surface), 98–99
Saturation embedding dimension, 85

SAW. *See* Self-avoiding walk (SAW)
Scale invariance, 139, 173, 323
   of curve of finite length, 147
   of DNA sequences, 15–27
   vs. self-similarity, 322
Scaling. *See also* Time scale
   anisotropic, 146, 147, 233
   crossover phenomena, 261, 262, 263
   determination of self-affine, 147–148
   and genome complexity with evolution, 21
   of heart interbeat interval, 10, 252, 260
      as adaptive behavior, 263
      as universal behavior, 262
   overview, 3–4, 7, 11, 38, 173, 206
   and protein chains, 72–73
   self-affine, 147–148, 149
   single exponential, 46, 48, 50, 51
   stretched exponential, 46
   testing for, 40–42
Scaling function, 40–41, 42, 43
Scaling law
   for correlation function in GPA, 286
   for survival probability in enzyme reactions, 109–110
Scaling region, 280, 287
Scaling relationships, 43–44, 48, 53, 66
Scattered colony, 133, 134
Scattering experiments, 339–340
Schizophrenia, 299
Scholl analyses, vs. time, 181
Schroeder, M., 173, 249
Screened-growth model, 97
Screening effect, 157
Sedimentation, 225
Segmentation, automatic, 217
Self-affine scaling exponents, 147–148, 149
Self-affinity, 57, 145–147, 207, 227, 233–234, 318, 326–329
   and dimension reduction, 238, 321
   interface between two bacterial colonies, 169
   of star polymers, 65
Self-avoiding invasion percolation, 218
Self-avoiding walk (SAW)
   of linear polymers, 63, 64, 67
   of proteins, 70
Self-exact fractals, 206, 207
Self-organized criticality, 16. *See also* Critical point
Self-similar dimension, 61
Self-similarity
   of branched polymers, 65
   of cardiovascular system, 249, 250, 253
   of cell types, 179
   connectivity of polymers, 62
   limitations to, in natural systems, 143–144, 175, 239–240
   number of parts to the reduction factor, 226
   overview, 4–5, 11, 38, 139–140, 206, 318, 319, 322, 323–326
   of proteins, 72, 82, 98–99
   statistical. *See* Scale invariance; Statistical self-similarity
   testing for, in ion channels, 39–40

of time scale, 10
of tumor shape, 215
Semilattice, 97
Semivariogram method, 236
Serine protease A, 94
Serratia marcescens, 129, 169
    colony morphology, 132, 135–138, 161, 162
        effect of serrawettin, 159, 163, 164
        effect of terrestrial factor, 160, 162
    growing tip, 161
    spreading growth, 152
Serrawettin, 159, 163, 164
    chemical structure, 160, 163
Shannon's redundancy, 24
Shape factor, 210
β-Sheets. *See* β-Pleated sheets
Shiff base, 114
Sholl analyses, 181
Sierpinski gasket, 6–7, 103, 108
    spectral dimension of, 61
Sierpinski pyramid, 4
Signal (EEG), 283, 285
    frequency resolution, 279, 286
    randomness, 278
    robustness, 273
    separation from noise, 272
    temporal resolution, 279
Single exponential scaling, 46, 48, 50, 51
Singularities, 59
Singularity strength, 60
Singular value decomposition, 271, 283–285
Size distribution, 232–233. *See also* Fractal fragmentation
Sleep, 69
Sliding box, 20
Slit island analysis, 230–231
Slope, of log-log plots. *See* Log-log plots, slope of
Sodium ion channel, 31, 49
Solvent accessible surface (SAS), 98–99
Spatial arrangement, in bacterial cultures, 157–158
Spectral density, 233
Spectral dimension. *See* Fracton dimension
Spectrum
    multifractal, 97
    power, 233, 234, 240, 257, 283
Spectrum, f-α, 96, 97, 119
Sperm, during fertilization, 189
Spin echo experiments, 54
Spurs, of bacterial colonies, 134, 135
Square lattice, 102
Stable distributions, 38–39
Staphylococcus epidermidis, colony morphology, 134
Star polymers, 64–65
State transitions, types of, 88–89
Stationarity, 272
Statistical self-similarity, 38, 40, 82. *See also* Self-similarity
    in human heartbeat dynamics, 251–252, 253
Stereology, 205
Stereospecificity, of enzyme catalysis, 114
Steroid-metalloporphyrin, 115–116
Stochastic phenomena, 225

Stretched exponential scaling, 46
Structural information index, 117–118
Structural interpretation, ion channel kinetics, 43, 44–50, 51
Structure factor, 340
Structure-function studies, 32
Substrate construction, 158
Substrate molecules
    circular microenvironment for reaction, 114
    concentration of, 102
    interaction of, 99, 107, 110–111
Subtilisin novo, 74
Succolarity, 238
Superoxide dismutase, 73, 74, 80, 97
Surface active exolipids, 159, 160
Surface corrugation, 62, 90, 93
    β-pleated sheets, 58, 73
Surface diffusion, 99–100, 101–102
Surface fractal dimension, 61, 66, 202, 227–228, 230
Surface tension, 143
Surface-to-volume ratio, 182
Surfactin, 159, 160
Survival probability, 106, 109
Survival rate, 41, 215
Swarming, 129, 149–151, 158
Switching
    in metabolite regulation, 90
    in proteins. *See* Conformational state change
Swollen conformation, in branched polymers, 65
Sympathetic stimulation, 253, 260

T

Takagi surfaces, 93
Takens, ART of, 275–278, 286
    an alternative to (MART), 285–286
    problems associated with, 289–290
Task workload, 302–303
Teeth
    grinding of, 224
    periodontal disease, 9–10, 221–222, 224, 225
Temperature. *See also* Thermodynamics; Vibration, atomic
    of cultivation, 129
    and fracton dimension of proteins, 70–71, 76
    random fluctuations in channel proteins, 31, 52
Temporal resolution, of EEG signals, 279
Terrestrial factor, of culture conditions, 159–160, 162
Tertiary structure of proteins, 58, 86
    fractal analysis of, 71–76
Tetanus, 136
Thermodynamics, of proteins, 84, 119. *See also* Temperature
Thermolysin, 74, 80
Thioredoxin, 74, 80
Three-point attachment theory, 114
Thresholding, 174
Thymine, 16, 17
Time-constant, 179–180

Time scales
  of conformational changes in proteins, 36
  dependence on ion concentration or voltage, 49
  and distributions, 48
  multiple, in heartbeat dynamics, 251–252, 253, 258–259
    clinical applications, 261–263
  of polymer entanglements, 63
  self-similarity at different, 10, 40
Time-scape, of heartbeat, 252, 254
Tissues, 189–242
  from chimeras, mosaic pattern in, 189–203
  embryonic differentiation, 189–191
  mechanisms of pattern formation, 191–193
Topological dimension, 4, 6, 278, 288
  non-integer. *See* Fractal dimension
Topological indices, 116–117
Tosylelastase, 74
Tracheo-bronchial tree, 250
Transient deterministic data, 292
Transition metal atoms, 113
Transport of ions. *See* Ion channel kinetics
Triangulation, 227
2,4,6-Trimethylphenol, 114
Triose phosphate isomerase, 74
Tritium, 34
Trophectoderm cells, 190
Trypsin, 74, 80
  active site probability distribution, 101
  surface mass exponent, 98
Trypsin inhibitor, 74, 80
Tryptic soy agar, 138
Tryptophan, 34, 58, 68
Tumors shape, 212, 214–215, 216, 217

U

Ulcers
  ameboid, 219
  corneal. *See* Cornea, herpes simplex virus ulcers
  dendritic, 219
Union principle of fractals, 112
University of North Texas, 268, 272, 292

V

Valent connectivity index, 116
Vascular bed, 184
Vasculogenesis, 217–218
Venous tree, 250, 251
Vertebrates, 26
Vibration, atomic, 68–69, 70, 119. *See also* Temperature
Vicsek's snowflake, 142

Videotaping
  of bacterial colony growth, 132, 149, 152
  in diagnosis, 206
  grey scale to binary images, 174, 239, 339
Virus coat protein, 74
Viscosity, of polymer solution, 65, 66
Viscous fingering, 225
Visual scanning task, and fractal dimension of EEG, 270, 292
Voltage
  across cell membrane, 49
  between two terminal bonds, 336–338
von Koch, Helge, 5

W

Walks. *See also* Lévy walk; Random walk model
  DNA walk, 17–18
  random self-avoiding walk, 80, 81, 82
  self-avoiding walk, 63, 64, 67, 70
Wavelet analysis, 20–21, 254
Wave-like row advancing
  in bacterial colony morphology, 149, 151
  microscopic generating process for, 152–153
Wavy expanding clusters, 133, 135
Weighing factors, 81, 116, 342
Weight indices of edges, 116
White noise, 235, 259, 263, 264. *See also* Noise
Wiener, N., 273
Woodcock-Johnson Revised Tests of Cognitive Ability, 292
Workload, 302–303

X

X-ray diffraction
  carbon backbone of proteins, 105
  measurement of motion/vibration, 35, 69

Y

Yardstick method, 198, 200, 207, 226–228
  interpolated, 228
  for tumors shape, 215, 216
Yeast
  DFA for, 19, 20
  division in, 225
  in glycolysis, 89
  linguistic features of DNA, 25
Yeast phosphoglycerate mutase, 74

Z

Zeroset, 239
Zinc, 115, 116, 117, 118
Zipf analysis, 15, 24, 26
Zona pellucida, 189, 194
Zygote, 189–190